Hartmut Zabel
Medical Physics

Also of Interest

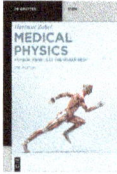

Medical Physics.
Volume 1: Physical Aspects of the Human Body
Hartmut Zabel, 2023
ISBN 978-3-11-075691-3, e-ISBN (PDF) 978-3-11-075695-1,
e-ISBN (EPUB) 978-3-11-075698-2
Volume 1, Volume 2 and Volume 3 also available as a set:
Set-ISBN 978-3-11-076102-3

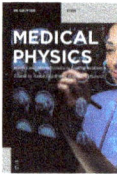

Medical Physics.
Volume 3: Physical Aspects of Therapeutics
Hartmut Zabel, 2023
ISBN 978-3-11-116867-8, e-ISBN (PDF) 978-3-11-116873-9,
e-ISBN (EPUB) 978-3-11-116906-4
Volume 1, Volume 2 and Volume 3 also available as a set:
Set-ISBN 978-3-11-076102-3

Medical Physics.
Models and Technologies in Cancer Research
Anna Bajek, Bartosz Tylkowski (Eds.), 2021
ISBN 978-3-11-066229-0, e-ISBN (PDF) 978-3-11-066230-6,
e-ISBN (EPUB) 978-3-11-066234-4

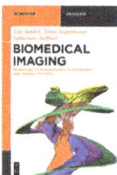

Biomedical Imaging.
Principles of Radiography, Tomography and Medical Physics
Tim Salditt, Timo Aspelmeier, Sebastian Aeffner, 2017
ISBN 978-3-11-042668-7, e-ISBN (PDF) 978-3-11-042669-4,
e-ISBN (EPUB) 978-3-11-042351-8

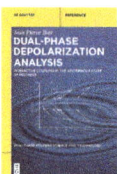

Dual-Phase Depolarization Analysis.
Interactive Coupling in the Amorphous State of Polymers
Jean Pierre Ibar, 2022
ISBN 978-3-11-075669-2, e-ISBN (PDF) 978-3-11-075674-6,
e-ISBN (EPUB) 978-3-11-075684-5

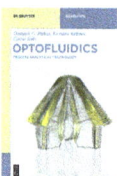

Optofluidics.
Process Analytical Technology
Dominik G. Rabus, Cinzia Sada, Karsten Rebner, 2018
ISBN 978-3-11-054614-9, e-ISBN (PDF) 978-3-11-054615-6,
e-ISBN (EPUB) 978-3-11-054622-4

Hartmut Zabel

Medical Physics

Volume 2: Physical Aspects of Diagnostics

2nd Edition

DE GRUYTER

Author
Prof. Dr. Dr. h. c. Hartmut Zabel
Ruhr-Universität Bochum
Fak. für Physik und Astronomie
44780 Bochum
hartmut.zabel@ruhr-uni-bochum.de

ISBN 978-3-11-075702-6
e-ISBN (PDF) 978-3-11-075709-5
e-ISBN (EPUB) 978-3-11-075712-5

Library of Congress Control Number: 2022946977

Bibliographic information published by the Deutsche Nationalbibliothek
The Deutsche Nationalbibliothek lists this publication in the Deutsche Nationalbibliografie;
detailed bibliographic data are available on the internet at http://dnb.dnb.de.

© 2023 Walter de Gruyter GmbH, Berlin/Boston
Cover image: andresr/E+/Getty Images
Typesetting: Integra Software Services Pvt. Ltd.
Printing and binding: CPI books GmbH, Leck

www.degruyter.com

Preface

The second edition of the textbook on Medical Physics is now published in three instead of two volumes. This is because the new edition is not only corrected and updated but also considerably expanded. The enhancement is primarily due to many new features designed to increase the usefulness of this textbook as a learning companion to regular courses. Each chapter concludes with a detailed summary, questions, exercises, and a self-assessment of the knowledge gained. All chapters contain info boxes for in-depth information on specific topics and math boxes for deriving central mathematical concepts. References to current literature and further reading recommendations lead to reviews in the subject area. Answers to questions and solutions to tasks can be found in the appendix.

The textbook is aimed at bachelor's or master's students in the first year of medical physics. It is recommended that students first acquire some basic knowledge of modern physics before proceeding with reading this Medical Physics textbook.

The range of topics in Medical Physics is extensive. For this reason, most textbooks in this field have been written by several authors. Covering all topics on your own is both a challenge and an opportunity to design the various chapters as coherently as possible and to relate them to one another. I would not have been able to meet this challenge without giving respective lectures.

Compared to the first edition, the chapters have been rearranged to allow a clear distinction between the physical aspects of the human body in the first volume, the physical aspects of diagnostics in the second volume, and the physical aspects of therapy in the third volume.

The first volume begins with classical mechanics, including forces, moments, and energy, concepts applied to the human body. The overarching importance of the action potential and signal transmission in the body's functioning is presented in chapters five and six, followed by a discussion of those organs, systems, and senses that have a clear connection to physics, such as the cardiovascular and the respiratory system. The last two chapters deal with the primary sensory organs of the body, the visual sense, and the auditory sense.

The second volume deals exclusively with imaging methods, distinguishing between those without ionizing radiation (ultrasound, endoscopy, magnetic resonance imaging) and those with ionizing radiation (X-rays, SPECT, PET). As an introduction to these radiological chapters, essential facts about X-ray and nuclear physics are presented, the interactions of radiation with matter are explained, and measures for radiation protection are discussed.

The third volume is devoted to the physical fundamentals of radiotherapeutic procedures using X-rays, protons, neutrons, and γ-rays. These chapters are introduced by first comparing the radiation response of benign cells and malignant tumor cells. The last two chapters deal with laser processes and highlight the physics of

https://doi.org/10.1515/9783110757095-202

nanoparticles in diagnostics and therapy. An additional chapter on medical statistics rounds off the third volume.

All these additions, guidance, and exercises will hopefully make studying this medical physics textbook a valuable and informative companion to your regular coursework. Questions can be directed to the editor or author and will be answered promptly. Corrections are very welcome and will be posted on the book's website.

Bochum, December 14, 2022

Acknowledgements

My first thanks go to all those who pointed out errors and made suggestions for improvement to the first edition. Constructive criticism is always very helpful in improving the text, correcting mistakes, and clearing up misunderstandings. I am very grateful to Dr. Alexey Saphoznik, who took the time to read the entire first volume and suggested many modifications. My special thanks go to Professor Birge Kollmeier (University of Oldenburg), who made valuable recommendations for improving the content, particularly concerning sound and sound perception. I would also like to acknowledge my ophthalmologist Dr. Elbracht-Hülseweh, from whom I learned a lot during many years of treatment. Thanks also go to my colleagues who helped and guided me during the preparation of the first edition.

My special thanks go to the editorial staff at de Gruyter Verlag and, in particular, to Kristin Berber-Nerlinger, who encouraged me in the first place to prepare a second edition and gave me valuable advice on the implementation of this project. Nadja Schedensack helped in all stages, and Kathleen Prüfer did an excellent editorial job. I am very grateful to the entire publishing team of de Gruyter.

Last but not least, I would like to thank my family, and in particular, Rosemarie, who accompanied this project with much patience, understanding, and encouragement.

https://doi.org/10.1515/9783110757095-203

Contents

Preface —— V

Acknowledgements —— VII

Part A: Diagnostics without ionizing radiation

1	Sonography —— 3	
1.1	Introduction and overview —— 3	
1.2	Ultrasound transducer —— 4	
1.2.1	Piezoelectric effect —— 4	
1.2.2	Ultrasonic head —— 6	
1.3	Reflection, transmission, and attenuation —— 9	
1.3.1	Reflection, transmission, and scattering —— 9	
1.3.2	Scattering and absorption —— 11	
1.4	Beam properties, pulsing, and focusing —— 13	
1.4.1	Pulse quality —— 13	
1.4.2	Time gain compensation —— 14	
1.4.3	Near field, far field, and focusing —— 15	
1.4.4	Physical parameters —— 18	
1.4.5	Safety issues —— 18	
1.5	Medical imaging —— 19	
1.5.1	A-mode scan —— 19	
1.5.2	B-mode scan —— 21	
1.5.3	C-scan —— 24	
1.5.4	M-mode —— 26	
1.5.5	Shear wave sonography —— 27	
1.6	Scan characteristics —— 28	
1.6.1	Dynamic focusing —— 28	
1.6.2	Line density —— 29	
1.6.3	Scan frequency —— 29	
1.6.4	Depth of view —— 30	
1.6.5	Penetration depth —— 30	
1.6.6	Spatial resolution —— 31	
1.6.7	Axial resolution —— 31	
1.6.8	Lateral resolution —— 31	
1.6.9	Artifacts —— 32	
1.7	Doppler method —— 33	
1.7.1	Doppler shift —— 33	
1.7.2	cw Doppler method —— 34	

1.7.3	Pulsed Doppler method (duplex mode) —— 38
1.7.4	Duplex scan of umbilical cord —— 40
1.8	Summary —— 42
	Exercises —— 43
	Suggestion for home experiments —— 45
	References —— 45
	Further reading —— 47
	Useful website —— 47

2	**Endoscopy** —— 48
2.1	Introduction —— 48
2.2	Standard uses of medical endoscopes —— 48
2.3	Fiber optics —— 50
2.4	Endoscope optics —— 54
2.5	Resolution and magnification —— 56
2.6	Specialized endoscopes —— 58
2.6.1	Narrowband imaging —— 58
2.6.2	Chromoendoscopy —— 60
2.6.3	Endomicroscopy —— 60
2.7	Confocal laser endoscopy —— 60
2.7.1	General working principle —— 60
2.7.2	Fiber-optic confocal reflectance microscope —— 62
2.8	Optical coherence tomography endoscopes —— 63
2.8.1	Basic principle of OCT —— 64
2.8.2	Resolution and scan range —— 66
2.8.3	Additional methods and applications —— 67
2.9	Capsule endoscopy —— 68
2.10	Future trends —— 69
2.11	Summary —— 71
	Suggestions for home experiment —— 72
	Exercises —— 73
	References —— 73
	Further reading —— 75
	Useful website —— 75

3	**Magnetic resonance imaging** —— 76
3.1	Introduction —— 77
3.2	Nuclear spin basics —— 77
3.3	Nuclear magnetic resonance basics —— 80
3.3.1	Zeeman splitting —— 80
3.3.2	Equation of motion —— 82
3.3.3	Magnetization of a two-level system —— 83

3.3.4	Toy model of magnetization relaxation —— 86	
3.3.5	Resonance absorption —— 88	
3.4	Spin-echo techniques —— 92	
3.5	Autocorrelation and spectral density (for experts) —— 96	
3.6	NMR and MRI procedures —— 101	
3.6.1	Saturation —— 101	
3.6.2	Chemical shift —— 101	
3.6.3	Standard nomenclature —— 103	
3.7	Contrast generation —— 105	
3.7.1	$T1$ contrast —— 105	
3.7.2	$T2$ contrast —— 107	
3.7.3	PD contrast —— 107	
3.7.4	Inversion recovery (IR) —— 110	
3.7.5	Short time inversion recovery (STIR) —— 111	
3.8	MR signal localization —— 112	
3.8.1	Slice encoding gradient —— 113	
3.8.2	Frequency encoding gradient (FEG) —— 114	
3.8.3	Phase encoding gradient (PEG) —— 115	
3.8.4	K-map —— 116	
3.8.5	Fourier transform —— 118	
3.8.6	Data acquisition —— 119	
3.9	Magnets and coils —— 120	
3.9.1	Main coil —— 121	
3.9.2	Gradient coils —— 122	
3.9.3	rf-coils —— 123	
3.9.4	MRI machine specifications —— 123	
3.10	Applications of MRI —— 126	
3.10.1	Joints —— 126	
3.10.2	Dynamic contrast enhancement (DCE) MRI —— 126	
3.10.3	Angio-MRI —— 130	
3.10.4	Diffusion-weighted imaging (DWI) —— 130	
3.10.5	Multiple parameter MRI (mpMRI) —— 133	
3.10.6	Functional MRI (fMRI) —— 133	
3.10.7	Real-time MRI —— 136	
3.11	Hyperpolarization MRI —— 137	
3.11.1	^{3}He-hMRI —— 139	
3.11.2	^{13}C-hMRI —— 140	
3.11.3	^{17}O-hMRI —— 141	
3.11.4	^{19}F-hMRI —— 141	
3.12	Further remarks —— 142	
3.12.1	New trends and comparisons —— 142	
3.12.2	Advantages–disadvantages and hazards —— 143	

3.13   Summary —— 145
              Exercises —— 148
              References —— 148
              Further reading —— 151
              Useful websites —— 151

Part B: **X-ray and nuclear methods**

4   **X-ray sources and generators** —— 155
4.1   Introduction —— 155
4.2   General components of x-ray tubes —— 156
4.3   Bremsstrahlung radiation —— 158
4.4   Characteristic radiation —— 160
4.4.1   Atomic transitions —— 160
4.4.2   Energy dispersive x-ray chemical analysis —— 164
4.4.3   Target material —— 164
4.5   X-ray generators —— 166
4.5.1   X-ray tubes for radiography —— 166
4.5.2   Linear accelerators for radiotherapy —— 168
4.5.3   Synchrotron radiation —— 171
4.6   Summary —— 174
              Exercises —— 176
              References —— 176
              Further reading —— 177
              Useful website —— 177

5   **Nuclei and isotopes** —— 178
5.1   Introduction —— 178
5.2   Isotopes —— 179
5.3   Atomic mass and atomic weight —— 181
5.4   Nuclear decay —— 183
5.4.1   Electron emission (β^-) —— 184
5.4.2   Positron emission (β^+) —— 184
5.4.3   Electron capture (EC) —— 186
5.4.4 α-Particle decay —— 187
5.4.5   Decay schemes —— 187
5.5   Radioactivity —— 189
5.5.1   Exponential decay law —— 189
5.5.2   Nuclear activity —— 192
5.5.3   Decay chains —— 194
5.6   Radioisotope production —— 195

5.6.1   Nuclear reactions —— 196
5.6.2   Isotope production via irradiation —— 196
5.6.3   Charge particle activation —— 198
5.6.4   Cyclotron isotope production —— 200
5.6.5   Radioisotope production by fission —— 205
5.6.6   Neutron activation —— 206
5.7   Summary —— 208
           Exercises —— 210
           References —— 212
           Further reading —— 212

6   Interaction of radiation with matter —— 214
6.1   Attenuation: Lambert-Beer law —— 214
6.2   Interaction of EM radiation with matter —— 216
6.2.1   Attenuation coefficient of photons —— 216
6.2.2   Mass attenuation coefficient of photons —— 218
6.2.3   Photoelectric effect —— 219
6.2.4   Compton scattering —— 221
6.2.5   Coherent scattering of x-rays —— 226
6.2.6   Pair production —— 229
6.2.7   Comparison of photon–electron interactions —— 230
6.3   Interaction of charged particles with matter —— 231
6.3.1   Alpha particles —— 231
6.3.2   Beta-particles —— 235
6.4   Interaction of neutrons with matter —— 237
6.5   Summary —— 240
           Suggestion for home experiment —— 241
           Exercises —— 242
           References —— 243
           Further reading —— 243
           Useful website —— 243

7   Dosimetry —— 244
7.1   Introduction —— 244
7.2   Definitions of dose and dose rate —— 245
7.3   Kerma —— 249
7.3.1   Flux and fluence —— 249
7.3.2   Energy fluence —— 249
7.3.3   Mass energy transfer coefficient —— 250
7.3.4   Mass energy absorption coefficient —— 250
7.3.5   Definition of kerma —— 251
7.3.6   Examples —— 253

7.4   Dosimeters and radiation monitors —— 255
7.4.1   Ionization chamber —— 256
7.4.2   Proportional counters —— 256
7.4.3   Geiger-Müller detectors —— 258
7.4.4   Dead time —— 258
7.5   Radiation exposure —— 259
7.6   Radiation protection —— 260
7.7   Summary —— 262
           Suggestions for home experiment —— 263
           Exercises —— 264
           References —— 265
           Further reading —— 265

Part C: **Radiography**

8   **X-ray radiography** —— 269
8.1   Introduction —— 269
8.2   Standard x-ray radiography —— 270
8.2.1   Beam delivery and beam hardening —— 270
8.2.2   Magnification and penumbra —— 272
8.2.3   Compton scattering and grids —— 273
8.3   X-ray attenuation and contrast —— 275
8.3.1   Contrast —— 275
8.3.2   Attenuation profile —— 277
8.4   X-ray recording —— 279
8.4.1   Film radiography —— 279
8.4.2   Fluoroscopy —— 282
8.4.3   Flat panel radiography —— 284
8.4.4   Comparison —— 286
8.5   Counting statistics, noise, quantum efficiency —— 287
8.5.1   Counting statistics —— 287
8.5.2   Noise and quantum efficiency —— 288
8.6   System integration —— 290
8.6.1   Projection radiography —— 290
8.6.2   Mammography —— 292
8.7   Attenuation contrast enhancement —— 292
8.7.1   Contrast agents —— 293
8.7.2   Digital subtraction angiography (DSA) —— 295
8.7.3   Dual-energy x-ray absorptiometry —— 295
8.8   Phase contrast imaging (PCI) —— 298
8.8.1   Physical background —— 298

8.8.2   Detection of phase contrast —— 300
8.9   Computed tomography (CT) —— 303
8.9.1   Overview —— 303
8.9.2   The Hounsfield scale —— 304
8.9.3   Specifications of CT scanners —— 306
8.9.4   Contrast enhancement —— 309
8.9.5   Radon transformation —— 310
8.9.6   Backprojection —— 313
8.9.7   Filter —— 314
8.10   Risks and comparisons —— 318
8.11   Summary —— 319
             Exercises —— 321
             References —— 323
             Further reading —— 324
             Useful website —— 324

9   Scintigraphy (SPE and SPECT) —— 325
9.1   Introduction —— 325
9.2   Collimators for scintigraphy —— 326
9.3   Detectors, counting, and artifacts —— 329
9.3.1   Photomultiplier tube —— 329
9.3.2   Anger counting —— 331
9.3.3   CZT detectors —— 331
9.3.4   Artifacts: Compton scattering —— 333
9.3.5   SNR and CNR —— 334
9.4   Isotopes for scintigraphy —— 335
9.4.1   Radioisotopes and radiopharmaceuticals —— 335
9.4.2   Isotope generators —— 337
9.5   Full body SPE scans —— 339
9.6   Single-photon emission computed tomography (SPECT) —— 341
9.6.1   SPECT systems and detectors —— 341
9.6.2   Clinical applications —— 343
9.6.3   SPECT image processing —— 345
9.7   Summary —— 346
             Exercises —— 347
             References —— 348
             Further reading —— 349

10   Positron emission tomography —— 350
10.1   Introduction —— 350
10.2   Basic principle of PET —— 351
10.2.1   Energy and momentum —— 351

10.2.2   Coincidence counting —— 352
10.2.3   Artifacts —— 353
10.2.4   Spatial resolution —— 354
10.2.5   TOF-PET —— 355
10.2.6   Ring designs —— 356
10.2.7   PET scanner and combinations —— 357
10.3   Data acquisition and image reconstruction —— 358
10.3.1   Detectors —— 358
10.3.2   Counting statistics —— 359
10.3.3   Image reconstruction —— 359
10.3.4   Standard uptake value —— 360
10.4   PET isotopes —— 361
10.4.1   General aspects —— 361
10.4.2 ^{18}F-decay —— 362
10.5   Clinical applications of PET —— 364
10.5.1   FDG-PET —— 364
10.5.2   FET-PET —— 366
10.5.3   Prostate-specific membrane antigen PET —— 367
10.6   Conclusion —— 369
10.7   Summary —— 370
            Exercises —— 372
            References —— 373
            Useful websites —— 374
            Further reading —— 374

Appendix

11   Answers to questions —— 377

12   Solutions to exercises —— 389

13   List of acronyms (used in all three volumes) —— 415

14   Selection of fundamental physical constants, conversions, and relationships —— 418

15   List of scientists named in this volume —— 419

16   Glossary —— 420

17   Index of terms —— 423

Part A: **Diagnostics without ionizing radiation**

1 Sonography

Physical parameters of sound and ultrasound	
Sound velocity in air	330 m/s
Sound velocity in water	1500 m/s
Sound velocity in tissue	1540 m/s
Sound velocity in bones	3600 m/s
Typical ultrasound frequency	10 MHz
Typical ultrasound wavelength	0.5 mm
Typical pulse repeat frequencies	1–5 kHz
Typical ultrasound half-value thickness in tissue	4 cm
Typical lateral resolution	1 mm
Typical axial resolution	0.5 mm
Typical frame rate	20–30 Hz
Typical power deposited	50 mW
Near field at 1 MHz	35 mm

1.1 Introduction and overview

Sonography works with ultrasound waves. Medical sonography is an imaging modality that uses ultrasound (US) for taking static images of organs and tissues, dynamic images of heart and lung movement, and kinetic images of blood flow. A well-known and common example of sonography is imaging fetuses as part of prenatal checkups. Sonography is not limited to medicine. Many other fields use US techniques such as submarine navigation, seafloor mapping, food control, security screening, surface cleaning, nondestructive material testing, and ultrasonic welding of plastics.

US imaging is much more practical to handle by a physician than any radiation-based imaging modality. It can be applied locally at the bedside or the site of an accident and does not require special safety procedures for the patient or the examining staff. Direct communication with the patient is possible during the examination, which is a significant advantage compared to other imaging modalities such as magnetic resonance imaging (MRI), computed tomography (CT), or positron emission tomography. Conversely, conventional US images have lower resolution and require considerable experience to interpret them correctly for useful diagnostics. US imaging for medical diagnostics are mainly in the following areas:

– Cardiovascular system
– Abdominal organs
– Urology/prostate
– Obstetrics/gynecology
– Ophthalmology
– Mammography

https://doi.org/10.1515/9783110757095-001

After discovering the piezoelectric effect by Paul-Jacques Curie[1] and Pierre Curie[2] in 1880, Langevin[3] first used it to produce ultrasonic waves, mainly for industrial and military applications. It was not until the 1940s to 1950s of the twentieth century that US was applied for medical investigations. In the 1960s, the first handheld contact B-mode scanner was produced and commercialized.

The basic properties and terms of sound waves are presented and defined in Section 12.2 of Volume 1: pressure amplitude, sound velocity, particle velocity, acoustic impedance, sound intensity, reflection, and transmission at interfaces. It is recommended to refresh these terms before continuing with the following discussions. Table 1.1 reviews the most important relationships. The frequencies used for sonography are much beyond audible sound, i.e., more than 20 kHz. In fact, most sonographic systems use frequencies in the range of 2–20 MHz. Although these have much higher frequencies than considered in Chapter 12 of Volume 1, the physics of sound waves is the same.

Tab. 1.1: Review of important relationships in sonography.

Sound velocity	$v_s = \dfrac{\omega}{k} = \sqrt{\dfrac{B}{\rho}}$
Wave amplitude	$\xi_0 = u_0/\omega$
Pressure amplitude	$p_0 = -B\xi_0 k$
Acoustic impedance	$Z = \rho v_s$
Particle velocity amplitude	$u_0 = p_0/Z$
Time average intensity	$\langle I \rangle = \frac{1}{2}Zu_0^2 = \frac{1}{2}\dfrac{p_0^2}{Z}$

1.2 Ultrasound transducer

1.2.1 Piezoelectric effect

For generating and detecting US waves, a piezoelectric head is used, also called a *transducer*. In general, transducers are devices that convert one form of energy into another. In the present case, US transducers convert electrical energy into vibrational energy of a crystal lattice using the piezoelectric effect. The exploitation of

---

1 Paul-Jacques Curie (1855–1941), French physicist.
2 Pierre Curie (1859–1906), French physicist, Nobel Prize in Physics 1903.
3 Paul Langevin (1872–1946), French physicist.

the piezoelectric effect requires a single crystal with appropriate properties. Piezo-electric crystals are ionic and insulating materials with high electrical polarizability that is strongly coupled to the crystal lattice. Mechanical compressive or tensile strain generates electrical potentials, and vice versa, electric potentials are converted into elastic strain. In either case, a polarization of electric dipole moments in the crystal is induced via strain or electrical potential. The direct piezoelectrical effect relates the polarization P of these electric dipole moments to the stress σ applied, as illustrated in Fig. 1.1:

$$P = g \cdot \sigma, \tag{1.1}$$

where g is the piezoelectric coefficient with the unit $[g] = m/V$. The converse piezo-electric effect relates the length change Δl of the crystal to the applied voltage change ΔU:

$$\Delta l = g \cdot \Delta U \tag{1.2}$$

Note that the expansion of the crystal does *not* depend on its size but only on the voltage applied. The magnitude of the piezoelectric coefficient g is on the order of 500 pC/N or 0.5 nm/V. The piezoelectric coefficient g is actually a tensor, but for simplicity, we keep it here as a scalar. A more complete discussion can be found in the review [1].

The piezoelectric transduction can be utilized in a static or dynamic mode up to high frequencies. Piezoelectric crystals have numerous applications in physics and technology as actuators, sensors, and for nano-positioning. For instance, the sensing head of scanning tunneling microscopes and atomic force microscopes is driven by piezoelectric crystals.

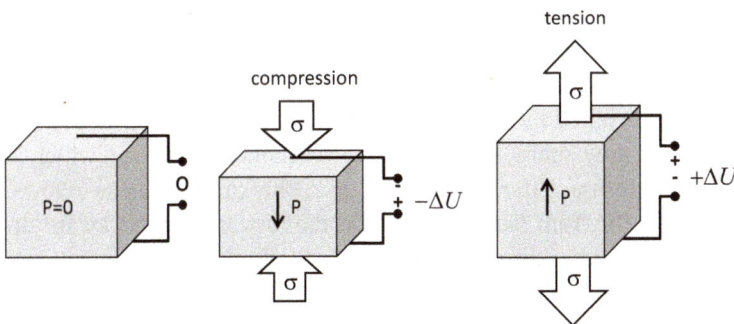

Fig. 1.1: Working principle of the direct piezoelectric effect. Application of compressive or tensile stress induces an electric polarization and a voltage change. Conversely, the application of a voltage changes the length of the piezoelectric crystal.

What are piezoelectric materials made of? There is a large variety of piezoelectric materials [1, 2]. The most common ones are crystals of insulating ceramic materials,

which lack inversion symmetry. The best-known example is the piezoelectric compound $PbTiO_3$ (short notation, PT) or the Zr-doped version $Pb(Zr_{1-x}Ti_x)O_3$ (short notation, PZT). The movement of Zr^{4+}/Ti^{4+} ions in and out of the oxygen plane (see
Fig. 1.2) by application of an electrical field or by stress induces an electrical dipole
moment. Therefore, these piezoelectric materials can either be used as actuators by
applying a voltage or as sensors of pressure or stress by measuring voltage changes.

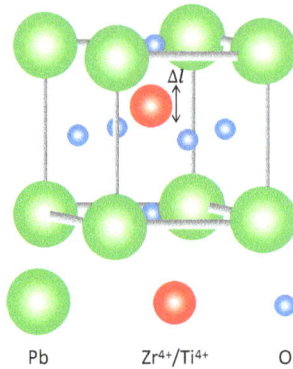

Pb Zr^{4+}/Ti^{4+} O

Fig. 1.2: Crystal structure of $Pb(Zr_{1-x}Ti_x)O_3$. By applying stress
or voltage, the center Ti^{4+} or Zr^{4+} ions move out of the oxygen
plane, creating an electric dipole moment.

1.2.2 Ultrasonic head

PZT crystals are the essential part of US transducers for US imaging and therefore
they are an integral part of US transducer heads. A schematic of such a US head is
shown in Fig. 1.3(a). In the transmitting mode, the energizing voltage is applied to the
backside of a piezoelectric disk covered by a thin metal film connected to a coaxial
cable while the front side is grounded. Furthermore, there is some porous damping
material on the backside of the piezo-crystal, ensuring that the backward traveling
sound wave is not being reflected forward and disturbing the main signal.

Transducers can be excited by an ac-voltage to emit sound waves at any frequency. But transducers are usually excited at their resonance frequency, which is
defined solely by the thickness of the piezoelectric disk. This can be seen as follows.
Sound waves emitted at the front face propagate in the forward and backward directions. The wave traveling backward is reflected on the backside of the disk and
overlaps with the sound wave emitted at the front side. For constructive interference of the waves from the front and the back, the path difference should be a multiple of the wavelength λ and therefore the thickness w of the disk should be $\lambda/2$.
This condition assumes that the front side and backside of the transducer are in
contact with materials characterized by lower impedances than the one of PZT.
Under these circumstances, the resonance condition is

$$f_0 = \frac{v_{PZT}}{\lambda_{res}} = n\frac{v_{PZT}}{2w} \quad (n = 1, 2, \ldots), \qquad (1.3)$$

where v_{PZT} is the sound velocity in the PZT disk and n is the order of interference. For a resonance frequency of 1 MHz, the thickness of the disk should be $w = 2$ mm. When used as a sensor, the sensitivity is highest at the resonance frequency. Changing the resonance frequency requires changing the disk's thickness.

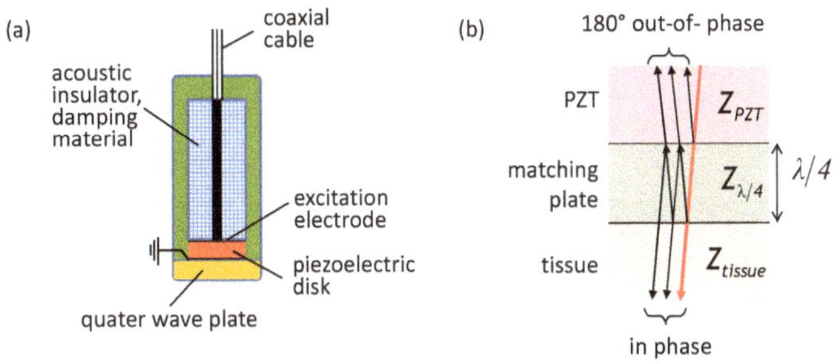

Fig. 1.3: (a) Section through an ultrasound transducer consisting of a piezoelectric disk connected to a coaxial cable, damping material on the backside, and quarter-wave plate as impedance bridge on the front side. (b) Reflection and transmission at the quarter-wave plate.

The front face of the piezoelectric disk is protected by a thin plastic cover of thickness $\lambda/4$, which serves as an impedance bridge to the body. The impedance of the plastic piece is chosen as the geometric mean between the value of the PZT disk ($Z_{PZT} = 30 \times 10^6$ Ns/m^3) and of soft tissue ($Z_{tissue} = 1.5 \times 10^6$ Ns/m^3):

$$Z_{\lambda/4} = \sqrt{Z_{PZT}Z_{tissue}} \qquad (1.4)$$

The thickness of the quarter-wave plate is fixed for the sound wavelength used; therefore, matching is only ideal for one particular wavelength. The quarter-wave plate enhances transmission of US in the forward direction and suppresses back-reflection into the PZT according to the principle illustrated in Fig. 1.3(b). Sound waves transmitted into the $\lambda/4$ plate reverberate back and forth at both interfaces. As the $\lambda/4$ plate has an impedance value in between PZT and tissue, the first reflection from the front interface is in phase, whereas the reflection on the backside suffers a phase jump by 180°. When the reflected wave arrives again at the front interface, it has a path length difference of $\lambda/2$ and is in phase compared to an incoming and straight-through US wave. Therefore, these waves enforce each other by constructive interference. On the other hand, the transmitted wave back into the PZT after it has been reflected at the interface to the tissue is 180° out of phase with the wave reflected at the interface of the PZT and matching plate. Therefore, they cancel each other by destructive interference.

The quarter-wave plate is covered with another plastic part that is impedance-matched to the tissue and acts as an acoustic lens; see the next section. Any air gap between the US head and the body must be strictly avoided because of the large impedance mismatch between the US head and air and at the air/body interface, which causes total reflection of the sound wave instead of transmission into the body. For this purpose, the body part to be scanned is covered with a gel and the US head is completely immersed in it. The gel has an impedance value similar to that of the plastic cover.

> **!** Piezoelectric materials are used to generate and detect sound waves. The resonance frequency of the piezoelectrical disk is given by the thickness of the disk.

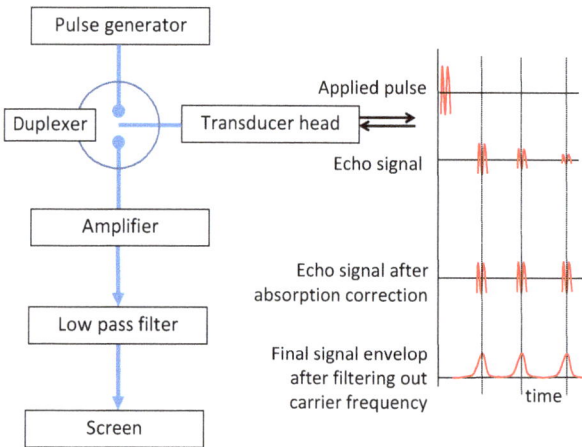

Fig. 1.4: Flowchart for application and signal processing of medical sonography.

In sonography, the back-reflected and scattered intensities are recorded by the same US transducer that also generates the sound wave in the first place. In order not to confuse the transmitted signal and the reflected echo signal, the standard operating mode is pulsed. First, the transducer generates a short high-frequency pulse that is transmitted into the body. The pulse repeat frequency (PRF) is set such that there is sufficient time between one pulse and the next to record a series of echo signal. To do so, a so-called duplexer quickly switches the transducer head from the transmitter to the receiver mode. The echo signals are then amplified and filtered for display. Figure 1.4 shows a flowchart of different steps, which are discussed in more detail further below. We will start with signal transmission and reflection and continue with attenuation effects, pulse shapes, and resolution.

1.3 Reflection, transmission, and attenuation

1.3.1 Reflection, transmission, and scattering

The back-reflected intensity ratio depends primarily on the acoustic contrast at the interface between two media expressed in terms of their acoustic impedances Z_1 and Z_2 and the angles α_1 and α_2 (see Fig. 1.5):

$$R = \frac{I_r}{I_i} = \left(\frac{Z_1 \cos \alpha_2 - Z_2 \cos \alpha_1}{Z_1 \cos \alpha_2 + Z_2 \cos \alpha_1} \right)^2, \tag{1.5}$$

where I_r is the back-reflected intensity and I_i is the incident intensity. The incident angle α_1 in medium 1 and the refracted angle α_2 in medium 2 are both measured with respect to the normal to the common interface (Fig. 1.5). For back-reflection, α_1 and α_2 should be within the opening angle of the detector.

The intensity I_t that is not reflected is transmitted across the interface to the next medium. The intensity ratio I_t/I_i is described by

$$T = \frac{I_t}{I_i} = \frac{4 Z_1 Z_2 \cos^2 \alpha_1}{(Z_1 \cos \alpha_2 + Z_2 \cos \alpha_1)^2}. \tag{1.6}$$

Energy conservation demands that

$$R + T = 1. \tag{1.7}$$

The same splitting between reflected and transmitted intensities occurs at all subsequent interfaces. Equations (1.6)–(1.8) are known in optics as Fresnel[4] equations for reflection and transmission of light at interfaces when replacing the acoustic impedances Z_i by the optical refractive indices n_i of the different materials [3].

Figure 1.4 illustrates schematically different idealized cases for reflection and scattering at interfaces. Case (a) shows a smooth interface perpendicular to the incoming sound wave and therefore has the highest back-reflected intensity. The same interface inclined by some angle as in case (b) will cause reflection and refraction, but the reflected beam may miss the detector. As in (c), rough interfaces cause diffuse scattering, which drastically diminishes the intensity in the backward direction. Curved surfaces like (d) widen the back-scattered beam, and only part of the intensity reaches the detector.

Table 1.2 lists some impedance values for tissues and bones. Aside from the lungs and bones, the impedance values for the other organs vary little. Two examples may illustrate the challenges of US imaging. Let us assume that we want to map the heart in the surrounding body tissue. We take the impedance of muscles

---

4 Augustin Jean Fresnel (1788–1827), French physicist and engineer.

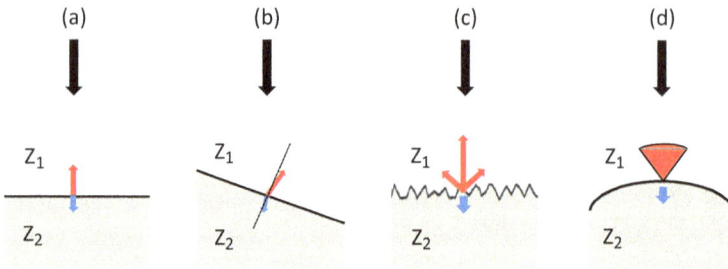

Fig. 1.5: Four different idealized cases for interfaces between tissues with different acoustic impedances Z_1 and Z_2: (a) smooth interface perpendicular to incoming sound wave (black arrow on top); (b) smooth and inclined interface; (c) rough interface; (d) curved interface. Black arrow: incident sound wave; Red arrows: reflected or scattered sound waves; blue arrows: transmitted sound waves; dashed line: normal to the interface.

Tab. 1.2: Sound velocities, densities, and impedance values for some characteristic materials of relevance for ultrasound imaging. PZT = $Pb(Zr_{1-x}Ti_x)O_3$.

	Sound velocity v (m/s)	Density ρ (kg/m³)	Acoustic impedance Z (10⁶ Ns/m³)	Attenuation (dB/cm/1 MHz)	Half-value layer thickness (cm)
Air	330	1.3	0.000430	7.5	0.4
Water	1500	998	1.5	0.0022	1360
Blood	1530	1000	1.62	0.15	20
Fatty tissue	1470	970	1.38	0.5–1.8	6–1.6
Soft tissue	1540	1050	1.6	0.75	4
Muscles	1570	1040	1.7	0.5	6
Bones	3600	1700	3–7	15	0.2
Lung	650–1160	300	0.3–0.4	40	0.075
PZT	4000	7500	30		

The fifth column lists attenuation values for a 1 MHz US source, and the sixth column gives the thickness at which the intensity drops to 50% of the incident value. Values are collected from [4–6] and calculated.

for the heart's myocardium and the impedance of water for the surrounding. Then, at normal incidence, the back-reflected intensity is 0.4% of the incident intensity. This estimate neglects all other effects, such as scattering from rough interfaces and absorption. In fact, an echo signal of about 0.1% is more realistic. The second example concerns the echo signal from bones. The back-reflected intensity is approx. 30% of the incident intensity on the front side and a further 30% of the transmitted intensity on the rear side, i.e., approx. 21% if scattering and absorption effects are neglected. To detect an echo signal from tissues behind bones, not only is the original intensity reduced by more than 50%, but the echo signal is again strongly reflected back by the bones. From this second example, we can conclude that all organs behind bones are essentially invisible to US imaging. Bones form sound walls and acoustic shadows; US imaging can only be applied to soft tissues, organs, and muscles that are in front of the bones. The same considerations apply to organs that are filled with air or gas, such as the lungs. Total reflection occurs at the air/tissue interface. As a result, US cannot penetrate the lungs, and the tissue behind the lungs remains invisible.

In sonography only the back-reflected intensity is detected. Back-reflection occurs only at inter- !
faces with sufficient impedance contrast between different organs.

1.3.2 Scattering and absorption

A propagating pressure wave traveling in a medium of impedance Z will be attenuated over distance x and time t. The pressure amplitude can be expressed as follows:

$$p(x,t) = p_0 \exp(-\mu x) \cos(kx - \omega t). \qquad (1.8)$$

Here μ is the *attenuation coefficient* with unit $[\mu] = \mathrm{m}^{-1}$; $k = 2\pi/\lambda$ is the wavenumber of the sound wave, unit $[k] = \mathrm{m}^{-1}$; all other symbols have their usual meaning. Attenuation of the pressure amplitude has two main contributions: absorption and scattering:

$$\mu = \mu_{abs} + \mu_{scatt}. \qquad (1.9)$$

Attenuation due to absorption is material specific and can be estimated. Damping due to scattering is interface specific and more difficult to estimate, although in most cases dominating.

Attenuation due to scattering can be visualized as follows. Sound propagating in the body usually hits an interface at an angle. Furthermore, interfaces of organs and bones are generally rough on the scale of US sound wavelengths, which are typically below 1 mm. Then at rough interfaces, the incident sound is partially transmitted from one medium in another, reflected in a different direction, and scattered in a wide angular range, as indicated in Fig. 1.6. For sonographic imaging, only the back-reflected and/or backscattered intensity (echo) is of interest and detected by the receiver.

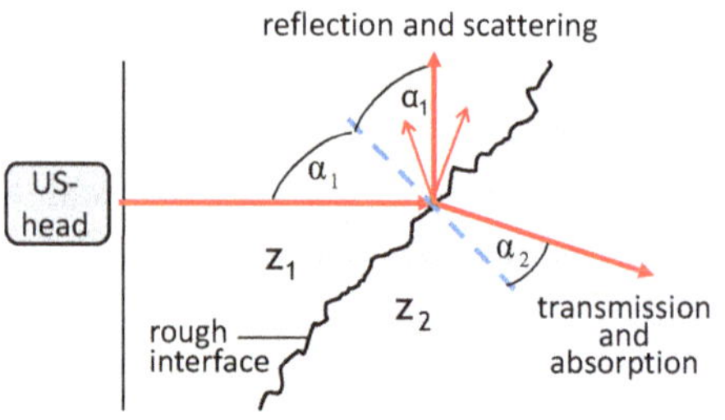

reflection and scattering

US-head

α_1

α_1

Z_1

Z_2

α_2

rough interface

transmission and absorption

Fig. 1.6: Reflection, transmission, and scattering at a rough interface between two materials characterized by different acoustic impedances Z_1 and Z_2.

Attenuation via absorption has two contributions, viscous damping and thermal conduction: $\mu_{abs} = \mu_{vis} + \mu_{therm}$. The "classical" result for both damping constants in gases and liquids is according to Kirchhoff[5] [7]:

$$\mu_{abs} = \frac{\omega^2}{2\rho v_s^3}\left(\frac{4}{3}\eta_s + \frac{(\gamma-1)\kappa}{C_V}\right), \tag{1.10}$$

where η_s is the viscosity of the media in which the sound wave propagates, κ is the heat conduction coefficient, C_V is the tissue-specific heat at constant volume (3.24 J/g K [8]), and γ is the adiabatic coefficient defined by the ratio of the specific heats at constant pressure and constant volume: $\gamma = C_p/C_V$. As the viscous and thermal damping both scale with ω^2, damping is more severe at high frequencies. However, thermal damping can be neglected because γ is essentially 1 for soft tissue. For a 1 MHz sound wave, the prefactor in eq. (1.10) has a value of 2 s/kg. Assuming a viscosity $\eta_s = 10$ Pa·s [9], we calculate a penetration depth of about 50 cm. For back-reflection, the attenuation path has to be taken twice, which reduces the penetration depth to 25 cm. At 10 MHz, the penetration depth is only 0.25 cm.

A simple thumb rule for damping of sound waves in biological tissue is as follows: 1 dB of damping occurs after traveling a distance z of 1 cm at a frequency f of 1 MHz, or

$$\frac{\mu}{dB} = \frac{z}{cm}\frac{f}{MHz}. \tag{1.11}$$

Since μ expressed in dB does not have a unit, z and f need to be divided by their units in order to get a simple number. The ratio μ/zf is roughly 1 for most organs and the brain, but it is about 2 for muscles and 0.5 for fatty tissue. An example of how to use eq. (1.5) is given in Exercise E1.2.

---

5 Gustav Robert Kirchhoff (1824–1887), German physicist.

Various intensity losses attenuate the echo signal: scattering at rough interfaces in off-specular directions, transmission, and absorption. Only a small fraction of intensity is reflected back and reaches the detector. US propagation is strongly damped due to the viscosity of the tissue.

1.4 Beam properties, pulsing, and focusing

This section examines additional conditions for pulse formation, pulse propagation, and focusing of sound waves from extended sources. These considerations are important in recognizing the limits of sonography on the one hand and optimizing the adjustment screws for best results on the other hand.

1.4.1 Pulse quality

Transducers for US imaging are used in pulse mode. Once excited, the duration of the US pulse depends on the damping constant. Resonators with a high-quality factor Q exhibit a lower damping. Therefore, they emit a longer pulse than resonators with low Q (for further information on the quality factor, see Infobox 12.1 in Chapter 12 of Volume 1). The quality factor Q is defined as the ratio of the resonance frequency f_0 to the bandwidth (full width at half maximum (FWHM)) of the frequency distribution Δf, which usually has a Gaussian form (Fig. 1.7):

$$Q = \frac{f}{\Delta f}. \tag{1.12}$$

Typically, the wave train length in a pulse comprises about three wavelengths, corresponding to a pulse duration (PD) of $3T = 3/f = 3$ μs at 1 MHz. The Fourier[6] transform[7], i.e., the freqency distribution of such a short pulse wave train is rather broad, typically on the order of 0.8 f_0, yielding a low Q factor of 1.25.

The PRF or probing frequency for US imaging typically is 1 kHz. Thus, there is almost a millisecond interval between any pulse of a few microseconds duration until the next pulse is fired. This rather long waiting time is utilized to switch the transducer from pulsing mode into receiver mode of echo signals after the time of echo TE. A duplexer performs the switching (see Fig. 1.4). Duplexers, not to be confused with duplex detection mode discussed in Section 1.7.3, are electronic devices that allow bidirectional communication via one joined signal path. Examples are an antenna for transmitter and receiver of microwaves in radio communication, or US

---

6 Joseph Fourier (1768–1830), French mathematician and physicist.
7 A Fourier transform is a mathematical procedure that decomposes waves into a distribution of frequencies, like a prism decomposes white light into a spectrum of monochromatic colors. More information can be found in Ref. [1.10].

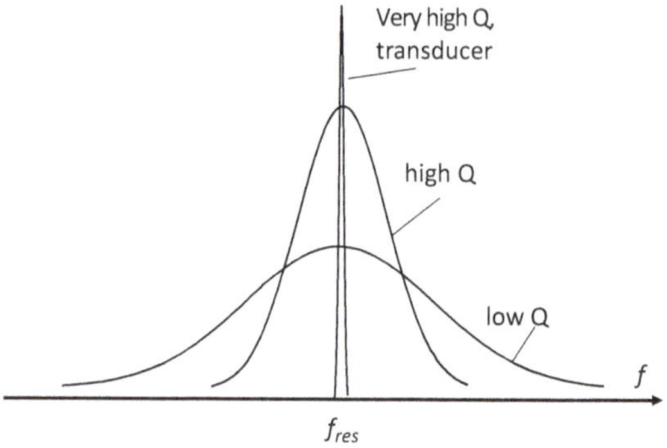

Fig. 1.7: Fourier transform of a US pulse for different quality factors of the resonator. The sharp line in the center corresponds to the intrinsic resonance frequency of the transducer.

head for transmitting and receiving of US waves. Duplexer may respond either to frequency, time, or amplitude. In the case of a US duplexer, switching is achieved by pulse amplitude, using nonlinear electronic devices (pn-junctions): high pulse amplitude for transmission and low pulse amplitude for detection.

1.4.2 Time gain compensation

The transducer in the receiver mode has to deal with much lower intensities than in the transmission mode. Furthermore, the echo signals decay over time and need to be corrected for attenuation and shaped to be useful for further signal processing. The time scales are shown in Fig. 1.8 for a US frequency of 1 MHz and a pulse repeat time (PRT) of 1000 µs, corresponding to a PRF of 1 kHz. As soon as the first echo pulse arrives after time TE, *time gain compensation* (TGC) is turned on, as indicated in the middle panel of Fig. 1.8. TGC amplifies successive echoes from deeper interfaces on a logarithmic scale. This is to assure that echoes from similar interfaces have equal signal amplitude on the screen even if they arrive later. TGC is linear on the log dB versus timescale in the simplest case but can be adapted to other special conditions. Echoes with lower intensity due to attenuation from interfaces with less contrast appear weaker even after amplification, illustrated by the third echo signal.

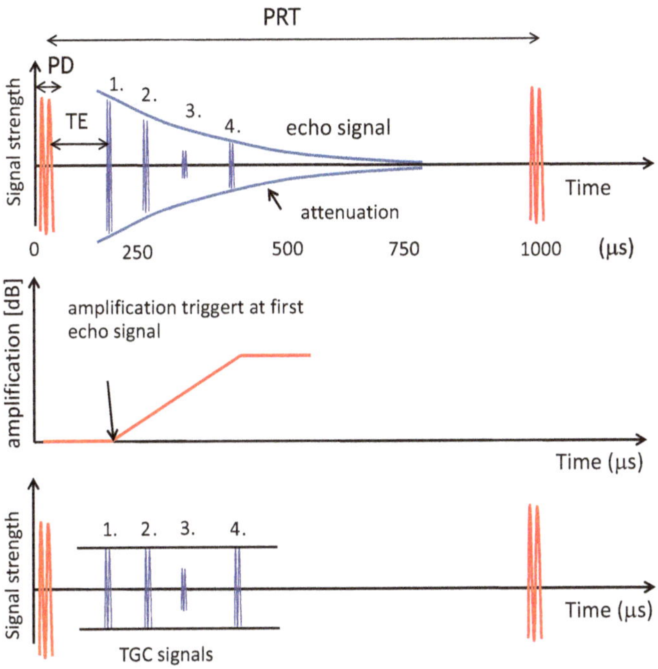

Fig. 1.8: Time sequence for pulsing and detecting of US. Top panel: US pulses are emitted in time increments of the pulse repeat time (PRT). The pulse duration (PD) is short compared to the echo time TE and the PRT. After receiving the first echo signal, second and third echo signals are detected with exponentially decaying amplitude. Middle panel: When the first echo signal has arrived, time gain compensation (TGC) is triggered to equalize the amplitude of subsequent echo signals. Bottom panel: The result is a roughly equal amplitude for all echo signals, as long as the original signal is still above the noise level. Typical values are for PRF = 1 kHz, i.e., PRT = 1000 μs; PD = 3 μs; TE = 100 μs.

> The common mode of operation for US imaging is a pulse mode. The bandwidth of the pulses is typically 80% of the resonance frequency. A sequence of echo signals is recorded between the repeat time of pulses. Attenuation effects of echo signals are electronically compensated.

1.4.3 Near field, far field, and focusing

Next, we consider the appropriate size, shape, and diameter of transducers. If the transducer head had the size in the order of emitting wavelength, the wave front would form a spherical surface. This is contrary to what is needed for US imaging. A planar wave front is preferred for US imaging with potential focusing options. This condition is met if the size of the emitter is much larger than the wavelength of the sound wave, at least 10 times as large. For a 1 MHz source, this requirement implies a head size of at least 15 mm diameter.

Figure 1.9 shows the near field and the far field of an extended sound wave source. In the near field, the main propagation direction is in the forward direction. All rays spreading out to the side are extinct by destructive interference. The near field extends up to a distance of [11]

$$N = \frac{D^2}{4\lambda_{\text{body}}} = \frac{D^2 f_0}{4v_{\text{body}}},$$ (1.13)

where D is the source size and v_{body} is the sound velocity in the body tissue. The wave front is nearly a plane wave within this distance, known as the Fresnel region, since strong interference effects eliminate all other spherical waves. In the far field, also known as Fraunhofer[8] region, interference effects are lost and the beam starts to diverge at larger distances. At the edges of the sound field, there are small areas of low intensity outside of the main beam, corresponding to the first side maxima of a diffraction pattern, called side lobes. These intensities may cause artifacts in US imaging. To give in numbers: for a 1 MHz source, a source size of 15 mm, the near field in soft tissue extends up to about 36 mm. With higher frequencies, the near field can be increased. The near-field region is the favorable distance for imaging organs, as we shall see later. In the far-field region, the angle of divergence is $\alpha = 2\lambda_{\text{body}}/D$. This shows that low frequencies (large wavelengths) diverge more than high frequencies.

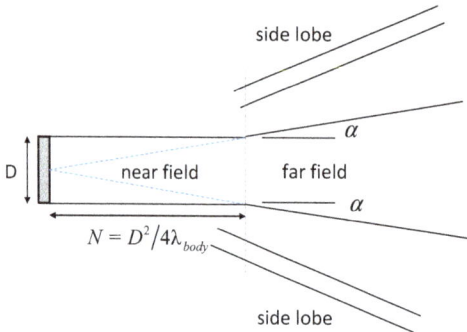

Fig. 1.9: Near field and far field of a sound wave emitter under the assumption that the source size $D \gg \lambda$. N is the near-field extension and α is the angle of divergence in the far field.

Obviously, it is favorable to focus the US field in the region of interest (ROI), i.e., at specific organs. Focusing enhances the intensity of the echo signal and increases

---

8 Joseph von Fraunhofer (1787–1826), German physicist and optical lens manufacturer.

the lateral resolution, which otherwise is given by the source size in the near field. Focusing can be achieved by several different methods; three of them are sketched in Fig. 1.10:

a. by a curved PZT transducer
b. by a plastic lens in front of the PZT crystal
c. by the signal arrival time

The third option, time focusing, can be realized if the transducer is subdivided into several smaller elements and the elements emit in time sequence: Element 1 emits first, followed by elements 2, 3, etc., at time increments corresponding to the path differences. Time focusing requires precise electronic timing. Changing the time increments between the elements changes the focal length. Therefore, time focusing is very flexible.

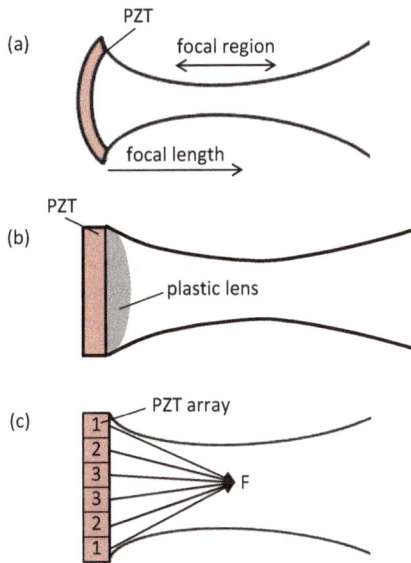

Fig. 1.10: Three different methods for focusing a US wave field: (a) focusing by a curved transducer; (b) focusing by a convex plastic lens in front of a flat transducer; and (c) electronically controlled time focusing.

The field of interest is positioned in the focus within the near-field area. Focusing of sound waves is achieved either by transducer curvature, lens application, or electronic time control.

1.4.4 Physical parameters

We summarize here the typical physical parameters for US imaging of body parts that we have discussed so far:

- *Sound velocity* (phase velocity) v_s in the soft matter is similar to water and is about $v_s \approx 1540$ m/s or 1.54 mm/μs. Throughout this text, we use this average velocity for various estimates. More precise tissue-specific values are listed in Tab. 1.2.
- *Echo signals* arrive after the time lap $\Delta t = 2L/v_s$, where L is the depth of a reflecting interface; 1 μs corresponds to 0.75 mm depth.
- *Wavelengths* stretch from $\lambda = v_s/f = 1500$ μm (1 MHz) to 40 μm (38.5 MHz).
- *Particle velocity* amplitude $u_0 = p_0/Z < 0.6$ m/s for $p_0 = 1$ MPa and $Z(\text{water}) = 1.5 \times 10^6$ Ns/m³.
- *Wave (displacement) amplitude* $\xi_0 = u_0/\omega = p_0/(Z\omega) = 30$ nm, for $f = 3$ MHz and $u_0 = 0.6$ m/s.
- *Pressure amplitude* $p_0 \leq 1$ MPa.

1.4.5 Safety issues

The safe operation of US imaging is an important issue. The (negative) peak pressure should not exceed 1 MPa = 10 bar to avoid any explosive rapture of internal cavities. At a pressure of 1 MPa, the particle velocity is accordingly 0.6 m/s and the intensity of the sound wave becomes

$$\langle I \rangle = \frac{1}{2}p_0 u_0 = \frac{1}{2}1 \text{ MPa} \cdot 0.6 \text{ m/s} = 0.3 \text{ MPa m/s} = 3 \times 10^5 \text{ W/m}^2 = 30 \text{ W/cm}^2$$

The calculated intensity derived from the quoted parameters above is by far too high for any safe medical application. The maximum power administered during sonographic imaging should not exceed 100 mW/cm². For estimating this limit, the following expression is used, known as *mechanical index (MI)* [12]:

$$\text{MI} = \frac{p_0^-/1 \text{ MPa}}{\sqrt{f/1 \text{ MHz}}}. \tag{1.14}$$

Here p_0^- is the negative peak pressure and f is the frequency of sound waves. The MI attempts to estimate the biomechanical effects of US to avoid formation and disruption of cavities (cavitation), for which only the negative or expansive pressure is responsible. The MI is found on most US display screens, along with other parameters. It is a dimensionless number and should be limited to values below 1.9 for safe operation of US. The MI suggests that increasing frequencies can compensate for higher pressures. But this does not guarantee safe operation. For instance, a pressure amplitude of 1 MPa and a frequency of 10 MHz yield an MI of 0.3. However, for this "safe," value, the sound intensity is, as already estimated, 30 W/cm^2, far more than what is considered safe. Therefore, one should be cautious when applying high intensities. Keeping the pressure level below 0.1 MPa is a safer bet.

For most of this chapter, we continue to consider only longitudinal waves. Although they exist in soft tissues and in bones, transverse waves are not helpful for imaging because of their high attenuation. However, transverse waves are used for special applications called elastography, to be presented later.

1.5 Medical imaging

As mentioned in the introduction, medical sonography is an imaging modality that takes static images of organs and tissues, real-time images of periodically moving organs (heart and lungs), and velocity measurements of blood flow. A quick overview of the different operational modes is listed here and presented in the subsequent paragraphs:

- **A-mode** is a single line scan through the body, recording the amplitudes of returning echoes from interfaces between tissues having different impedances as a function of time.
- **B-mode** scans provide two-dimensional (2D) images, representing changes in acoustic impedance of the tissue within one section.
- **C-mode** scans produce 2D images formed by taking a sequence of slices in B-mode, repeated in a direction normal to B-mode images at a constant depth.
- **M-mode** or motion mode allows imaging of moving organs by A- or B-scans but with higher PRFs, adequate for recording videos.
- **Doppler mode** makes use of the frequency shift by moving reflectors, allowing visualization of blood flow.
- **Pulse inversion mode** uses two successive pulses with opposite signs, and the difference of which displays body parts with nonlinear compressibilities.

1.5.1 A-mode scan

During an A-scan (= echo ranging in amplitude mode), the US head is held stationary at a certain point. US pulses are transmitted into the body with the help of a gel between the US head and skin in order to avoid any reflection at this first interface. Therefore, an echo signal from position "0" in Fig. 1.11 is not to be expected. As soon as the sound wave penetrates further into the body and hits an organ at depth L (position 1) with impedance Z_2, in contrast to the surrounding tissue with Z_1, part of the sound wave is reflected and the echo signal arrives after a transit time

$$\Delta t = \frac{2L}{\langle v \rangle_{body}} \tag{1.15}$$

at the detector.[9] On the backside of the organ (position 2), the sound wave is again reflected, but the echo signal is weaker than from the top interface because of attenuation within the organ. Finally, the echo signal arrives from the backside of the body at position 3. These three signals can be seen as "blips" on the screen of an oscilloscope, where the horizontal axis is the time axis (equivalent to the depth of the reflected sound wave), and the vertical axis is the voltage output after signal amplification, proportional to the echo signal amplitude. Because of this display mode, the A-scan is also known as *amplitude mode*.

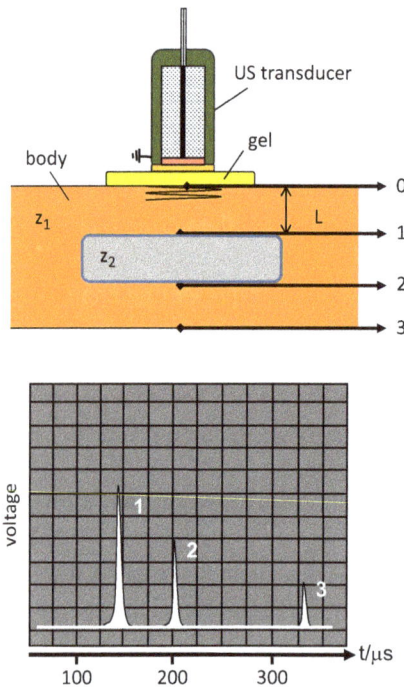

Fig. 1.11: Ultrasound transducer as source of sound wave pulses and sensor of echo signals from the interfaces at positions 1, 2, and 3. In the lower panel, the echo signal is displayed on an oscilloscope screen.

In A-scans, no further signal processing is performed. These unprocessed signals help locate a particular organ and optimize the echo signal concerning pulse length, frequency, appropriate focal length, etc. Obviously, the A-scan has no lateral resolution, but it has depth (axial) resolution coupled to the time resolution. The time resolution depends on the FWHM of the pulse length or the bandwidth Δf

---

9 Although the velocity varies between different body parts, for signal processing an average and constant sound velocity in the body of $\langle v \rangle_{body} = 1540\,m/s$ is assumed.

of the pulse. For standard pulse signals, the bandwidth is usually about 0.8 f_0, where f_0 is the resonance frequency. From this, an axial resolution ΔL can be estimated according to

$$\Delta L = \frac{\langle v \rangle_{body}}{2\Delta f} = \frac{\langle v \rangle_{body}}{2 \times 0.8 f_0}. \tag{1.16}$$

Signals that are received within a period of 1 µs (FWHM of the bandwidth) come from interfaces that are 750 µm apart, which corresponds to the specified (axial) depth resolution. A higher depth resolution can be achieved with shorter pulses. However, since shorter pulses require higher frequencies, the attenuation increases. Therefore, a compromise between resolution and signal strength has to be found for each application.

1.5.2 B-mode scan

The most frequently used US imaging mode is the B-mode or brightness mode scan. The B-scan can be regarded as a sequence of A-scans in time and space that together represents a slice through the body. Three basic types of B-scanners are known: sector scanner, array scanner, and phase array scanner.

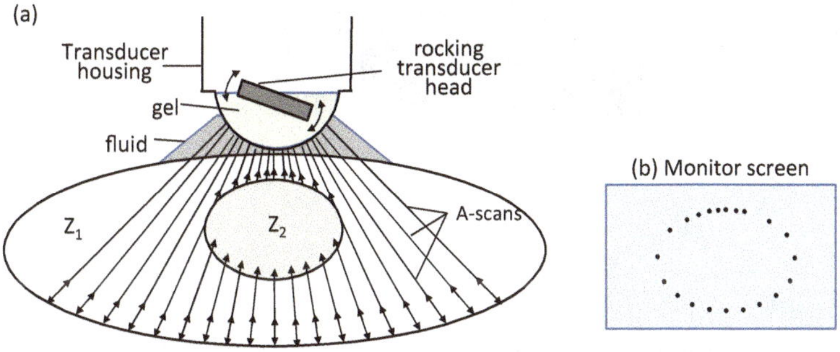

Fig. 1.12: Panel (a): Sector scan of an organ, consisting of a sequence of A-scans. The transducer head is rocked in an impedance matching fluid, and sound waves are emitted perpendicular to the transducer surface. Echo signals from the contours of an organ at one particular slice are shown in panel (b).

1.5.2.1 Sector scanner

The basic working principle of a sector array B-scanner is shown in Fig. 1.12. In this example, a transducer head is rocked back and forth in an impedance-matching fluid, emitting pulses perpendicular to its surface at different angles. The backscattered and back-reflected waves from interfaces between the organ (Z_2) and the surrounding tissue (Z_1) are detected by the transducer and are stitched together, yielding the contours of the imaged organ on a monitor screen as indicated in panel (b) by the dotted line. The grayscale of each dot represents the signal strength of the reflected sound wave at the interfaces. Back-reflection from the far end of the body is usually so weak by absorption and scattering that it can be safely neglected. Here and in all other B-scans, TGC is turned on for amplifying the signal. With TGC on, variations of echo signal amplitudes are ideally only due to variations of acoustic impedance and not due to depth. However, applying TGC also increases the noise, and the signal-to-noise ratio (SNR) may worsen. Nevertheless, TGC provides a more balanced image. A typical example of a B-scan of the heart is shown in Fig. 1.13.

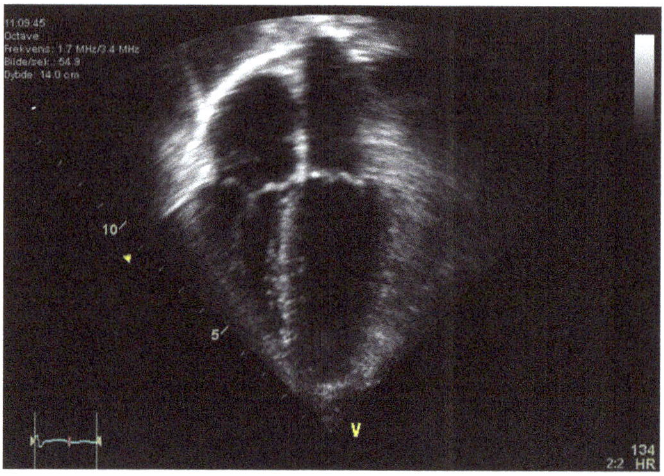

Fig. 1.13: Typical US image in B-mode of the heart chambers (reproduced from https://openi.nlm.nih.gov/).

1.5.2.2 Array scanner

Instead of sweeping the transducer head mechanically, other emitter/receiver designs and electronic sweep control arrangements are available in modern US heads. One such arrangement is sketched in Fig. 1.14. An extended transducer is subdivided into an array of narrow stripes, each one about the width of a wavelength. A single stripe would not produce a near field but a diverging far field, opposite to the intention. The single elements are therefore activated in groups. Time focusing is applied to define a focal length in point P within a group. For this, elements 3 and 5 in the graph of Fig. 1.14 are activated first, followed by element 4. After a short time delay, groups 4–6, then 5–7, etc., are successively activated until a sweep is completed. In reality, the number of elements in the array is much larger than indicated in the schematics.

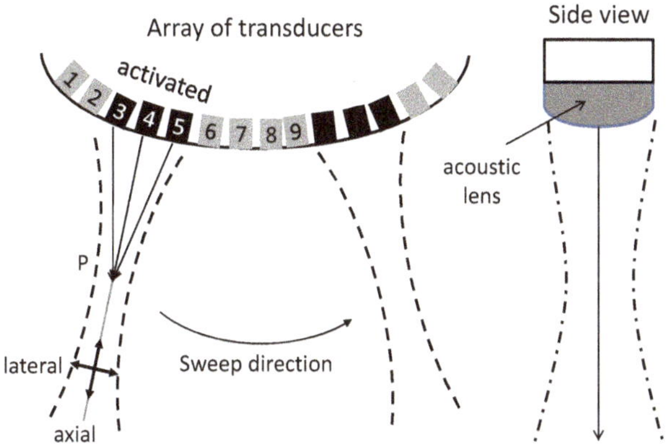

Fig. 1.14: Left panel: Array scanner for an electronic sweep of the US across an organ. Each transducer element is much smaller than in the section scanner. Right panel: Side view also features an acoustic lens for defining the width of the imaged slice.

As mentioned earlier, the focal length in the sweep direction (lateral or azimuthal plane) is adjusted by time focusing and can be electronically controlled. A larger time delay will result in a shorter distance of the focal point P. Focusing in the perpendicular direction is achieved by an acoustic lens (right panel in Fig. 1.14) defining the thickness of the slice imaged.

1.5.2.3 Phase array scanner

Similar to the array scanner, the phase array scanner consists of many transducers arranged in a linear array (Fig. 1.15). But there are two important modifications compared to the previous scanners. Each individual element is smaller than in the array scanner and there are fewer of them. All elements are excited almost simultaneously, but with a small time delay between neighboring elements. The time delay determines the phase difference and the extent of the destructive and constructive interference between neighboring elements. The phase difference ultimately controls the sweep direction and the time focus point. The sweep covers a sector field similar to the sector scanner, but without moving parts.

Different types of scanners used for US imaging are shown in Fig. 1.16. The field of view is the portion of organs or tissues that are intersected by the scanned slice. Depending on the probe used, the shape of this field can be a sector, a rectangle, a trapezoid, or a convex field.

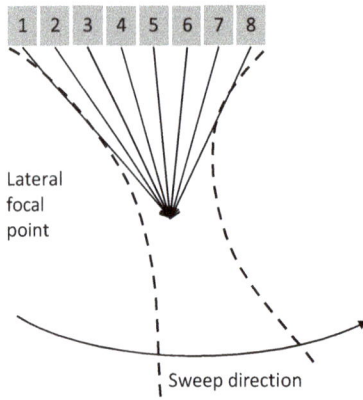

Fig. 1.15: Phase array scanners consist of a linear array of small transducer elements activated simultaneously with minute time delays between neighboring elements defining sweep direction and focal length.

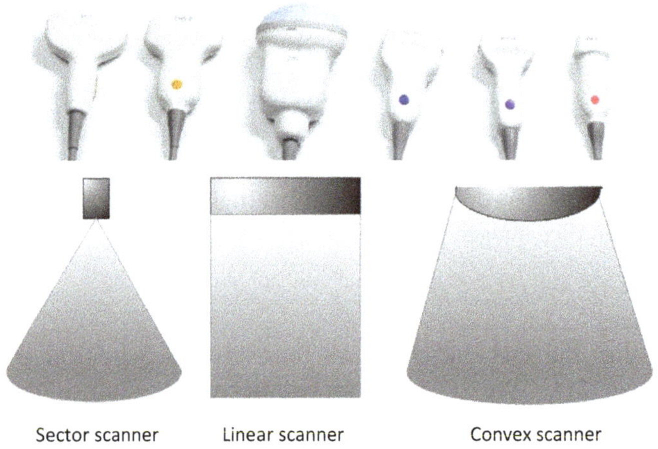

Fig. 1.16: Upper panel: Various types of scanner heads used for US imaging. Lower panel: Depending on the scanner type, the imaged slice may have a different shape.

1.5.3 C-scan

C-mode scans are collections of B-mode scans for reconstructing 2D images of inner body organs at a constant depth. All B-mode scans discussed so far take one slice of the body at a time. In C-mode, a sequence of B-mode scans is acquired in a direction normal to B-mode slices (Fig. 1.17). Then the pictures are stitched together, and a plane of constant depth is selected, indicated by the red-bordered area in Fig. 1.17. Three different scan types can be distinguished as sketched in Fig. 1.17: linear, sweep, or rotational. The rotational scan requires a transducer at the end of a stick

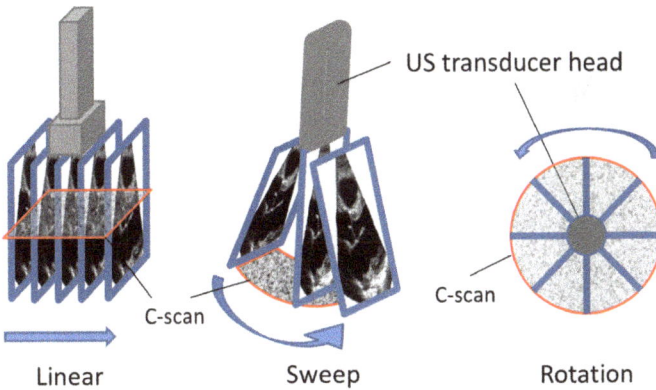

US transducer head

C-scan

C-scan

Linear   Sweep   Rotation

Fig. 1.17: Perpendicular cuts through B-mode scans at constant depth form two-dimensional images. The red-bordered areas correspond to C-mode scans.

that can be inserted in an opening, such as for rectal or vaginal screenings. In all cases, C-scans add information on slices perpendicular to the scan direction.

If adequately stitched together, either B-mode or C-mode scans have the potential of generating tomographic images. However, in contrast to MRI and CT, the coordinate system is not fixed in C-mode scans. The manually held US head has to touch the skin and follow the contours of the body surface. This makes it much more challenging to reconstruct 3D images. In the past, practitioners had to mentally integrate sequences of pictures to get a feeling of the size and shape of an organ in the lateral direction. In recent years, different methods have been proposed and are being used, all relying on fast data acquisition and computer processing. The simplest version is a handheld scan of predefined mode (linear, sweep, or rotation) and preselected scan speed. The resulting image will give a 3D impression, although the geometry may not be precise but sufficient for a diagnosis. Sweep scans are easier to perform manually than linear scans since only a rotation at a constant position is required. However, the penetration depth varies with rotation angle because of the changing distance between transducer head and organ. Manual scans can be improved by tracking the US head with a receiver attached to it in a gradient magnetic field surrounding the patient, similar to techniques used for MRI imaging (see Chapter 2). Furthermore, linear scans can be acquired by mechanically scanning the transducer head using stepper motors, while the contour of the body surface is monitored capacitively and fed into a dc-motor that adjusts the elevation of the transducer. An overview on methods and techniques for 3D US imaging is provided in [13]. Impressive high-quality 3D topography images can nowadays be taken, such as the one shown in Fig. 1.18 of a fetal face.

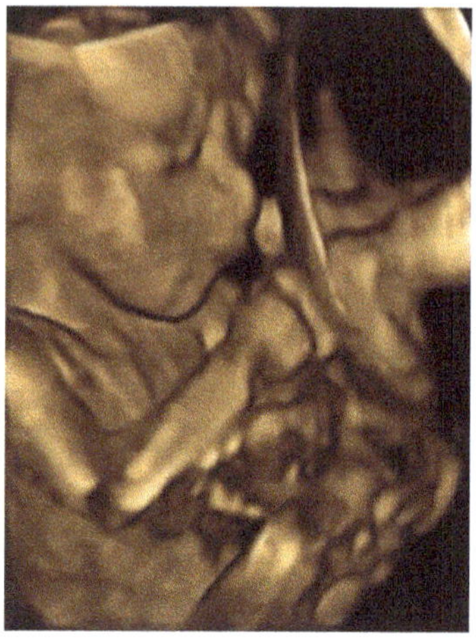

Fig. 1.18: Surface-rendered 3D image of a twin fetus (head of one and foot of the other) taken by a 3D echo scan (private communication).

1.5.4 M-mode

M-mode or motion mode provides information about variations in signal amplitude versus time due to object motion. The transducer head acquires data in an A-mode scan at a fixed position. The data are displayed as a series of dots or pixels with brightness levels representing the intensity of the echoes. The signal intensity variation is displayed on a horizontal time axis in repeating A-lines at the fixed position. At the same time, the vertical axis indicates the distance of the echo from the transducer. An example of an M-mode image from the beating heart is shown in Fig. 1.19.

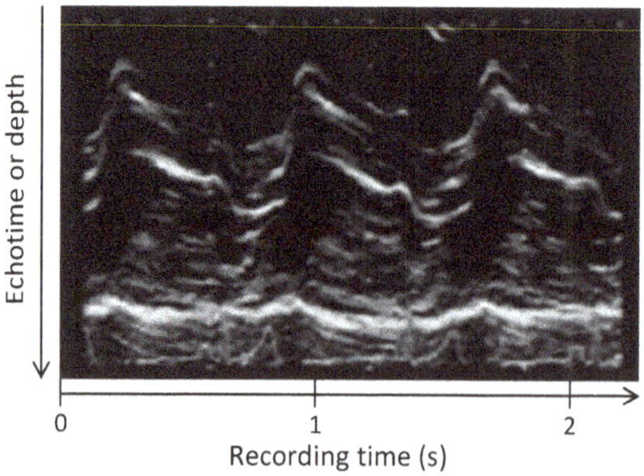

Fig. 1.19: M-mode display of mitral valve leaflet of a beating heart (reproduced from https://openi.nlm.nih.gov/).

Similar imaging concepts are utilized to take videos of a single A-line and a full sector scan taken in B-mode. Current technologies enable to record about 50 frames per second (FPS) with sufficient depth information and resolution. This is an adequate temporal resolution for visualization of regular heart action with a beat frequency of 70 per minute. Three-dimensional image echocardiogram of a beating heart can be viewed on the webpage [14], where 3D refers to a 2D spatial image plus time resolution. Even 4D images (3D spatial plus time dimension) are now available and mainly used for perinatal diagnosis [15].

1.5.5 Shear wave sonography

Standard US sonography uses the echo signal from longitudinal compression waves. Although shear waves propagate in soft tissues and bones, their attenuation is high and penetration is much more limited compared to compression waves. Nevertheless, shear wave sonography, or elastography (SWE), was developed over the last 20 years and is now available for studying the shear elastic properties of organs and tissues [16, 17].

Compressional and shear waves are compared in Fig. 1.20. Compressional waves are longitudinal waves: the propagation velocity $v_{\text{sound}}^{\text{comp}}$, the particle velocity u_0, the pressure amplitude p_0, and the particle displacement amplitude ξ_0 all point in the same propagation direction. In contrast, shear waves are transverse waves, meaning that the amplitudes of particle displacement, particle velocity, and pressure are perpendicular to the sound wave's propagation direction. Compressional waves change the volume elements periodically via compression and expansion. Shear waves do not change the volume but change the shape of volume elements. Therefore, shear waves probe the shear elastic modulus G. The velocity of shear waves is

$$v_{\text{sound}}^{\text{shear}} = \sqrt{G/\rho}. \tag{1.17}$$

Knowing the density derived from compressional sonography, the shear elastic modulus G can be determined by measuring the shear velocity. On average, in tissue $v_{\text{sound}}^{\text{shear}}$ is by a factor of 100 lower than compressional velocities, and the corresponding shear elastic constant G is lower than the bulk modulus B by a factor of 10^{-4} to 10^{-5}. Furthermore, transverse waves are polarized: they feature two main polarization directions. If the wave propagates along the z-direction, the displacement amplitude may be oriented in the x- or y-directions. Compressional waves are excited by placing a piezoelectrical crystal together with the electrodes face on, as indicated in Fig. 1.20(a). For excitation of shear waves, the electrodes are placed on the sides of the piezoelectric disk, as sketched in panel (b).

SWE analyzes and judges tissues that display anisotropic properties like the striated skeletal muscles (Chapter 2 of Volume 1). Another area of increasing activity is the elastographic analysis of the liver. The liver lies close to the skin and is located within the shear wave penetration depth range. Liver fibrosis is characterized by increasing porosity and change of stiffness, which is detectable by SWE. An overview is given in [18]. SWE is optionally provided and integrated in B-mode scanning heads in commercial US equipment.

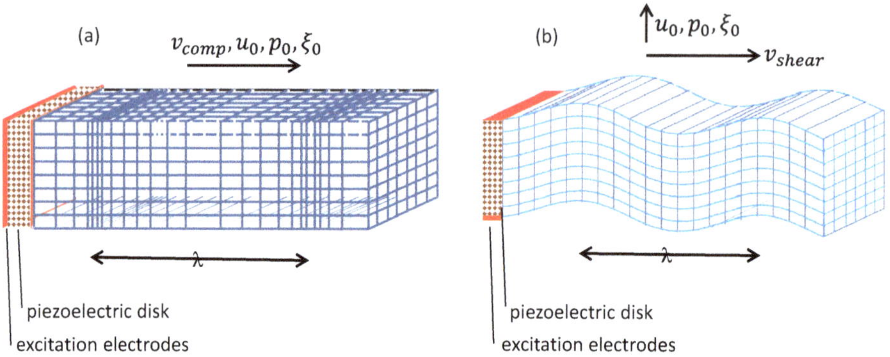

Fig. 1.20: Comparison of compressional wave propagation (a) and shear wave (b). Note that the compressional wave is a longitudinal wave, whereas the shear wave is a transverse wave with polarization perpendicular to the propagation direction.

Four scan modes (A, B, C, M) deliver gray scale images of internal organs with adjustable depth and resolution. A-, B-, and C-scans are static, and M-scans yield dynamic information. SWE probes the elastic properties of tissues.

1.6 Scan characteristics

After introducing the general concepts of A-, B-, C-, and M-scans, a few additional considerations about focusing, imaging, and resolution are discussed that apply to all scan modes.

1.6.1 Dynamic focusing

Focusing improves the spatial resolution and image quality of the objects within the focal depth but also degrades the image quality. To overcome this conflict, a new multizone focusing technique was introduced [19]. A focus per pulse emission is applied along each sweep, which is fixed but can be adapted to the depth of view (DOV). These pulses are emitted with the usual pulse repetition frequency (PRF). Dynamic focusing is then used to receive the echo signal, called *dynamic receive focusing*. This is achieved through continuous phase control of the receiving elements within a phase array scanner. Starting with the near-field area and continuing to deeper fields, the number of detector elements increases each time to receive echoes from the corresponding focal zones. This is schematically illustrated in Fig. 1.21. For receiving signals from shallow regions, three detector elements (−1, 0, 1) in an array of 15 are activated; for receiving signals from deeper regions, further elements from −3 to 3 are added, etc.

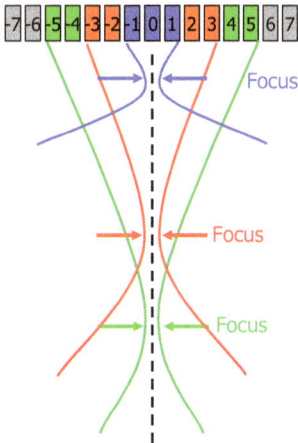

Fig. 1.21: Dynamic focusing for depth-dependent detection of echo signals (adapted with permission from [19]).

1.6.2 Line density

Whatever scan technique is used, the tissue of patients is always sampled by a number of line scans for completing one slice (section). The spatial resolution in the sweep direction is intrinsically limited to about 1 mm. Therefore, for a 10 cm wide slice, 100 lines are necessary for acquiring a high-quality and high-resolution image.

1.6.3 Scan frequency

Moving tissues require a high PRF to follow the movement. The PRF depends on the number of lines per frame and the frame rate:

$$PRF = lines\,per\,frame \times frame\,rate.$$

Typically, 30–50 frames/s are taken with 100 lines/frame, which yields a PRF = 3 kHz.

1.6.4 Depth of view

To image tissue at a particular depth, the sound pulse needs to have sufficient time to travel forth and back before the next pulse is emitted. This limits the PRF. When increasing the DOV, the PRF has to be reduced correspondingly:

$$\text{DOV} = \frac{v_{\text{sound}}}{2 \times \text{PRF}}. \tag{1.18}$$

From this consideration, we conclude that it is not possible to take high-resolution images of moving tissues at large depth. A compromise is necessary either with respect to resolution or depth. From the above relations follow the criteria:

$$\text{DOV} \times \text{PRF} \leq \frac{1}{2} v_{\text{sound}} = \text{constant}. \tag{1.19}$$

Any choice of parameters has to obey this boundary condition. These considerations are independent of the actual operating frequency used. The choice of US frequency has, in addition, a big effect on the penetration depth of US waves, focus, and resolution.

1.6.5 Penetration depth

The penetration depth of US waves in the body is a complex function of frequency. It depends on viscous damping and heat conduction, scattering at small particles and rough interfaces, and already back-reflected intensity at interfaces. When US waves hit bones, water, or gas-filled tissue, penetration into tissue on the backside drops dramatically. Because of this complexity, it is difficult to give a general expression covering all eventualities. However, the half-intensity depth $L_{1/2}$, i.e., the depth at which the intensity has dropped to half of its original value, as a function of frequency f can be estimated with the help of an empirical expression:

$$L_{1/2} = \frac{C}{f}. \tag{1.20}$$

The constant is $C \cong 50$ cm MHz. Some examples are listed in Tab. 1.3. One method to circumvent the attenuation problem is by using (higher) *harmonic imaging:* the beam is transmitted at a fixed fundamental frequency f_0, whereas the received signal is analyzed at higher harmonics of $2f_0$ or $3f_0$. This increases the SNR of reflected signals, particularly from the deepest parts of images, without compromising resolution.

1.6.6 Spatial resolution

Spatial resolution is the ability to observe two objects A and B, separated by a certain distance δ in space, as two distinct images A' and B' on a screen. If objects A and B are too close together, the echo pulses from A and B will overlap and will not be distinguishable. In US imaging, the axial or depth resolution δ_{axial} is distinctly different from the lateral resolution δ_{lat} for intrinsic physical reasons. Therefore, we need to separate the discussion of both.

1.6.7 Axial resolution

Axial resolution is a question of pulse length. The higher the frequency of sound waves, the shorter the pulse length that can be chosen and the better the axial resolution that can be achieved. The axial resolution is the one already quoted for A-scans in eq. (1.16):

$$\delta_{axial} = \langle v \rangle_{body}/2\Delta f, \tag{1.21}$$

where Δf is the pulse bandwidth. Examples of axial resolutions for different frequencies are listed in Tab. 1.3. They range from 150 to 800 µm.

1.6.8 Lateral resolution

The lateral resolution δ_{lat} is given by the width of the focal point. The width of the focal point follows from the axial depth of the focal point L, the wavelength λ, and the width of the active aperture length D:

$$\delta_{lat} = \frac{L\lambda}{D} = \frac{\lambda}{2\alpha} = \frac{1}{2\alpha}\frac{v}{f}. \tag{1.22}$$

Tab. 1.3: Penetration depth, lateral resolution, and axial resolution for different US frequencies or wavelengths in body tissue with an average sound velocity of 1540 m/s (from PD Dr. Marc Kachelrieß, Erlangen).

f (MHz)	λ (nm)	Depth (cm)	Lateral resolution (mm)	Axial resolution (mm)
2.0	0.78	25	3.0	0.80
3.5	0.44	14	1.7	0.50
5.0	0.31	10	1.2	0.35
7.5	0.21	6.7	0.8	0.25
10	0.16	5.0	0.6	0.20
15	0.10	3.3	0.4	0.15

Axial depth L and angle α are defined in Fig. 1.10. The last equation implies that the lateral resolution depends linearly on the US wavelength, which is confirmed by the values listed in Tab. 1.3. The lateral resolution is on the average about 1–2 mm.

In general, the following conclusions can be drawn, which are tradeoffs between penetration depth and spatial resolution:

Low-frequency US waves offer a high penetration depth at the expense of low axial and lateral resolution; in contrast, *high-frequency* US waves have a lower penetration depth but provide higher axial and lateral resolution.

1.6.9 Artifacts

There are numerous artifacts to consider in US imaging. These make the correct interpretation of the US images quite difficult. The practitioner must be aware of these pitfalls to avoid a potentially incorrect diagnosis. The main artifacts are:

– **Speckle images** are due to objects smaller than the length of sound waves, giving rise to strong scattering in all directions (4π), also known as Rayleigh scattering.
– **Reverberation** occurs when the US wave is scattered several times between different organs before being received in the detector. The detector shows a series of delayed echoes, which appear as spurious objects in the distance.
– **Double reflection** may occur at two parallel interfaces like the diaphragm. Structures in the liver can appear to lie in the lung, when reflected twice.
– **Acoustic shadowing** occurs behind strongly reflecting objects such as bones and gas-filled hollow spaces.
– **Acoustic enhancement** occurs when sound attenuation is less than in normal tissue, such as in liquid-filled spaces like the bladder. Then the tissue behind such organs appears brighter than expected.
– **Refraction** occurs at inclined flat interfaces, which makes objects appear closer than they are, like the refraction effect of light observed in swimming pools.

Higher frequency sound waves have smaller beam divergence and narrower beamwidth, resulting in higher lateral resolution. But higher frequencies also lead to higher attenuation and lower penetration depth.

Low frequencies lead to a high penetration depth at the expense of a lower lateral resolution.
Artifacts disturb the US image and yield potentially wrong information.

1.7 Doppler method

1.7.1 Doppler shift

The Doppler[10] effect is well known in physics and describes frequency changes if either source or observer moves with a speed v_{source} or v_{obs}, respectively. The frequency shift occurs due to an artificial wavelength change whenever the source of sound or the receiver of sound has a finite velocity projection in the direction of source and receiver. The wavelength change can be translated into a frequency change as recognized by the observer. The most general expression for the Doppler effect is [20]

$$f_{obs} = f_{source} \frac{v_{sound} \pm v_{obs}}{v_{sound} \pm v_{source}}, \tag{1.23}$$

where f_{obs} is the frequency received by the observer, f_{source} is the frequency emitted by the source, v_{sound} is the sound velocity in the respective medium, and v_{source} is the velocity of the moving source. The sign convention is such that $v_{obs} \geq 0$, if the observer moves toward the source, and $v_{source} \geq 0$ if the source moves away from the observer.

In medicine, the Doppler effect is used to determine the blood flow velocity in blood vessels or organs like the heart and the kidneys. The velocity is determined by emitting US sound waves from a steady source ($v_{source} = 0$) into tissues containing moving blood cells (erythrocytes). Then in a first step, the erythrocytes receive Doppler-shifted US waves. The frequency shift is

$$f_{source} - f_{obs} = \Delta f = f_{source} \frac{v_{blood}}{v_{sound}}. \tag{1.24}$$

In this first step, the blood cells take on the role of an observer. In a second step, the blood cells become a moving source of frequency-shifted US frequencies by scattering the US waves back to the detector at rest. For this second step, the same equation applies again so that the total observed frequency shift by the transducer is

$$\Delta f = 2f_{source} \frac{v_{blood}}{v_{sound}}. \tag{1.25}$$

Finally, we need to consider that the transducer as source and receiver of US waves encloses an angle θ against the blood flow direction (Fig. 1.21) and only the flow projection is detected. Therefore, we have for the frequency shift:

---

10 Christian Andreas Doppler (1803–1853), Austrian physicist and mathematician.

$$\Delta f = 2 f_{\text{source}} \frac{v_{\text{blood}}}{v_{\text{sound}}} \cos \theta. \tag{1.26}$$

Notice that the shift Δf increases with the blood velocity and with the frequency of US waves, and that Δf is largest for an angle $\theta = 0$ but zero at perpendicular orientation. The sign of Δf tells the direction of flow, toward or away from the transducer. The sign is often color coded on displays, red for flow toward the transducer and blue for flow away.

1.7.2 cw Doppler method

According to the probing depth, the operating frequencies for US Doppler measurements are between 2 and 10 MHz (see further). For precise measurements of frequency shifts, a high quality factor Q of the transducer is required. Therefore, the transducer is continuously energized at the resonance frequency, and the usual damping block on the backside of the reducer is removed. In fact, the continuous operation requires two transducers, one for continuously emitting US waves and the other one for continuously receiving the US echo signal. Usually, both transducers sit in the same housing with little separation between them, as indicated in Fig. 1.22. The arrangement is similar to an A-scan with the important difference that Doppler application requires an inclination angle $\theta \neq 90°$.

A quick estimate tells us that the frequency shift is not big. Using a 5 MHz US source, an average sound velocity of 1540 m/s in tissue, and assuming a blood velocity of 2 m/s, the frequency shift expected at angle $\theta = 0$ is only

$$\Delta f = 2 \cdot 5 \, \text{MHz} \cdot \frac{2 \, \text{m/s}}{1540 \, \text{m/s}} = 13 \, \text{kHz}.$$

The frequency shift of US waves due to blood flow is 0.2% of the source frequency, and the difference is in the audible frequency range! This low-frequency shift is filtered by *fast Fourier transform* [10]. First, the US wave with the source frequency f_{source} and amplitude A_1 (Fig. 1.22(a)) is multiplied by the wave detected at the receiver having a much smaller amplitude A_2 and a Doppler-shifted frequency $f_{\text{source}} \pm \Delta f$ (Fig. 1.22(b). The result is shown in Fig. 1.22(c). After filtering the high-frequency component, the wave with the Doppler beat frequency Δf is obtained (Fig. 1.22(d)). This frequency can be made audible by a loudspeaker. The higher the pitch, the higher is the velocity. However, it cannot distinguish between positive or negative frequency changes, i.e., between flow directions toward the receiver or away from the receiver.

For converting Doppler shifts into velocities, the inclination angle θ must be known. This is shown in a screenshot in Fig. 1.23 for the case of a blood vessel. The operator needs to adjust a cursor on the display parallel to the flow direction, and

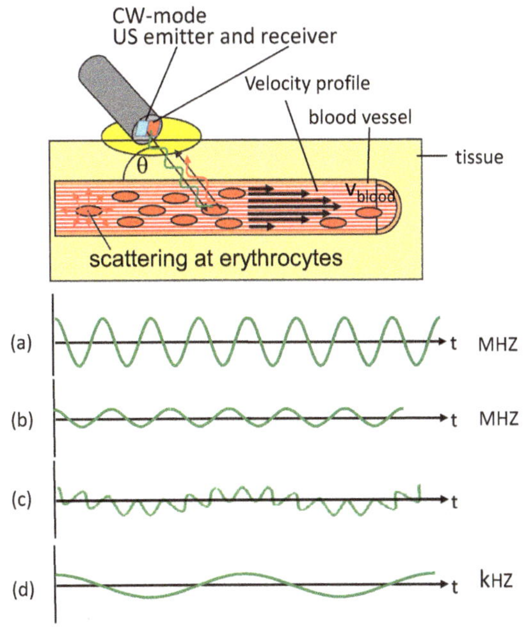

CW-mode
US emitter and receiver

Velocity profile

blood vessel

tissue

θ

v_{blood}

scattering at erythrocytes

(a)   t   MHZ

(b)   t   MHZ

(c)   t

(d)   t   kHZ

Fig. 1.22: The top panel shows the usual arrangement with a US wave emitter on the skin coupling a continuous wave (cw) into the tissue that contains blood vessels. The MHz US wave is scattered by erythrocytes. The scattered and Doppler-shifted wave at angle θ is detected in the receiver. Emitter and receiver are split. (a) Emitted US frequency; (b) Doppler-shifted frequency received from backscattering; (c) product of original wave and Doppler shifted wave; and (d) Doppler frequency in the kHz regime after filtering.

the operational system calculates the angle θ and the blood velocity. In the lower part, the Doppler frequency is plotted versus time, which can be converted to velocity versus time via

$$|v_{blood}| = \frac{\Delta f}{2f_{sound}} \frac{v_{sound}}{\cos \theta}. \tag{1.27}$$

The time structure expresses the pulsing frequency/velocity in the rhythm of the heartbeat. The peak's maximum corresponds to the peak systolic velocity (PSV) and the minimum at the end of the tail corresponds to the end diastole velocity (EDV). Different quantities can be derived from these two values, characterizing proper blood flow versus potential stenosis independent of the inclination angle. The resistance index RI is defined as the ratio:

$$RI = (PSV - EDV)/PSV. \tag{1.28}$$

The pulsatile index PI is defined as follows (see also eq. (8.40) of Volume I):

$$PI = (PSV - EDV)/MV, \tag{1.29}$$

where MV is the mean velocity. PI is more difficult to evaluate than the RI, because the former index involves the shape of the velocity profile and requires an integration of the profile. Often, simply the ratio PSV/EDV is taken. For these indices, reference values are known and tabulated for different blood vessels and organs, such as the carotid artery, the kidneys, the umbilical cord, and others.

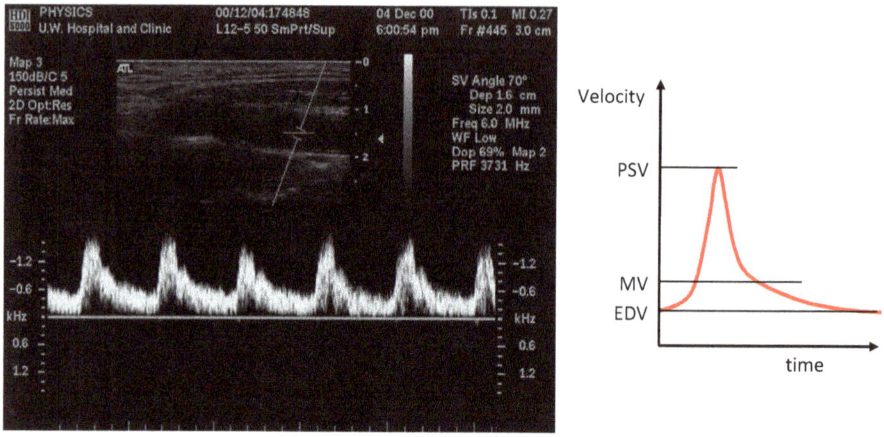

Fig. 1.23: Left panel: Plotted on the screen is the source frequency versus time. The upper part shows a US image of the blood vessel taken in B-mode. The fine line inclined against the horizontal indicates the detector angle (70°). The fine horizontal line crossing the inclined line in the middle of the blood vessel indicates where the blood velocity is measured. Right panel: Schematics of the velocity profile with definitions of different velocities. PSV, peak systolic velocity; EDV, end diastolic velocity; MV, mean velocity.

Actually, the blood velocity not only changes in time according to the heartbeat. In addition, there is a velocity profile across the diameter of the blood vessel. Assuming laminar blood flow, the velocity is highest in the center and drops to almost zero toward the walls. Therefore, it is necessary to define a sample volume length over which the velocity is determined. The velocity spectrum spreads out if the volume is too large, while the maximum velocity is still clearly visible for laminar flow. However, a better signal is obtained if the sample volume length is reduced to a narrow ROI, filtering out velocities close to the walls. Since blood is a non-Newton-type fluid (see Section 8.5 of Volume 1), the velocity profile is flatter in the center than expected from a parabolic profile. This helps focusing on the center, and small deviations are not so severe.

By the same means, the Doppler method can distinguish between laminar flow and turbulent flow. Turbulent flow may occur either due to a large diameter artery such as the carotid artery or local constrictions. Deviations from laminar flow can be detected by the frequency or velocity distribution as function of time and illustrated in Fig. 1.24. In case of turbulence, both the velocity profile and the Doppler angle show a wide distribution.

Doppler sonography in continuous wave mode allows to determine the blood velocity and to distinguish between laminar and turbulent flow. !

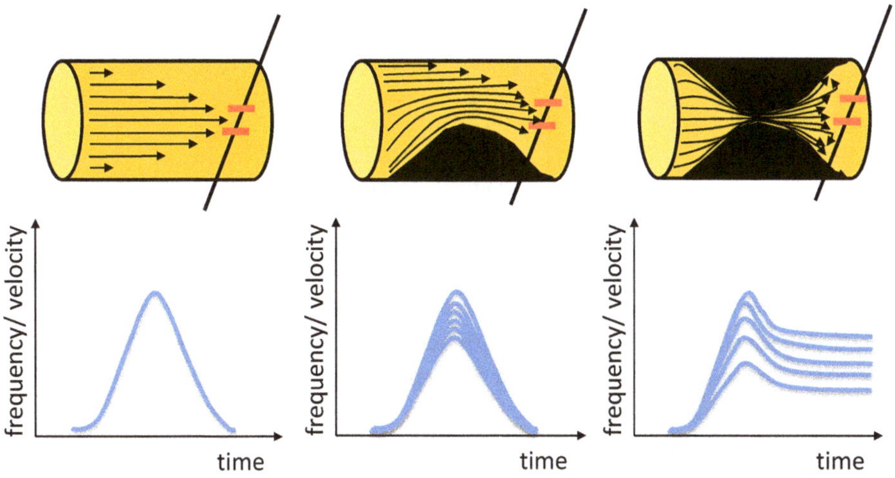

Fig. 1.24: Velocity distribution for laminar flow (left panel), some constriction (middle panel), and more severe constriction (right panel), causing turbulent flow. The region of interest, selected by the detecting system, is indicated by red bars.

A typical application is a checkup of the carotid artery blood supply to the brain. The artery goes through the base of the skull to the brain. Carotid means "providing sleep." Since every small circulatory disorder can impair consciousness and promote a stroke, examining the blood flow using the Doppler effect can indicate a calcification of the carotid artery at an early stage.

A typical Doppler test of the carotid artery is demonstrated in Fig. 1.25 [21]. The Doppler velocity test can even be combined with a simultaneous electrocardiogram to determine the phase relationship between systolic ejection and peak velocity at the carotid artery from which the pulse wave velocity is determined (see Section 8.5 of Volume 1 for more details).

Before closing this section, we discuss the optimum frequency for US Doppler shift application. The scattering of US at the erythrocytes is of the Rayleigh type, since the extension of blood cells (6–8 μm) is much smaller than the wavelength of US (300 μm at 5 MHz). Rayleigh scattering is an isotropic scattering in a solid angle of 4π, which scales with the fourth power of the frequency (f^4). Therefore, it would be advantageous to use higher US frequencies to increase the intensity in the receiver.

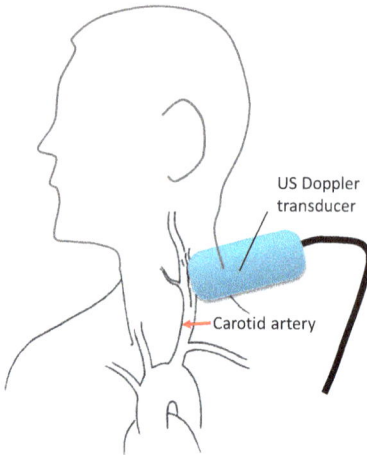

US Doppler
transducer

Carotid artery

Fig. 1.25: Test of the blood flow through the carotid artery with the help of a US Doppler transducer.

However, higher frequencies also experience higher damping, since viscous damping increases with f^2 (eq. (1.4)), and therefore, the penetration depth becomes much reduced. A tradeoff between signal strength and penetration depth has to be made. For blood vessels close to the surface, high frequencies can be used, but for deeper vessels lower frequencies are to be preferred.

In cw operational mode, it is impossible to locate the moving source or distinguish between flow in overlapping blood vessels at different depths. On the other hand, it is impossible to get accurate Doppler flow information with short pulses used for imaging. If information on both location and velocity are required, a compromise between accuracy of depth information (short pulse) and accuracy of velocity (cw) is required. Rephrasing, axial resolution requires large signal bandwidth, whereas velocity resolution requires signal duration. A compromise is a longer signal duration (longer wave train and shorter bandwidth) at higher PRF. This method is known as pulsed Doppler method or *duplex mode,* presented in the next section.

1.7.3 Pulsed Doppler method (duplex mode)

In pulsed mode, the Doppler method is combined with B-mode scans. This combination is known as the *duplex mode,* which stands for B-mode plus pulse Doppler mode. Some transducers in the array used for B-mode scanning are spared for Doppler shift detection in duplex mode. The duplex mode has the advantage that focusing can be applied like in normal B-mode with inherent depth information. Furthermore, the ROI is adjustable for avoiding integration over all blood vessels in the beam, as would be the case in cw mode. Furthermore, the angle of inclination can be determined directly, as shown in Fig. 1.24. The price to pay for the extra depth information is a wider frequency distribution in pulsed mode and, therefore, a lower resolution for Doppler shifts compared to cw operation. To partially overcome this conflict, a pulse length of about 10 times the wavelength is used, much longer than in normal B-mode operation, where the pulse lengths are about 2–3 times the wavelengths. Another significant difference concerns the pulse rate frequency (PRF). The PRF is in the order of 1 kHz for normal B-mode operation, but must be chosen much higher in duplex mode, as we will see below.

In addition to blood velocity measurements, in the duplex mode, the blood flow direction can be determined via the sign of the Doppler shift $\pm \Delta f$. The sign is usually displayed in color on the screen for so-called color Doppler imaging. The color code is as follows: blue for negative $-\Delta f$ (flow away from the transducer) and red for positive Δf (flow toward the transducer) (Fig. 1.26).

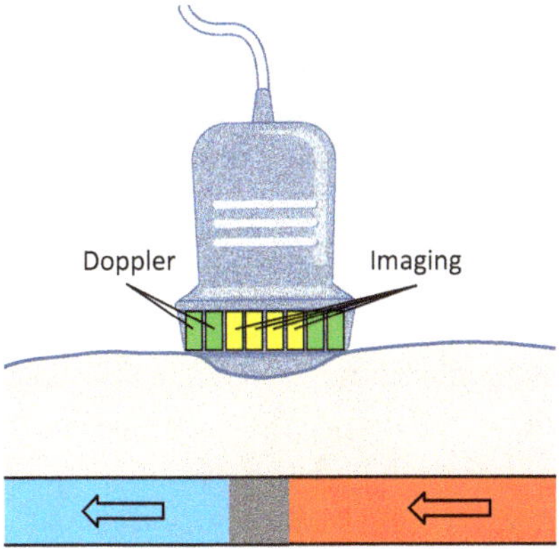

Doppler   Imaging

Fig. 1.26: Color code chosen for the Doppler shift. Flow toward the transducer with positive Δf is coded red, and flow away from transducer with negative Δf is coded blue. In the gray area under the transducer, there is no Doppler shift detected. The scanner contains separate elements for Doppler shift detection (marked green) and for imaging (marked yellow).

In pulse mode, some boundary conditions need to be considered. First, a new pulse cannot be emitted before the previous one is scattered back and received by the detector. The travel time for the pulse is $\Delta t = 2L/v_{\text{sound}}$, where L is the depth of the blood vessel. Thus, the maximum PRF is $f_{\text{PRF}}^{\max} = 1/\Delta t = v_{\text{sound}}/2L$. For a depth of 5 cm, the maximum PRF is 15.4 kHz. Second, the minimum PRF depends on the Doppler shift frequency Δf. If the PRF is less than $2\Delta f$, *aliasing effects* occur. Then the detected frequency in the receiver will show wrong results. The aliasing effect is demonstrated in Fig. 1.27. If the probing frequency f_{probe} is lower than twice the source frequency f_{source}, the frequency of the source may be represented by a frequency that is too low. Thus, the criterion for a correct representation of waves is $f_{\text{probe}} > 2f_{\text{source}}$, which is known as the *Nyquist*[11] *criterion* [22].

If, for instance, the Doppler frequency shift is $\Delta f = 4\,\text{kHz}$, corresponding to a velocity of 0.6 m/s at a transducer frequency of 5 MHz, the sampling PRF should be at least 8 kHz. The *aliasing effect*, therefore, sets a lower bound to the PRF:

---

11 Harry Nyquist (1889–1976), Swedish-American electrical engineer and information scientist.

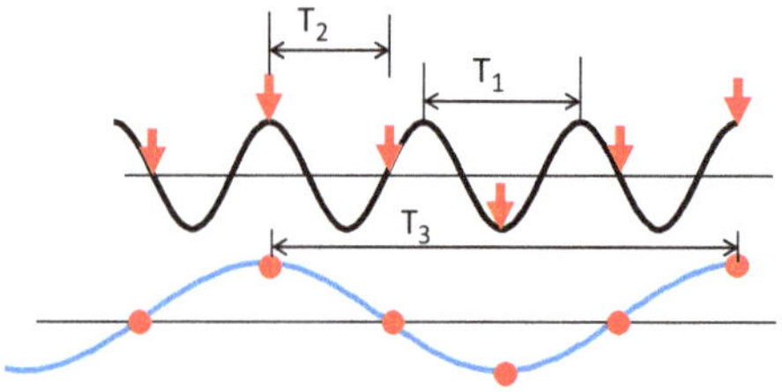

Fig. 1.27: Aliasing effect: The black line is the original wave with a period T_1 and red arrows are probing events in intervals of $T_2 < T_1$. The blue line results from a probing measurement, yielding a period $T_3 \gg T_1$ and T_2. Only for $T_2 \leq T_1/2$ or for frequencies $f_{probe} > 2f_{source}$, the correct frequency of the original wave is represented.

$f_{PRF}^{min} = 2\Delta f$, where Δf depends on the blood flow velocity: $\Delta f = 2f_{source} \cdot (v_{blood}/v_{sound})$, neglecting the inclination angle. For this example, the boundary condition is 8 kHz ≤ PRF ≤ 15.4 kHz. Setting these two limits equal, we obtain for the maximum blood velocity that can be detected at a depth L without aliasing effects:

$$v_{blood}^{max} = \frac{v_{sound}^2}{8 f_{source} L}. \tag{1.30}$$

For a source frequency of 5 MHz, a depth of 5 cm, and a sound velocity of 1540 m/s, we find for the maximum blood velocity that can be detected $v_{blood}^{max} = 1.2$ m/s. Higher velocities would be displayed with false colors. In some blood vessels, the velocity can be much higher. In order to measure these high velocities, the source frequency has to be reduced. Alternatively, one may conclude that high velocities are unsuitable for detection in duplex mode. Low PRF in duplex mode and high blood velocities do not match. This immediately leads to *aliasing effects* and false color coding. The pulse mode obviously has a number of pitfalls, and the operator needs to be very careful in choosing the proper parameters.

1.7.4 Duplex scan of umbilical cord

Finally, we discuss an example of duplex scans from the umbilical cord of a fetus. The umbilical cord that connects the fetus with the placenta is an intertwined cord of one vein and two arteries shown schematically in the inset of Fig. 1.28(a) with equal but antiparallel blood flow velocities. In panel (a) of Fig. 1.28, the focus is on the blue area at a depth of 8.4 cm, indicated by the dashed vertical line and two short horizontal lines. In this area, the velocity is negative with umbilical peak systolic velocity (PS$_{Umb}$) of −26.7 cm/s and umbilical end diastolic velocity (ED$_{Umb}$) of −10.7 cm/s. It is per se not possible to decide whether this signal is from the vein or the arteries. However, as the arteries come in a pair, and if two blood vessels next to each other have the same color, they must originate from the arteries. In the present case, the blue color is indeed from the arteries. The velocity profile is shown in the lower part of the same panel. In panel (b) of Fig. 1.28 the focus is on a red part of the cord (vein) at a depth of 15.2 cm. Here we observe a positive velocity profile with $PS_{Umb} = 33.7$ cm/s and $ED_{Umb} = 12.7$ cm/s. The ratio $PS_{Umb}/ED_{Umb} = 2.65$. In analogy to eqs. (1.28) and (1.29), the umbilical RI is defined as follows:

$$RI_{Umb} = \frac{PS_{Umb} - ED_{Umb}}{PS_{Umb}}. \tag{1.31}$$

Here $PS_{Umb} = 0.62$. The umbilical PI is defined by:

$$PI_{Umb} = \frac{PS_{Umb} - ED_{Umb}}{M}, \tag{1.32}$$

where $M = \langle v_{mean} \rangle$ is the mean velocity. As the pulse shape is asymmetric, the mean velocity has to be determined graphically. In panel (b) the umbilical PI is 0.96. Note that the nomenclature for the velocities is similar to the one on the screens, but different from the standard notation. More examples of umbilical cord Doppler imaging can be found in Ref. [23].

In pulse mode, Doppler sonography yields information on the blood flow velocity and direction, for instance, in the umbilical cord.

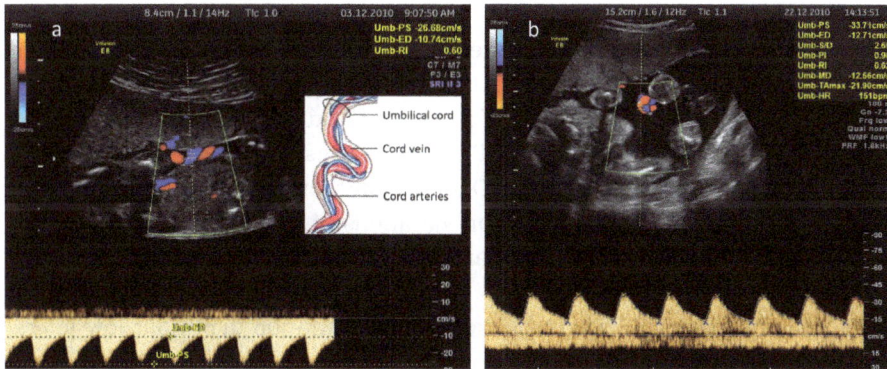

Fig. 1.28: Duplex mode imaging of the umbilical cord. The inset in (a) shows schematics of the umbilical cord with intertwined vein and artery blood vessels. Panels (a) and (b) show simultaneous depth resolution and frequency shifts. The frequency shift from antiparallel velocities in veins and arteries in the umbilical cord are color-coded. The velocity profiles are shown at the bottom of the respective panels. In (a), the frequency shift is negative (blue color) and originates from the arteries. In (b), the frequency shift is positive (red color) and comes from the vein. The region of interest within which the velocity is measured is indicated by the vertical dashed line crossing two horizontal bars (adapted and reproduced from https://openi.nlm.nih.gov/ US National Library of Medicine).

1.8 Summary

S1.1   Medical sonography has three main applications: (1) static imaging of tissues and organs, (2) dynamic imaging of the heart, and (3) measurement of blood flow velocity and direction.

S1.2   Ultrasonic sonography uses sound waves with frequencies in the range of 2–8 MHz.

S1.3   Ultrasonic waves are produced by piezoelectric transducers.

S1.4   The transducer also acts as receiver of sound waves.

S1.5   In medical sonography, six modes can be distinguished: A-mode, B-mode, C-mode, M-mode, cw Doppler mode, and pulse Doppler mode.

S1.6   MI is a criterion for the safe application of US to patients.

S1.7   US propagating into tissue is partially reflected at interfaces separating tissues with different acoustic impedances.

S1.8   The back-reflected echo signal is used for imaging organs with different impedances compared to the surrounding.

S1.9   Blood, air chambers, and bones strongly reflect US waves.

S1.10   US waves propagating into tissue are attenuated due to dissipation in viscous media and due to thermal conduction.

S1.11   TGC alleviates attenuation effects of echo signals.

S1.12   The bandwidth of pulses is typically 80% of the resonance frequency.

S1.13   PRF is the probing frequency for US imaging, typically 1 kHz.

S1.14   Imaging is usually performed in the near-field or Fresnel regime.

S1.15   A-scan is a line scan that probes the depth of reflecting interfaces.

S1.16   B-scan is a sector scan composed of an angular sweep of A-scans. B-scans probe slices of the tissue within the probing depth of US.

S1.17   C-scans consist of sequences of B-scans probed in the lateral direction normal to the B sectors providing 2D images of constant depth.

S1.18   By stitching together C-scans, 3D topographic images can be formed.

S1.19   M-scans provide information on variations of signal amplitude to record moving objects, such as the heart. M-scan technique is also used for taking movies of the cardiac activity.

S1.20   Low-frequency US waves provide high penetration depth but low lateral resolution.

S1.21   High-frequency US waves have low penetration depth, but high axial and lateral resolution.

S1.22   US scanning suffers from many artifacts due to double reflections and shadowing.

S1.23   Using Doppler effect, the flow velocity of blood can be determined.

S1.24   Pulsed Doppler wave methods add depth resolution and directional information on the blood flow.

Questions

Q1.1 In which fields of science and technology is US used?

Q1.2 What is US used for in medicine?

Q1.3 Which frequencies are typically used for US imaging?

Q1.4 What is the definition of acoustic impedance Z?

Q1.5 What is meant by the term "impedance mismatch"?

Q1.6 Sound waves are attenuated in matter. What are the physical reasons?

Q1.7 At interfaces between tissues of different acoustic impedance, the sound wave is either _____or_____.

Q1.8 What is required for receiving a strongly reflected echo signal?

Q1.9 What is a transducer?

Q1.10 How is US generated?

Q1.11 What are the essential parts of a transducer?

Q1.12 Why is a gel required between transducer head and skin?

Q1.13 Why is US transmitted into tissue in pulsed form instead of continuously?

Q1.14 What is TGC?

Q1.15 What distinguishes the near-field region and the far-field region of an extended sound wave source?

Q1.16 Is the focusing of US achieved in the near-field or far-field region?

Q1.17 How is focusing achieved?

Q1.18 The echo time, i.e., the time between the emission of a sound wave and receiving the echo signal, is proportional to the distance between transducer and interface. What determines the axial resolution of the reflecting interface?

Q1.19 What is a B-scan?

Q1.20 What type of scanners are used for B-scans?

Q1.21 What is the PRF and how should it be chosen?

Q1.22 What is a C-scan?

Q1.23 What kind of artifacts may occur during US imaging?

Q1.24 What do you understand by the terms (a) acoustic shadowing, (b) acoustic enhancement, and (c) acoustic reverberation?

Q1.25 How can the fluid velocity of blood be determined?

Q1.26 Is the Doppler frequency shift recorded in pulse or continuous mode?

Q1.27 When flow velocity and flow direction are to be combined, which method is used and, what are the limiting considerations?

Q1.28 What are the boundary conditions for the PRF when applying pulsed Doppler methods?

Q1.29 How are the RI and the PI defined?

Exercises

E1.1 **Liver screening:** During an US screening of the liver, no usable picture can be produced. The examiner notices that there is an air-filled intestine in the path between the transducer and the liver. Explain why this may cause a problem.

E1.2 **Half-value thickness:** At what thickness is the intensity of a 1 MHz and 5 MHz sound wave in soft tissue reduced to 50% of its incident value? Use the attenuation values from Tab. 1.2.

E1.3 **1/e Attenuation:** What is the 1/e attenuation depth of a 1 MHz US wave in soft tissue?

E1.4 **Reflection and transmission:** Assume flat interfaces and normal incidence of the sound wave at two interfaces, separating areas with impedances Z_1, Z_2, and Z_3. Furthermore, assume that the intensity is not attenuated by absorption or scattering. The surface at the top surface is impedance match such that no reflection occurs at the surface.
    a. Draw all reflected and transmitted intensities.
    b. Calculate the back-reflected intensities received by the transducer in terms of equations.
    c. Calculate the reflected and transmitted intensities, assuming the following impedances: $Z_1 = 1.7 \times 10^6$ Ns/m^3 (muscles); $Z_2 = 1.5 \times 10^6$ Ns/m^3 (water); $Z_3 = 0.3 \times 10^6$ Ns/m^3 (lung).
    d. Which intensities are received by the transducer and which interface contributes the most?

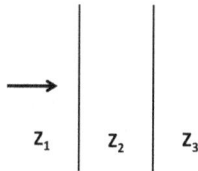

$$Z_1 \quad Z_2 \quad Z_3$$

E1.5 **Examination at high resolution:** An abscess 10 cm below the skin and 2 mm wide has to be examined by applying US with high resolution. What strategy will you follow for a successful image production? Specify the
    a. position of the field of interest;
    b. frequency;
    c. lateral resolution;
    d. axial resolution;
    e. PRT and PRF.

Justify your selections and always give reasons for your choice.

E1.6 **Bulk modulus:** In a setting like the one shown in Fig. 1.12, the first echo signal is received after $\Delta t = 68$ µs from an interface that is $L = 5$ cm below the skin. Assume that the tissue within the 5 cm thickness is homogeneous.
    a. What is the sound velocity in the tissue?
    b. With the help of Tab. 1.2, determine the kind of tissue in which the sound wave travels.
    c. Using again values from Tab. 1.2, determine the bulk modulus B of the tissue.

E1.7 **Doppler sonography:** An RI of 60% and a PSV of 5 m/s were determined in a Doppler sonographical checkup at 7 MHz. Calculate the EDV and the Doppler frequency change for the difference PSV – EDV at an angle $\theta = 0$.

Suggestion for home experiments

HE1.1 Try to observe echo signals in the mountains in front of a steep cliff. It works best if the cliff is more or less smooth.

HE1.2 Estimate the speed of a fire engine by guessing the siren frequency shift before and after passing by.

HE1.3 Estimate the distance of lightning by counting the time difference between the thunderbolt and the thunder. Note that sound travels 330 m/s.

Attained competence checker	+	0	–	⚡
I know what sound waves are.				
I know how to characterize sound waves with respect to amplitude, velocity, and intensity.				
I know how acoustic impedance is defined.				
I know how sound waves behave at interfaces with different impedance values.				
I know that sound waves are attenuated in media and at interfaces.				
I know how sound waves are generated in the MHz frequency regime.				
I can distinguish between different US scanning modes.				
I know what TGC refers to and how it works.				
I know how bandwidth and quality factors are related.				
I know what the MI is and how it is used.				
I can name possible artifacts that may occur during US scanning.				
I can distinguish between axial and lateral resolutions.				
I know the criteria to distinguish between near field and far field.				
I know what PRF means and how PRF and DOV are related.				
I know how the blood flow velocity can be measured.				
I know what the boundary conditions are for PRF in pulsed Doppler mode.				

References

[1] Rathod VT. A review of acoustic impedance matching. Techniques for piezoelectric sensors and transducers. Sensors. 2020; 20: 4051, p. 1–64.
[2] Richard MM. Piezoelectricity. Phys Rev B. 1972; 5: 1607–1613.
[3] Klein MV, Furtak TE. Optics. 2nd edition. New York: John Wiley & Sons; 1986.

[4] Bushberg JT, Seibert JA, Leidholdt EM Jr, Boone JM. The essential physics of medical imaging. 3rd edition. Philadelphia, New York, London: Lippincott Williams & Wilkins, Wolters Kluwer; 2012.

[5] Ziskin M. Fundamental physics of ultrasound and its propagation in tissue. Radiographics. 1993; 13: 705–709.

[6] Riley KF, Hobson MP, Bence SJ. Mathematical methods for physics and engineering: A comprehensive guide. London, New York: Cambridge University Press; 2006.

[7] Kirchhoff G. Über den Einfluss der Warmeleitung in einen Gase auf die Schallbewegung. Ann Phys Chem. 1868; 134: 177–193. English translation: On the influence of thermal conduction in a gas on sound propagation, in Physical Acoustics, ed. by Lindsay R.B. (Dowden, Hutchinson & Ross, 1974). pp. 7–19.

[8] Popovic ME, Minceva M. Thermodynamic properties of human tissues. Therm Sci. 2020; 24: 4115–4133.

[9] Chen S, Urban MW, Pislaru C, Kinnick R, Zheng Y, Yao A, Greenleaf JF. Shearwave dispersion ultrasound vibrometry (SDUV) for measuring tissue elasticity and viscosity. IEEE Trans Ultrason Ferroelectr Freq Control. 2009; 56: 55–62.

[10] Brown A. A guide to Fourier transform and fast Fourier transform. London, New York: Cambridge Paperback; 2019.

[11] Krautkrämer J, Krautkrämer H. Werkstoffprüfung mit Ultraschall. Berlin Heidelberg: Springer-Verlag; 1980.

[12] de Jong N. Mechanical index. Eur J Echocardiogr. 2002; 3: 73–74.

[13] Fenster A, Parraga G, Bax J. Three-dimensional ultrasound scanning. Interface Focus. 2011; 1: 503–519.

[14] https://en.wikipedia.org/wiki/Echocardiography

[15] Kurjak A. 3D/4D Sonography. J Perinat Med. 2017; 45(6): 639–641.

[16] Chen S, Urban MW, Pislaru C, Kinnick R, Zheng Y, Yao A, Greenleaf JF. Shearwave dispersion ultrasound vibrometry (SDUV) for measuring tissue elasticity and viscosity. IEEE Trans Ultrason Ferroelectr Freq Control. 2009 Jan; 56(1): 55–62.

[17] Taljanovic MS, Gimber LH, Becker GW, Latt LD, Klauser AS, Melville DM, Gao L, Witte RS. Shear-wave elastography: Basic physics and musculoskeletal applications. Radiographics. 2017; 37: 855–870.

[18] Sarvazyan AP, Urban MW, Greenleaf JF. Acoustic waves in medical imaging and diagnostics. Ultrasound Med Biol. 2013; 39: 1133–1146.

[19] Ermert H, Hansen C. Ultraschall. In: Dössel O, Buzug TM, eds. Medizinische Bildgebung. Vol. 7. Berlin, Munich: de Gruyter; 2014, 217–326.

[20] Halliday D, Resnik R, Walker J. Fundamentals of physics. 10th edition. New York, London, Sydney, Toronto: Wiley and Sons; 2007.

[21] Hetzel G, Lang W, Strobel D. Physical principles of Doppler and color Doppler ultrasound. In: Iro H, Bozzato A, Zenk J, eds. Atlas of head and neck ultrasound. Stuttgart, New York, Delhi, Rio: Thieme Verlagsgruppe; 2013, P. 10–15.

[22] Salditt T, Aspelmeier T, Aeffner S. Biomedical imaging. principles of radiography, tomography, and medical physics. Berlin, Munich: De Gruyter. Graduate Text; 2017.

[23] https://sonoworld.com/Client/Fetus/html/doppler/capitulos-html/chapter_01.htm

Further reading

Jenderka KV, Delome S. Diagnostischer Ultraschall. In: Schlegel W, Karger CP, Jäkel
    O. Medizinische Physik. Berlin, Heidelberg, New York: Springer Spektrum; 2018; 285–305.
Ermert H, Hansen C. Ultraschall. In: Dössel O, Buzug TM, eds. Medizinische Bildgebung. Vol. 7.
    Berling Munich: de Gruyter; 2014, 217–326.
Dhawan AP. Medical image analysis. 2nd edition. New York, London, Sydney, Toronto: Wiley-IEEE
    Press; 2011.
Bushberg JT, Seibert JA, Leidholdt EM Jr, Boone JM. The essential physics of medical imaging. 3rd
    edition. Philadelphia, Baltimore, New York, London: Lippincott Williams & Wilkins, Wolters
    Kluwer; 2012.
Kremkau FW. Sonography principles and instruments. 9th edition. Amsterdam, Boston, Heidelberg,
    London, New York, Oxford, Paris, San Diego — San Francisco — Singapore — Sydney — Tokyo:
    Elsevier, 2015.
Hoskins P, Martin K, Thrush A, eds. Diagnostic ultrasound – Physics and equipment. 2nd edition.
    London, New York: Cambridge University Press; 2010.
Allisy-Roberts PJ, Farr RF. Physics for medical imaging. Philadelphia, Pennsylvania: Saunders
    Elsevier; 2008.

Useful website

Image gallery of US scans: www.medison.ru/uzi/eng/all/27.11.2022 (accessed on 24 January 2023)

2 Endoscopy

Physical parameters of endoscopes	
Lateral resolution	100–150 µm
Working distance	5–10 mm
Depth of field	5–100 mm
Spatial resolution	100 µm
Subsurface information depth	5 µm
Magnification	Up to 150
Field of view	140°
Confocal laser endoscopy	Subcellular resolution
Optical coherence tomography depth resolution	3–4 µm
Capsule endoscope transit time	80 min

2.1 Introduction

Endoscopy is a imaging modality that enables the examination of dark cavities with visible light. The word "endoscopy" is derived from Greek and consists of the prefix "endo" for "within" and the verb "skopein" for "to see" or "observe." Endoscopy is one of the earliest medical imaging methods, having been invented well before x-ray imaging. Nevertheless, it played a minor role until recently, when significant advances in fiber optics, light sources, and charge-coupled device (CCD) sensors made this technique indispensable in everyday medical practice. Nowadays, endoscopes are used in clinics not only for the optical inspection of surface structures in cavities in search of malignant tissues but also to support minimally invasive surgery, for taking biopsies, and for microscopic and spectroscopic *in-vivo* examinations. Endoscopy is exceptionally versatile. The latest breakthroughs in micro-optics, light sources, and light sensors have already revolutionized endoscope technology, and many more inventions are on the horizon. Following a brief overview of their main areas of application, this chapter explains the physical basics of endoscopes and their various further developments.

2.2 Standard uses of medical endoscopes

Basic versions of medical endoscopes conduct light from an external white light source through a bundle of glass fibers to hollow spaces inside the body; a second bundle of glass fibers receives and transmits the backscattered light from the tissue to a light-sensitive detector. Both bundles are wrapped into a flexible tube that is

https://doi.org/10.1515/9783110757095-002

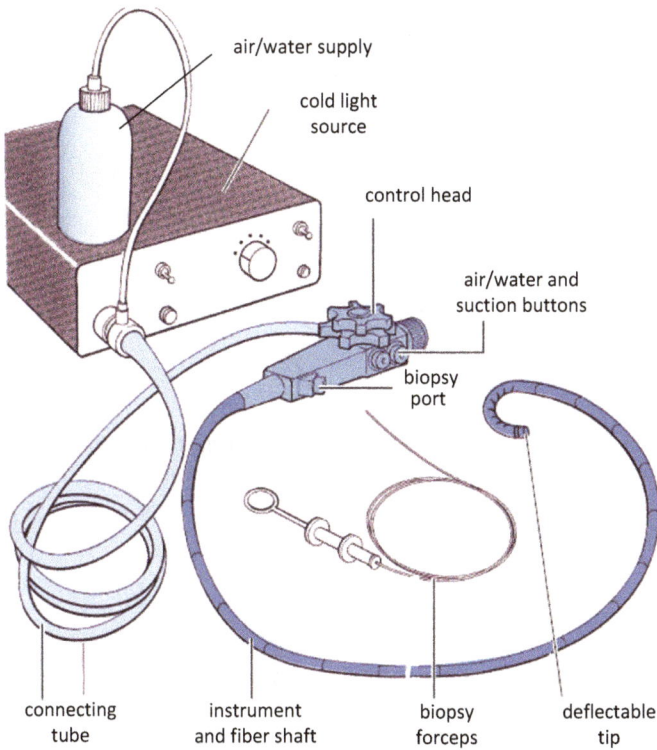

Fig. 2.1: Sketch of a traditional endoscope featuring the main components: light source, control head and tube containing the fiberglass bundles for illumination and imaging, air/water duct, and biopsy channel (adapted from [1] by permission of John Wiley and Sons Inc.).

inserted into open ducts of the body. The main features of a standard traditional endoscope are sketched in Fig. 2.1.

Endoscopes provide visual evidence of problem zones such as ulceration, inflammation, and cancerous tissue when inserted into the body through open channels (esophagus, rectum, urethra, and vagina). The main applications of endoscopes are *gastroscopy* through the esophagus and *colonoscopy* through the rectum. They are sometimes also referred to as upper endoscopy and lower endoscopy, respectively (Fig. 2.2). Endoscopes can also be used for taking tissue samples for further laboratory examination (biopsy), removing polyps, lumps, etc., and for local application of pharmaceuticals. Furthermore, endoscopes support minimally invasive surgery (*laparoscopy*), such as hand, knee, and gall bladder. Endoscopes are equipped with extra channels for all these additional applications to insert and maneuver special instruments like snares and biopsy forceps, as indicated in Fig. 2.1. Other channels supply air or other gases and suck off fluids (blood and various debris).

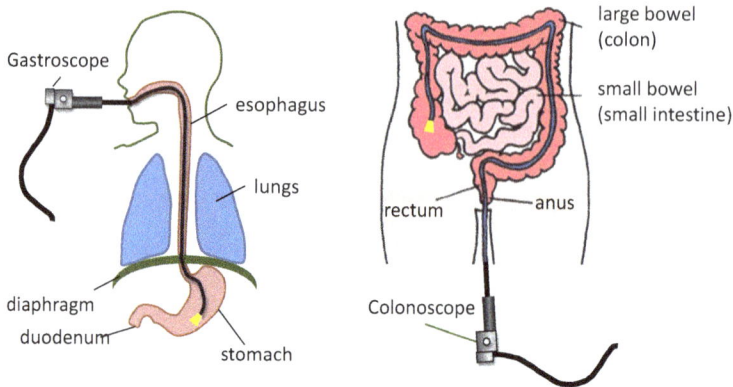

Fig. 2.2: The most common applications of endoscopy are gastroscopy through the esophagus (left panel) and colonoscopy through the rectum (right panel). Both applications are used for early cancer recognition and removal of precancerous tissues.

2.3 Fiber optics

Endoscopes, as we know them today, use a point-by-point imaging method of an object. This is an entirely different optical imaging scheme than the usual pinhole-type optics. The endoscopic view is similar to viewing through an assembly of straight drinking straws. Each straw has a tiny *field of view* (FOV). But the collection of all overlapping FOVs yields a pixelated picture that is upright and not enlarged, in contrast to images taken with a pinhole camera. Both schemes are compared in Fig. 2.3.

Early endoscopes were straight like straws. But with the introduction of thin *glass fibers*, endoscopes can be bent and light can be made to go around "corners," just like the flexible fiber itself sketched in Fig. 2.4. Assuming that the fibers are well ordered in a coherent fashion rather than scrambled up, the image corresponds pixel by pixel to the object, where the diameter of single fiberglasses gives the pixel size. Kapany[1] is recognized as the inventor of fiber optics and endoscopic applications.

Guidance of light through glass fibers is achieved by *total reflection* of light when it passes from a transparent matter of higher optical density characterized by the refractive index n_1 to lower optical density n_0 ($n_0 = 1$ for air). Figure 2.5(a) shows the case when light first enters from air into the fiber from the front end at an angle α. Then the light ray is refracted toward the optical axis of the fiber according to Snell's law:

---

1 Narinder Singh Kapany (1926–2020), Indian-American physicist.

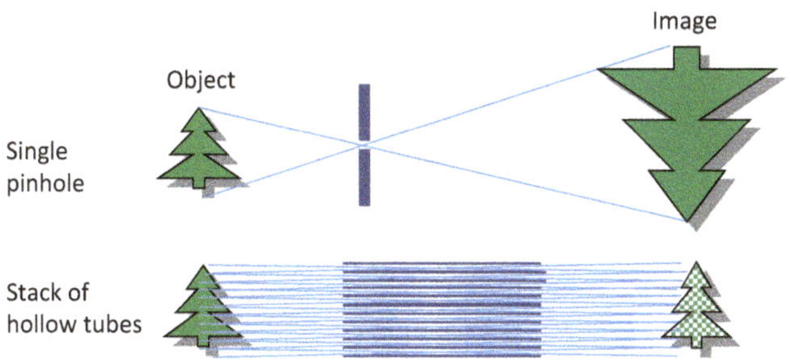

Fig. 2.3: Upper panel: image generated by a pinhole camera; lower panel: image produced by a stack of hollow tubes.

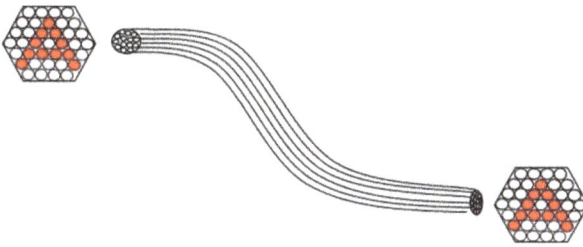

Fig. 2.4: Point-to-point imaging of objects by a close-packed bundle of fiberglass.

$$n_0 \sin \alpha = n_1 \sin \beta, \tag{2.1}$$

and reflected at an angle γ at the interface to air. Total reflection occurs for an angle γ_c fulfilling the condition:

$$\cos \beta_c = \sin \gamma_c = n_0/n_1. \tag{2.2}$$

For all angles $\gamma \geq \gamma_c$, light rays are reflected back from the fiber/air interface to the inside; incident light rays at angles $\gamma \leq \gamma_c$ will pass through the fiber wall to the surrounding. The FOV, for which total reflection occurs, is 2α and depends on the refractive index of the fiber according to

$$2\alpha_c = 2\sin^{-1}\left(\frac{1}{n_0}\sqrt{n_1^2 - n_0^2}\right). \tag{2.3}$$

The aperture of the fiber then follows from:

$$A_{\text{fiber}} = n_0 \sin \alpha_c = \sqrt{n_1^2 - n_0^2}. \tag{2.4}$$

Fibers are always coated with a cladding material characterized by a refractive index $n_2 < n_1$ (Fig. 2.5(b)). The cladding has the purpose of suppressing cross-communication between fibers. Total reflection then occurs at the core/cladding interface. In endoscopes, fibers are, in fact, embedded in a cladding material shared by all fibers, as indicated in Fig. 2.6. With cladding, the condition for total reflection at the internal interface is modified to

$$A_{\text{fiber}} = \sin \alpha_c = \sqrt{n_1^2 - n_2^2}. \tag{2.5}$$

Thus, the FOV and the aperture increase with an increasing difference in the refractive indices $\Delta = n_1^2 - n_2^2$.

Fig. 2.5: Refraction and reflection in fiberglass. (a) Simple fiberglass without coating; (b) fiberglass with a core and a cladding material; and (c) constructive interference in a waveguide.

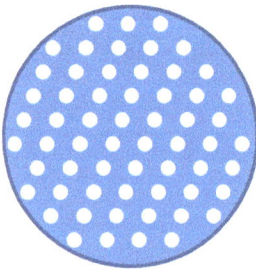

Fig. 2.6: Optical fibers embedded in a cladding material.

Fiberglasses are optical waveguides. They use the total reflection of light at smooth interfaces to optically less dense materials.

Endoscopes typically have an outer diameter ranging from 0.5 mm for very narrow channels up to 9 mm for wider channels like the esophagus and the colon. As already stated, endoscopes house two bundles of optical fibers: one for illumination at the distal end of an endoscope, and one for guiding the scattered light back to the proximal end of the instrument. Both fiber bundles contain $20 \cdot 000$–$40 \cdot 000$ fine glass fibers, each about 5–$10 \cdot \mu m$ thick.

Endoscopes have a conflicting design problem to solve. The outside diameter should be small, yet the number of fibers inside should be as large as possible for high resolution images. When the coherence length of light, i.e., the length of the wave train, becomes larger than the diameter of a single fiber, then wave optics has to be taken into account. This implies that after two reflections at the fiber/cladding interface the wave front of the light has to interfere constructively with the part of the same wave train that has not yet been reflected. In Fig. 2.5(c) the incoming wave travels the distance AB, while the reflected wave travels the distance AC. After reflection, both waves should match in phase. Therefore, the difference in path length $AC - AB$ should be a multiple of the wavelength λ/n_1 in the medium with refractive index n_1. With $AC = d/\sin\beta$ and $AB = AC\cos(2\beta)$, the condition for constructive interference is

$$\sin\beta = \frac{m\lambda}{2n_1 d},$$

(2.6)

where m is the order of interference. Phase jumps at the core/cladding interface do not occur at the boundary to a medium with lower refractive index. In terms of the aperture, the constructive interference condition then reads:

$$n_1 \sin\beta = \frac{m\lambda}{2d} \leq A_{\text{fiber}} = \sqrt{n_1^2 - n_2^2}$$

(2.7)

and

$$m \leq \frac{2d}{\lambda}\sqrt{n_1^2 - n_2^2}.$$

(2.8)

The largest number $m_{\max}$ that fulfills the condition of constructive interference before reaching the critical angle for total reflection defines the number of allowed light ray directions. If $m_{\max} = 0$, the waveguide is called a *single mode guide* allowing only the fundamental mode to go through. Single-mode fibers (SMF) and *multimode fibers* are compared in Fig. 2.7. Narrow fibers with high packing density for high-resolution images are usually SMFs.

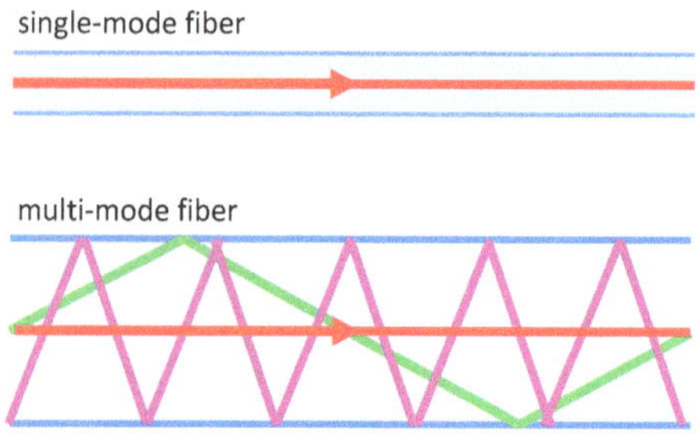

single-mode fiber

multi-mode fiber

Fig. 2.7: Single- and multimode fibers. Note that the colors do not indicate different wavelengths but different light beam directions for the same wavelength.

! Most fiberglass endoscopes use SMF for illumination and imaging.

2.4 Endoscope optics

In the early endoscopy days, the examiner watched the backscattered light directly through an ocular eyepiece. However, modern technology has replaced direct viewing with a light-sensitive chip, such as a CCD or a complementary metal-oxide-semiconductor (CMOS) sensor for higher image quality, schematically shown in Fig. 2.8(a). The electrical output is fed into a PC for image processing and displayed on an LCD screen. The CCD chip measures only light intensity but is not color sensitive. Three color filters cover the sensor for generating color images according to the RGB color code like in digital cameras. A lens in front of the sensor may either focus the light on the chip or magnify the pixel picture at the proximal end of the fiber bundle. Endoscopes designed according to this scheme are known as *fiberoptic endoscopes*.

Viewing with a CCD chip has the advantage that more than one person can simultaneously watch images and videos taken during an examination and even remote viewing is possible. Furthermore, the pixel density of a modern CCD chip matches well with the density of fiber bundles. Therefore, there is no loss of information or effect on the high definition (HD) of the recorded image.

While keeping the functionalities described earlier, the simple optics of a standard endoscope has room for improvement. Although endoscopes do not feature an intrinsic focus, modern micro-optics allow collecting the rays on the object side and focus them on the fiber bundle. By inserting an aperture and an objective lens on the distal end, the *focal length* and the *depth of field* can be adjusted (Fig. 2.8(b)). The depth of field is the distance range measured from the objective lens to the tissue over which images remain in focus. This is typically 5–100 mm in modern video endoscopes presented further.

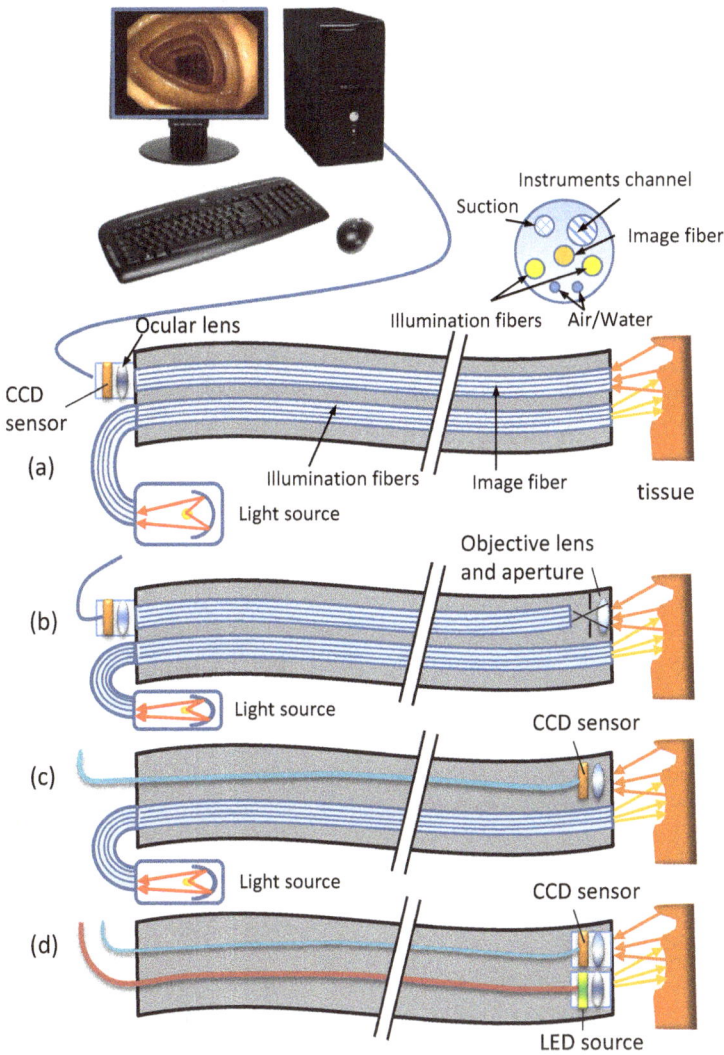

Fig. 2.8: (a) Essential parts of endoscopes consisting of fiberglass bundles for illumination and transmitting backscattered light. A cross section of the endoscope features channels for surgical instruments and air/water inlets; (b) lens and aperture at the distal end in front of the fiberglass bundles improves the optics; (c) a CCD chip at the distal end completely replaces the fiber bundle for imaging; and (d) a LED light source replaces the illuminating fiber bundle.

With the miniaturization of CCD chips, the receiving fiberglass bundle could be entirely removed and replaced by a CCD sensor at the distal end while retaining the illuminating optics (panel (c)), referred to as *"chip in the tip."* This design, known as *video endoscope*, offers flexibility for positioning the CCD chip either at the distal end (distal tip) or on the side of the tube. It has been shown that imaging with distal sensor endoscopes provides better images concerning resolution, contrast, and color discrimination than standard ones. Most endoscopes currently used in clinics have this version of video endoscope [2].

The next step of endoscopic development is the substitution of the illuminating fiberglass bundle by a tiny but very bright solid-state light source such as a light-emitting diode (LED), indicated in panel (d). In this design, endoscopes are realized without any fiber bundles and without the need of a very expensive xenon white light source. In the past, the development of fiberglass technology has paved the way to the successful use of endoscopes. But simultaneous advances in semiconductor technology and microelectronics have made optical fiber bundles obsolete. Aside from better illumination and higher image quality, another important improvement is the reduction of weight. The lighter endoscopes are, the better the additional instruments can be manipulated. Furthermore, the optical fiber bundles can no longer break by over-bending. These developments are still in progress [3].

Although endoscopy is doubtless an enormously successful procedure for visual inspection and diagnoses of early stages of cancer, some limitations should be addressed. Endoscopy can only be executed in empty hollow spaces, where any liquids including blood are removed. In case of minimal invasive surgery with the help of endoscopes, this constitutes additional difficulties. Often body parts to be examined or treated need first to be inflated with air or carbon dioxide. Additional fluids and debris have to be sucked off. Hygiene is another issue of concern. All equipment parts that come in contact with patients must be sterilized. Alternatively, but more costly than reconditioning are endoscopes with one-time-use components that are delivered sterilized by the manufacturer. For further practical procedures of endoscopy we refer to [1].

2.5 Resolution and magnification

Resolution is the ability to distinguish two closely spaced objects. In the diffraction limit, the resolution is defined by the Abbé[2] limit or Rayleigh[3] criterion and is mainly determined by the wavelength used. However, for endoscopes, which operate still far away from the diffraction limit, the resolution is defined by the ratio of illuminated area $A = \pi r^2 = \pi(d \, \tan(\alpha))^2$ provided by the FOV (see Fig. 2.9) divided by the number N of fibers (fiber endoscopes) or the number N of pixels (video endoscopes):

$$R = \sqrt{\frac{\pi(d \, \tan(\alpha))^2}{N}}. \tag{2.9}$$

---

2 Ernst Karl Abbe (1840–1905), German physicist and optical scientist.
3 John William Rayleigh (1842 –1919), British mathematician and scientist, Nobel Prize 1904.

With a working distance of 10 mm and an opening angle $\alpha = 70°$, assuming $N = 10^5$, a resolution of 0.15 mm is achieved. For even higher resolution and lower signal-to-noise ratio (SNR), more costly CMOS sensors are used instead of CCD chips [4].

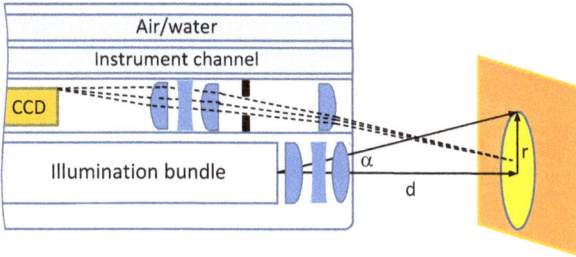

Fig. 2.9: Distal end of a video endoscope. The surface of the illuminated tissue is at a distance d and the field of view has a radius of r.

Currently, *HD video endoscopes* with 9 mm outer tube diameter, suitable for gastroscopy and colonoscopy, contain sensors allowing images with 1–2 megapixels to be taken and provide a spatial resolution of about 100 µm at a working distance of about 10 mm [5]. The spatial resolution decreases with increasing working distance.

With higher resolution, more details are visible. In contrast, *magnification* enlarges the images without improving the resolution. Standard instruments magnify up to about ×35. High-magnification instruments using special optics reach magnifications of about ×150. This high magnification is achieved by using movable (zoom) lenses in the tip of the endoscope, as indicated in Fig. 2.9. Figure 2.10 shows an example of a high-resolution and high-magnification endoscopic image of the colon.

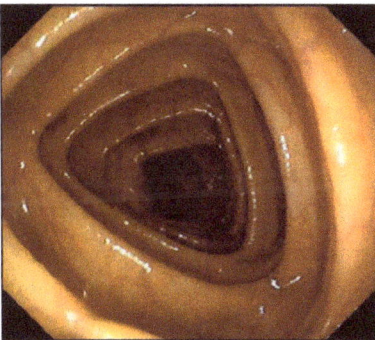

Fig. 2.10: High-resolution image of the colon taken with a modern endoscope (reproduced from http://www.gastrolab.net/ni.htm).

To summarize, standard but state-of-the-art video endoscopes have an FOV of about 140°, depth of field variable between 5 and 100 mm, a spatial resolution of about 100 µm at a working distance of 5–10 mm, and in some cases a magnification up to

150. The information depth of the backscattered light spans from the surface to about 5 μm depth into the skin, depending on the wavelength used. The physical length is about 1 m for gastroscopy and 1.6 m for colonoscopy. These lengths are sufficient for reaching the stomach on one side and the full length of the colon on the other side, respectively. But it is not sufficient for imaging the small intestines. For the latter examination, different methods are applied, such as the capsule endoscope presented further.

> ! Video endoscopes are the most frequently used endoscopes in clinical applications. Video endoscopes use illuminating fibers and a CCD camera at the distal end of the scope.

2.6 Specialized endoscopes

In medicine, endoscopes are used to visualize and diagnose inner hollow body parts and support of minimally invasive surgery. In addition, some specialized endoscopes have been developed for particular tasks such as microscopy and spectroscopy and to overcome some limitations of standard endoscopy. Some of the developments are briefly presented here. Lasers, used in most of these advanced endoscopes, are described in Chapter 6 in Volume 3.

2.6.1 Narrowband imaging

Narrowband imaging (NBI) refers to endoscopic imaging techniques that uses only a narrowband of wavelengths for imaging and testing [6]. Standard endoscopes use the full wavelength band of visible light from 450 to 650 nm. The color of an object is then a question of absorption versus scattering. Wavelengths that are scattered mix and yield the specific color of that body part. For instance, if the wavelength bands for blue (~440–460 nm) and green (~540–560 nm) are absorbed, but the wavelength band for red (~580–600 nm) is scattered, the color of this body part will appear red. However, if a red body is illuminated with blue light, it will appear black. In standard endoscopy, blood vessels give images a reddish color since hemoglobin has absorption bands in the blue and green region but scatters red light. Blood vessels appear dark if a filtered light source blocks out the red wavelength band. This allows concentrating on other surface structures, such as the mucosa in the stomach or colon, which then stands out in greenish color when illuminated with a green filter. A comparison of a normal image taken with a white light source and one with a filter inserted is schematically shown in Fig. 2.11. Detection of fluorescent light is also possible if the tissue has been stained with an appropriate fluorophore.[4] Alternatively, instead of using an optical transmission filter for one particular color, one may read out just one of the three-pixel sets of the CCD sensor. Conversely, for illumination one may also use LEDs with preselected wavelength bands.

Using a narrow color band for illumination has another advantage. The penetration depth of light into tissue depends on the wavelength. Wavelengths in the blue color regime absorb and reflect near the surface, whereas the red color band penetrates deeper into the tissue and reflects from subsurface structures. NBI is an endoscopic option that can be easily added to any type of endoscope as it requires only one or two color filters. A recent review is given in [7].

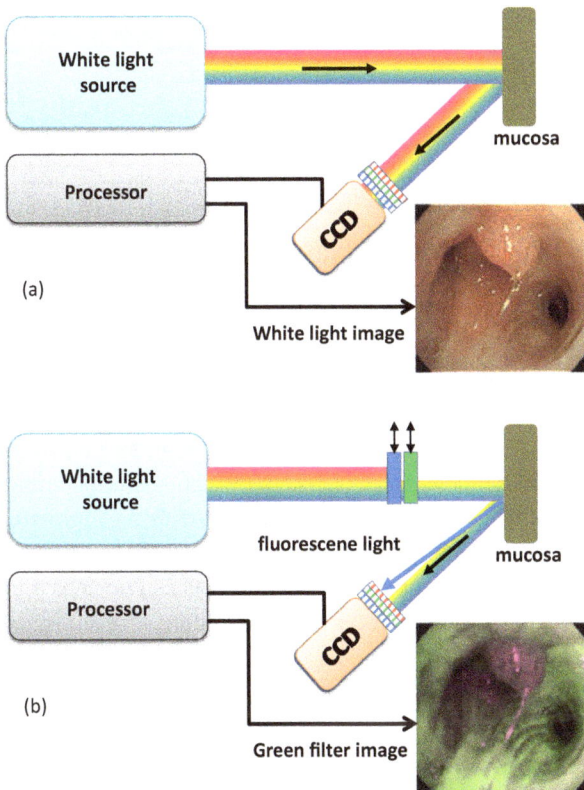

Fig. 2.11: Comparison of endoscopic imaging with white light (a) and with filtered light (b). In both cases, the backscattered light is detected by a CCD sensor. When a green filter blocks the red light, the backscattered light contains mainly information from structures scattering green light. The same holds also for a blue filter. At the same time, fluorescence light becomes visible.

---

[4] Fluophores are organic chemicals that absorb photons of wavelength λ_1 and re-emit longer wavelength photons $\lambda_2 > \lambda_1$ after a short time delay.

2.6.2 Chromoendoscopy

Chromoendoscopy is an alternative endoscopic procedure where dyes or stains are used to enhance contrast [8]. The chemicals are infused into the gastrointestinal tract just before inspection with an endoscope. The dye enhances characteristic features of the tissue and facilitates the identification of different tissue types or pathologies based on the pattern recognized. Several specific dyes are available to identify different diseases.

> **!** A number of endoscopes have been developed for special tasks beyond the imaging capabilities of standard endoscopes. These include NBI, chromoendoscopy, fluorescence endoscopy, Raman endoscopy, confocal laser endoscopy (CLE), and optical coherence tomographic endoscopy.

2.6.3 Endomicroscopy

Endomicroscopy is a technique that enlarges the image as the name suggests and obtains histology-like images from inside the human body in real time, a process known as *optical biopsy*. In medicine, biopsy implies the removal of some tissue for external histological examination. Endomicroscopy allows histological tests in vivo and in real time. The magnification is achieved either by *confocal microscopy* or *optical coherence tomography* (OCT, for explanations see further). Clinical endomicroscopy achieves a resolution on the order of 1 µm which is 100 times higher than for HD video endoscopes. At this high resolution, the FOV is reduced to several hundred micrometers. Endomicroscopy is mainly used for imaging the gastrointestinal tract and, in particular, for diagnosing and characterizing Barrett's esophagus (heartburn) and other precancerous lesions.

One of the characteristics of precancer conditions is an enlarged cell nucleus and an increased size ratio of the nucleus to the cell body [9]. Therefore, the ability to image nuclear cell morphology in vivo represents a major step toward detecting epithelial precancers. Epithelial cells cover the inner surface of all tubes and ducts of the body. They are the equivalent of the epidermis for inner body surfaces. Early observation of structural changes in the epithelial structure can be treated and save lives.

2.7 Confocal laser endoscopy

2.7.1 General working principle

Confocal laser endoscopy (CLE) uses techniques developed for optical microscopy in biology and medicine over the past 50 years, which has been adapted to the special requirements of endoscopes. Confocal microscopes use three characteristic features that distinguish them from standard microscopes [10]:

a. The objective lens produces an illuminated focus with a small FOV.
b. A scanner system moves the focus across the sample.
c. A pinhole in front of the detector blocks all light rays not originating from the focal spot.

The focal spot of the objective lens and the pinhole work together to yield a confocal "image" in the detector. However, there is no image in the usual sense that can be seen through an ocular lens. Instead, the image is generated point by point by scanning the focal point across the sample. Confocal microscopy aims at improving the contrast by reducing background intensity from scattered light. The resolution is not dramatically increased compared to a conventional microscope, but the sharpness is considerably enhanced [11].

After these general remarks, we want to look at the ray tracing of a confocal microscope. Panel (a) in Fig. 2.12 gives a simplified overview of the optical path in a confocal microscope and panel (b) shows some more details when used in the NBI mode. The light source can be a white source, but a broadband laser is more adequate. The light is reflected by a beam splitter and passed through an objective lens onto the sample surface. The highest resolution is used to create a tiny illuminated spot on the sample surface and within the penetration depth of the light. The incident and reflected light pass an XY scanner, which moves the illuminated spot over the sample surface. The XY scanner can be a mirror scanner or a plate scanner containing an array of pinholes. The reflected light then goes back through the beam splitter and the confocal pinhole before entering the detector. The confocal pinhole blocks all light that is not reflected from the illuminated spot. Overall, compared to a standard microscope, this method results in images with more contrast and less blurring due to scattered light. The highest reported lateral resolution is about 0.2 μm, slightly higher than achievable with conventional optical microscopes [12, 13].

Confocal microscopes are mainly used in biology and medicine for an additional aspect, which is explained in panel (b) of Fig. 2.12. Certain tissue parts can be made visible by staining with fluorophores. Let us say that the selected fluorophore absorbs blue light and emits green light. We then use either a blue laser or a white source with a blue filter as the light source. The beam splitter is replaced by a dichroic mirror that only reflects blue light. The blue light excites the stained parts within the focal spot, and green light is emitted, which passes through the dichroic mirror and the confocal pinhole. Other colors or rays from other (deeper) areas will not pass through the mirror and not through the pinhole, as indicated by the "red" ray in panel (b). This procedure produces a "weighted" high-resolution histological image of specific parts in the tissue.

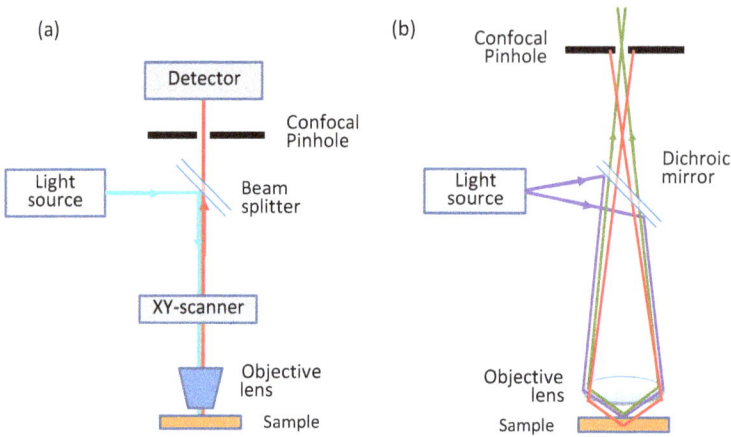

Fig. 2.12: Confocal microscopy. (a) The general ray path of a confocal microscope. Note that the focus of the incident light is scanned across the sample surface. The reflected light passes a confocal pinhole instead of an ocular lens. (b) Dichroic version of a confocal microscope. The shorter exciting wavelength (blue) is scanned across the surface, and the longer wavelength (green) is detected after passing the confocal pinhole, whereas the red light is blocked.

2.7.2 Fiber-optic confocal reflectance microscope

Now we return to the endoscopic version of confocal microscopy. Figure 2.13 schematically outlines a *fiber-optic confocal reflectance microscope* (CRM) [10]. A fiber-optic bundle is placed between the scanning mirrors and the objective lens. At the proximal end of the fiber bundle, the illuminating laser beam couples light into only one fiber at a time using an XY mirror scanner. At the distal end, each fiber serves both as a point light source and as a detection pinhole. The illuminated fiber passes the light onto the tissue through a miniature lens; light scattered back from the tissue is imaged by the objective lens and passed on through the fibers to a beam splitter and photodetector. Backscattered light from out-of-focus areas in the tissue is distributed over several fibers and mostly rejected by a pinhole in front of the detector.

Confocal microscopes work in standard reflection or in fluorescence mode. On the proximal side, the FOV is essentially identical to the scanning area of the fiber bundle. Typical FOV is 150 µm for the lateral scan area with a resolution of 1.5 µm. The tissue must be stained with a suitable fluorophore for fluorescence imaging.

If not used internally in hollow spaces, fiber optics is not required. The objective lens of the CRM can then directly be focused on the area of interest. This is indeed done in skin oncology [10]. There are two versions of CRM: fixed and movable. In the fixed version, the laser beam is scanned over the area of interest with a lateral resolution of about 1 µm and a depth resolution of 3–5 µm up to a depth of about 250–300 µm. A sequence of scans in steps of 5 µm allows an in vivo "optical biopsy," meaning that the scanner provides an analysis of the epidermal structure at nearly histologic resolution, including images of melanoma in the skin. The movable CRM version can be used to probe parts of the skin that are otherwise difficult to reach and allows a simple and fast full-body skin examination. The handheld device is used like a US scanner and the image can be seen live on the screen.

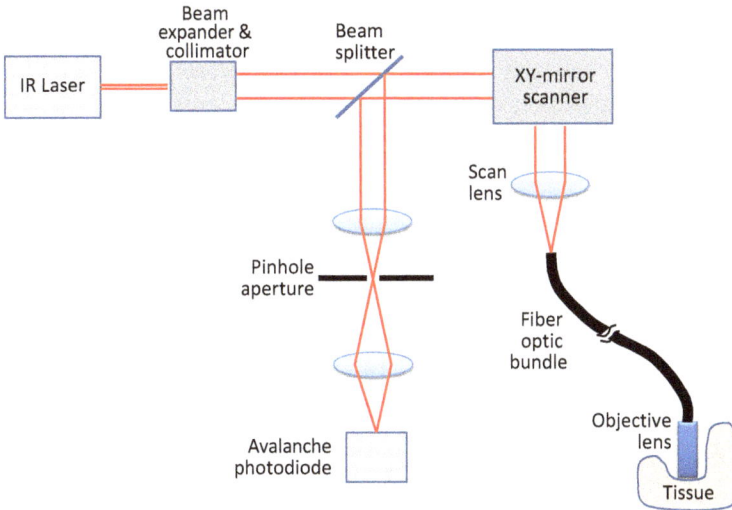

Fig. 2.13: Schematic diagram of the fiber-optic confocal reflectance endoscope (adapted from Ref. [10]).

Confocal endoscopes combine fiber optics with a confocal microscope. The surface area to be inspected is illuminated line by line with an XY scanner. The back-reflected light has to pass a pinhole aperture that is scanned synchronously with the illuminating scanner.

2.8 Optical coherence tomography endoscopes

OCT endoscopes add the third dimension to confocal microscopy. OCT works as an optical biopsy tool for in vivo microstructural information of tissue in three dimensions (3D) [14]. It has a lateral resolution comparable to CLE and enables cross-sectional depth-resolved views in the third dimension by using interferometry.

Infobox I: Microscopy and tomography: What is the difference? A microscope takes an enlarged image of a surface area. Tomographs image structures below the surface. The depth of information is adjustable and depends on the attenuation of the probing beam. If the surface image is combined with images from different depths, a 3D tomographic image is constructed.

OCT is an optical analog to sonographic B-mode imaging, discussed in Section 1.5.2. Like the B-scan, with OCT, the information depth is determined by the echo time delay of back-reflected light. However, since the speed of light is 200 000 times faster than the speed of sound in soft tissue, the echo time delay of light cannot be measured directly. Instead, the delay time is measured interferometrically with the aid of a reference beam in two arms of a Michelson–Morley interferometer. The Infobox II outlines the basic principle of such an interferometer.

2.8.1 Basic principle of OCT

In contrast to usual Michelson[5]–Morley[6] interferometers, for OCT, the light source must have a short coherence length, as shown in Fig. 2.14 top left inset (a). A light source of short coherent light features a central wavelength and a wave train over a short distance. The Fourier transform of such a short wave train shows a Gaussian distribution of wavelengths with a full width half maximum $\Delta\lambda$, centered around the central wavelength λ_0; the coherence length of the wave train is determined by [15, 16]

$$L_c = \frac{2\ln 2}{\pi}\frac{\lambda^2}{\Delta\lambda}. \tag{2.10}$$

A heuristic derivation of eq. (2.10) is given in Exercise E2.4. An ideal coherent monochromatic light source has, in comparison, only one wavelength and an infinite long wave train, indicated in panel (d) of Fig. 2.14. But such a light source would not provide any depth resolution. Similar to sonography, the shorter the pulse of the light source, the higher the depth resolution.

---

5 Albert A. Michelson (1852–1931), American physicist, Nobel Prize in Physics 1907.
6 Edward W. Morley (1838–1923), American chemist.

Infobox II: Michelson–Morley interferometer

Monochromatic light from the laser source is split into two parts via a half transparent mirror. Half of the beam travels from the laser source to the fixed mirror and back, while the other travels through the beam splitter to the scanning mirror and back. When combined in the return path, intensity oscillations are observed in the detector due to constructive interference of both beams. Constructive interference occurs whenever the total path of the partial beams is a multiple of their wavelength:

$$2L_{\text{fixed}} - 2L_{\text{scan}} = c(t_1 - t_2) = c\Delta t = m\lambda. \tag{2.11}$$

where L_{fixed} and L_{scan} are the respective distances between the beam splitter and the mirrors, Δt is the time difference or delay time, and c is the speed of light.

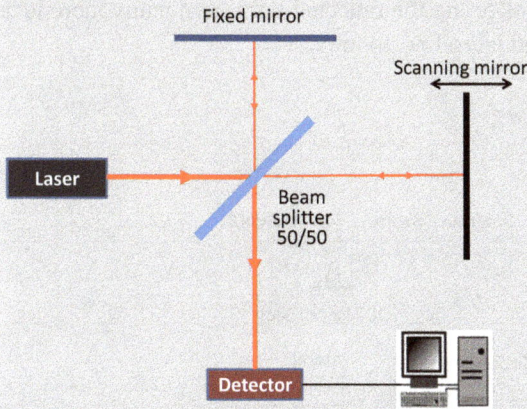

Basic design of a Michelson interferometer

OCT uses a Michelson–Morley interferometery for imaging, schematically shown in Fig. 2.14 panel (b). First, light pulses of short coherence length are funneled simultaneously through the optic fiber to the probe arm and the interferometer's reference arm. The probe arm contains a scanning system that scans the light beam over the sample surface (lateral scan in Fig. 2.14(b)) and collects the backscattered light. The scanning system determines the FOV and the lateral resolution. The reference arm of the interferometer has a scanning path delay u that is translated into the desired imaging depth z. Optical interference between the light from the sample and the reference beam occurs on the return path in the fiber coupler only when the optical delays match within the coherence length of the light. Because of the low coherence length, the signal falls off rapidly with delay mismatch. This is the reason for using a short coherence length light source. If the light source had a longer coherence length, constructive–destructive interference would be possible over a long scanning distance and, therefore, would lose depth resolution. For a light pulse of short coherence length, the reference mirror probes different depths z_n, i.e., the reference mirror scanning range u translates the delay path into a depth z_n. Both beams are fed into the fiber coupler for interference. At the photodetector, the constructive inference intensity pattern is modulated by the reflected intensity from the specific depth. The spectral interferogram $I_D(k)$. can be expressed as [15, 16]:

$$I_D(k) \propto S(k) \sum_{n-1}^{N} \sqrt{R_n R_R} \cos(2kz_n), \qquad (2.12)$$

where $S(k)$ is the optical power density of the light source, $k = 2\pi/\lambda$ is the wavenumber of the light source, R_R is the reflectivity of the reference mirror, and R_n refers to the reflectivity of the sample at depth z_n. The cos-term in eq. (2.12) describes the interference pattern. Figure 2.14(c) shows exemplarily the signals from two different depths, z_1 and z_2. After processing, the signals $A_1(z_1)$ and $A_2(z_2)$ are recorded. A complete image is formed when collecting the reflected light from many more locations as a function of depths z_n and lateral position (x_n, y_n).

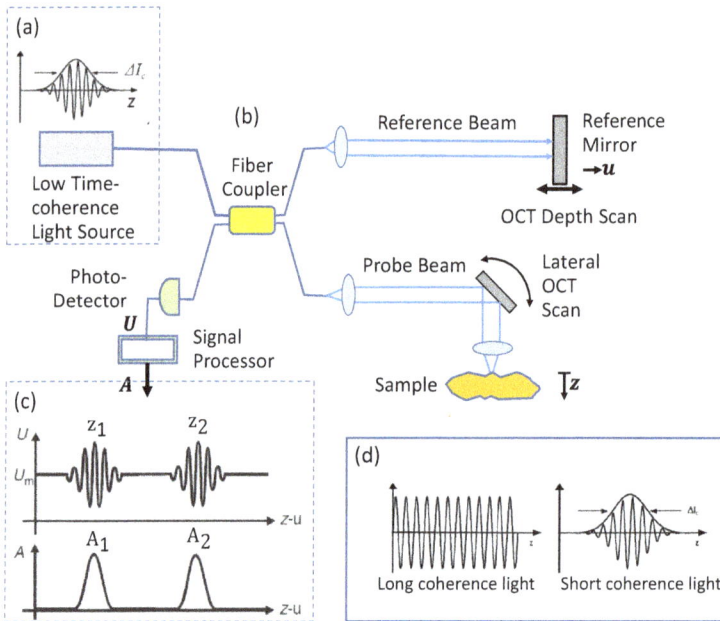

Fig. 2.14: Schematic block diagram of a typical OCT system using light with a short coherence length (adapted from [15, 18]). (a) Light source with short coherence length; (b) Michelson interferometer; (c) interferometry signal before and after processing and for two different depths; and (d) comparison of light sources with infinite long coherence length and short coherence length.

2.8.2 Resolution and scan range

The axial resolution dz is identical to the roundtrip coherence length L_c defined in eq. (2.10): $dz = L_c \approx \lambda/(\Delta\lambda/\lambda)$, where $\Delta\lambda$ is the bandwidth of the light source. With a bandwidth $\Delta\lambda$ of 10% of the wavelength λ, the coherence length comprises five wavelengths: $L_c = \lambda/0.2$. With $\lambda = 1000$ nm, a depth resolution of 3–4 µm can be achieved. This depth resolution is sufficient to identify details in the retina, which has a thickness of about 330 μm [17].

Ophthalmology is one of the main applications of OCT. Another question of interest is to what depth OCT can scan. The axial imaging depth follows from the maximum fringe frequency N, which is achievable within the round trip coherence length: $z = N \times L_c$. In our example, $N=10$. With a wave train of 5 wavelengths, 10 interference fringes can be generated, and the scanning depth is therefore about 50 μm. The coherence length has to be extended to increase the depth range, which, however, decreases the resolution. As in confocal microscopy, the lateral resolution is a question of the spot size. For further information and the theoretical background of OCT, we refer to the reviews in [15, 16].

2.8.3 Additional methods and applications

The OCT imaging method sketched here is the so-called time-domain version (TD-OCT) because the delay time is scanned [9]. In most applications, a much faster swept Fourier domain version is applied, also known as optical frequency domain imaging (OFDI) or simply spectral-domain (SD-OCT) imaging, the description of which goes beyond the scope of this introduction but can be found in [15, 16, 18]. Using SD-OCT, a sweep rate of up to 10 MHz has been reported, more than 1000 times faster than TD-OCT [19].

OCT devices use low-power infrared light with a 750–1300 nm wavelength, which penetrates deeper into the tissue than visible light. The depth of penetration is typically 1–3 mm, depending on the tissue structure and the depth of focus. The light is either from a laser source with an artificially shortened coherence length or an LED with an intrinsically short coherence length. When used in combination with an endoscope, the lateral scanning unit can either be placed on the endoscope's proximal or distal end [21, 22]. Proximal scanning is done with a relay lens across the fiber bundle, and distal scanning is performed by rotation of the laser beam sideways about the optical axis or by moving mirrors for scanning. Figure 2.15(a) shows the cross section of an OCT endoscope with a side-viewing diffractive scanner, and panel (b) reproduces an in vivo image of the esophagus of a guinea pig with an axial resolution of 2.7 μm [22].

3D-OCT has successfully been applied to tissues of inner organs and for scanning the outer skin. Another extensive and successful application of OCT is in ophthalmology, where the thickness of the retina and in particular of the macula can be determined, including imaging of any lesions [17, 18]. Reviews of the OCT technique and its various clinical uses can be found in [20–22].

> OCT in combination with an endoscope provides 3D tomographic images of internal tissues. OCT, in general, is a highly successful imaging modality, used in many medical areas and in particular in ophthalmology.

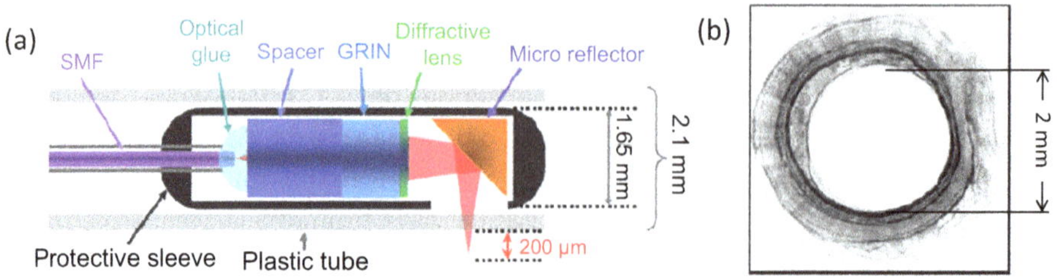

Fig. 2.15: (a) Schematic cross section of an OCT endoscope with a distal diffractive scanning unit for side viewing. (b) Image taken from the esophagus of a guinea pig at one fixed lateral position with an axial resolution of 2.7 µm using SD-OCT (adapted from [22]). SMF, single-mode fiber; GRIN, gradient index lens.

2.9 Capsule endoscopy

Capsule endoscopy is a method of taking images of the digestive tract that cannot be obtained with standard endoscopes [23, 24]. The capsule is the size and shape of a pill and contains a tiny light source, camera, video transmitter, and battery to power the devices. After the patient has swallowed the capsule, it records images of the interior of the gastrointestinal tract at constant time intervals. Typically, two images per second are taken, recorded, and stored by a recording device worn externally on the patient's body. During the examination time, the patients can go about their usual activities. The capsule moves in the intestine through normal contraction of the surrounding muscles. The average transit time through an empty gastrointestinal tract is about 80 min before the capsule is expelled through the anus. At present, neither the capsule's speed nor orientation can be controlled from the outside. The primary use of capsule endoscopy is to examine the small intestine, which is inaccessible to standard endoscopes. In the future, it will be possible to maneuver the capsule externally by adding a small magnet or other robotic functions. This would also open the door for the transition from passive video recording to more active but minimally invasive procedures. The capsule can also be equipped with additional sensors for scanning pressure, temperature, and pH values. The size of a capsule endoscope and its relationship to the small intestine are shown in Fig. 2.16.

An interesting new development in capsule endoscopy is one that is attached to a thin tether with optical fibers that connect the tiny light sensors to an imaging console [23]. After swallowing, the capsule enters the stomach through the esophagus and can be pulled out again after taking pictures. This procedure takes about 2 min and is mainly used to examine Barrett's esophagus syndrome, which is the gastrointestinal reflux of acidic fluid from the stomach into the esophagus (heartburn), a major precursor of esophageal cancer. A rapidly rotating laser tip emits near-

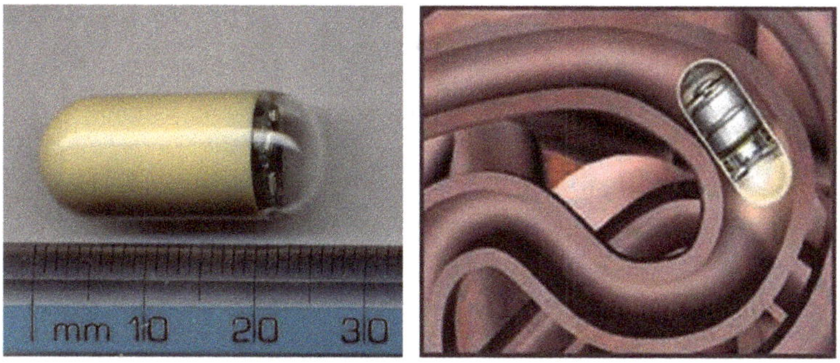

Fig. 2.16: Pill size capsule for endoscopy of the intestinal tract. The capsule contains a light source and a camera. Recorded images are sent to a storage device worn by the patient. The capsule moves by the intestinal muscle activity (reproduced from https://en.wikipedia.org/wiki/Capsule_endoscopy, © creative commons).

infrared light (Fig. 2.17(a)). At the same time, tiny sensors record the light reflected from the lining of the esophagus, which is sent back through the fiber bundle to the image processor. Very sharp microscopic images that were generated with the OFDI-OCT method were recorded. The images reveal structures below the surface that are not easily recognizable with standard endoscopy [24]. Figure 2.17(a) shows the cross section of the tethered capsule endoscope and (b) a picture of manufactured capsule attached to a tether.

2.10 Future trends

The various types of endoscopy have benefitted strongly from rapid developments in nanotechnology, particularly in optics, including lasers and electronics [25]. This trend will most likely continue in the future, although the lateral and in-depth resolution achieved today is already at its optimum for in vivo diagnostics. At present, the focus is on high-resolution imaging for detecting pathological anomalies. For enhanced diagnostics, spectroscopic capabilities may be added to the arsenal of endoscopic tools. Further advances may be seen in the decrease of size (microendoscopy), such that ducts not accessible presently can be examined in the future. Miniaturized capsules may be designed using biodegradable materials. Those could be injected into the vasculature for patrolling potential problem areas like arteriosclerosis.

Furthermore, the capsule could be equipped with tools that remove calcifications and provide drugs to lesions. Miniaturized capsules may be equipped with self-propelling or self-steering capabilities or controlled by an outside console. In general, one may foresee an endoscopic development that carries imaging and diagnostics forward to more spectroscopy and treatment in more narrow and localized areas. Another area of potential development is the combination of different imaging techniques or

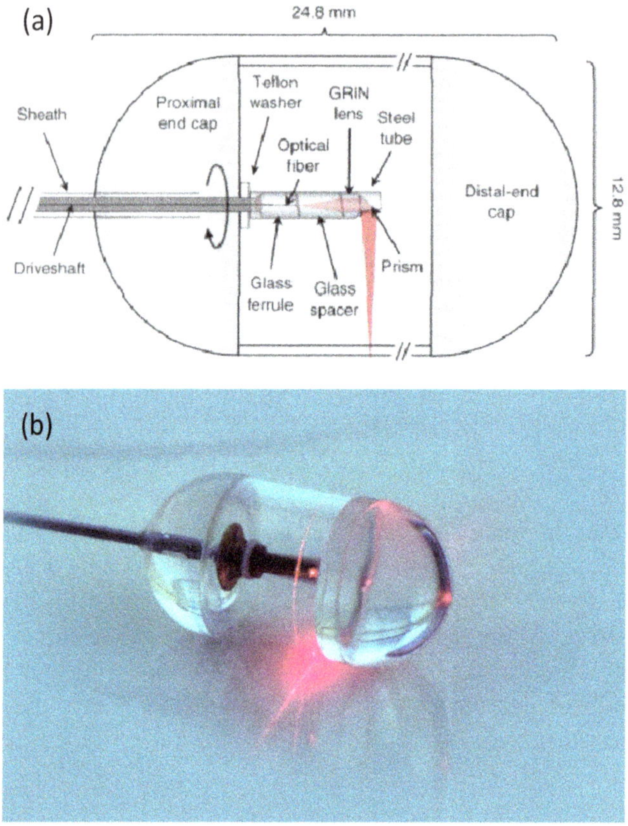

(a)

(b)

Fig. 2.17: (a) Cross section of capsule endoscope, showing the rotatable scanning unit of the infrared laser light and optical sensor. (b) Tethered inch-long endomicroscopy capsule (courtesy Michalina Gora, PhD, and Guillermo Tearney, MD, PhD, Wellman Center for Photomedicine, Massachusetts General Hospital).

medical treatments. For instance, an endoscope may carry a transducer for US imaging, or may be loaded with radioisotopes for local exposure like in brachytherapy (see Chapter 5, Volume 3). A third emerging field is the combination of endoscopy with surgical robotics, where the demands for HD and fast imaging are particularly stringent.

In Tab. 2.1 different imaging methods are compared, which have been discussed in this chapter, including their pros and cons. The last line contains a comparison with sonographic techniques.

Tab. 2.1: Comparison of different imaging modalities.

Imaging modality	Depth sensitivity range	Depth resolution	Pros	Cons
CE	N/A	N/A	Enhanced contrast	Superficial imaging Use of dyes
NBI	N/A	N/A	Enhanced contrast – no dye required	Superficial imaging
CLE	<250 μm	1–5 μm	Cellular resolution	Limited imaging depth Limited FOV
LCI	~250 μm	N/A		Limited detection depth
OCT	3 mm	5–30 μm	3D imaging No contrast agent	No fluorescence imaging, 3D not in real time
US	1–10 cm	50 μm	Broad field of view	Low resolution

CE, chromoendoscopy; NBI, narrowband imaging; CLE, confocal laser endoscopy; LCI, low coherence interferometry; OCT, optical coherence tomography; US, ultrasound; 3D = three dimensional.

2.11 Summary

S2.1 Endoscopy is a method for remote optical inspection of hollow dark spaces in the body.

S2.2 Traditionally, endoscopes consist of a flexible tube that contains two bundles of fiber-optical glass, one for illumination and one for guiding scattered light back to the observer.

S2.3 Endoscopes can be diagnostic for observation only or operative, having channels for irrigation, suction, and the insertion of accessory instruments for surgical procedures.

S2.4 The most common applications of endoscopy are gastroscopy through the esophagus and colonoscopy through the anus. In both cases, endoscopy supports early cancer recognition.

S2.5 Fiber optics is based on the principle of total reflection. For all angles of incidence, which are larger than a critical angle of incidence, the light is reflected into the material with a higher refractive index.

S2.6 In modern endoscopes, a CCD chip at the distal end replaces the fiber-optic bundle for back-reflected light.

S2.7 Endoscopes have been developed for special tasks beyond the capabilities of standard endoscopes. These include imaging with one color (NBI), imaging of stained tissue (chromoendoscopy), taking microscopic pictures (endomicroscopy), for high-resolution confocal imaging (confocal laser endoscopy), and for 3D volume imaging (OCT endoscope).

S2.8   Confocal and OCT endomicroscopies provide morphological information with subcellular resolution for in vivo and real-time biopsy.

S2.9   Fluorescence spectroscopy using fluorophores reveals molecular information.

S2.10 NBI visualizes the upper epithelium and mucosa by blocking out scattered light from blood vessels.

S2.11 Chromoendoscopy is sensitive to infused dyes and stains.

S2.12 Capsule endoscopes contain miniaturized optics for illumination and image recording. The capsule is swallowed, and images from the gastrointestinal tract are transmitted to an external recording device.

S2.13 One of the main advantages of endoscopy is its versatility.

? Questions

Q2.1   What is an endoscope in simple terms?

Q2.2   What are the main components of an endoscope?

Q2.3   Traditionally endoscopes use glass fibers for illumination and for image formation. Which optical boundary condition has to be considered?

Q2.4   What are the most common endoscopic applications in medicine?

Q2.5   What is the typical lateral resolution of an endoscope?

Q2.6   What is the probing depth of an endoscope?

Q2.7   How does a confocal endoscope work?

Q2.8   What are the main features of a confocal microscope? Name at least two.

Q2.9   What is meant by an OCT endoscope?

Q2.10 How can the small intestines be examined?

Q2.11 What is a tethered endoscopic capsule?

Suggestions for home experiment

HE2.1. Have a look through a small round hole in an opaque board and estimate your FOV as a function of distance between eye and board.

HE2.2. Repeat this experiment by looking through a bundle of drinking straws hold together in front of one of your eyes.

⚡ Attained competence checker

	+	0	–
I know what the two main uses of endoscopes are.			
I can name the essential parts of an endoscope.			
I know that fiber optics works differently compared to lens optics.			
I realize that endoscopes are not only used for imaging but also for laparoscopy.			

I realize that there are many modifications of the basic endoscope design for special applications.

I know what narrowband endoscopy means and what it does.

I appreciate the capabilities that confocal microscopy adds to a standard endoscope.

I can describe the essential working principle of an OCT.

I realize that OCT is used in many areas of medical diagnostics.

Exercises

E2.1 **Optic fiber:** Show that eq. (2.3) for the maximum opening angle of an optic fiber is correct.

E2.2 **Refractive index:** Consider a single fiberglass in air. What is the maximum refractive index n_1 of the fiberglass that allows light to enter from the front side?

E2.3 **Order of interference:** Consider a 1 µm diameter fiber, a 500 nm light source, and a fiberglass with a refractive index $n_1. = 1.5$ embedded in a matrix with $n_2=1.2$. What is the maximum order of interference m? How does m change when using a 1000 nm light source?

E2.4 **Longitudinal coherence length:** The longitudinal coherence length L_c is determined by the monochromaticity of the wave train. Consider two monochromatic waves with wavelengths λ and $\lambda + \Delta\lambda$.
(a) What is the number of wavelengths N until both waves interfere destructively?
(b) Then determine the coherence length $L_c = N\lambda$.
(c) Assuming $\lambda = 1000$ nm, and $\Delta\lambda = 0.1\lambda$, what is the coherence length?

E2.5 **OCT scanner:** With OCT, a tissue of 200 µm depth should be probed, using a light source with mean wavelength $\lambda = 1000$ nm. What is the optimum coherence length, resolution, and the number of fringes to be scanned if the required resolution is set to 5% of the scanning depth?

References

[1] Cotton PB, Williams CB. Practical gastrointestinal endoscopy: The fundamentals. 6th. Oxford: – Wiley and Sons, Blackwell Publishing; 2008.
[2] Tsunoda K, Tsunoda A, Ishimoto S, Kimura S. Clinical applications of commercially available video recording and monitoring systems: Inexpensive, high-quality video recording and monitoring systems for endoscopy and microsurgery. Surg Technol Int. 2006; 15: 41–43.
[3] Teh JL, Shabbir A, Yuen S, So JB. Recent advances in diagnostic upper endoscopy. World J Gastroenterol. 2020; 26(4): 433–447.
[4] Aoki H, Yamashita H, Mori T, Fukuyo T, Chiba T. Ultrahigh sensitivity endoscopic camera using a new CMOS image sensor: Providing with clear images under low illumination in addition to fluorescent images. Surg Endosc. 2014; 28: 3240–3248.
[5] Bhat YM, Dayyeh BK, Chauhan SS, Gottlieb KT, Hwang JH, Komanduri S, Konda V, Lo SK, Manfredi MA, Maple JT, Murad FM, Siddiqui UD, Banerjee S, Wallace MB. High-definition and high-magnification endoscopes. Gastrointest Endosc. 2014; 80: 919–927.

[6]   Emura F, Saito Y, Ikematsu H. Narrow-band imaging optical chromocolonoscopy: Advantages and limitations. World J Gastroenterol. 2008; 14: 4867–4872.

[7]   Kurumi H, Nonaka K, Ikebuchi Y, et al. Fundamentals, diagnostic capabilities and perspective of narrow band imaging for early gastric cancer. J Clin Med. 2021; 10: 2918.

[8]   Singh R, Chiam KH, Leiria F, Pu LZCT, Choi KC, Militz M. Chromoendoscopy: Role in modern endoscopic imaging. Transl Gastroenterol Hepatol. 2020; 5: 39.

[9]   Fischer EG. Nuclear morphology and the biology of cancer cells. Acta Cytol. 2020; 64: 511–519.

[10]   Sung KB, Richards-Kortum R, Follen M, Malpica A, Liang C, Descour MR. Fiber optic confocal reflectance microscopy: A new real-time technique to view nuclear morphology in cervical squamous epithelium in vivo. Opt Express. 2003; 11: 3171.

[11]   Meschieri A, Pupelli G, Pellacani G, Rajadhyaksha M, Longo C. Reflectance confocal microscopy: A new tool in skin oncology. Photon Lasers Med. 2013; 2: 277–285.

[12]   Paddock SW. Principles and practices of laser scanning confocal microscopy. Mol Biotechnol. 2000; 16: 127–149.

[13]   St. Croix CM, Shand SH, Watkins SC. Confocal microscopy: Comparisons, applications, and problems. BioTechniques. 2005; 39: S2–S5.

[14]   Huang D, Swanson EA, Lin CP, et al. Optical coherence tomography. Science. 1991; 254: 1178–1181.

[15]   Fercher AF, Drexler W, Hitzenberger CK, Lasser T. Optical coherence tomography – Principles and applications. Rep Prog Phys. 2003; 66: 239–303.

[16]   Izatt JA, Choma MA. Theory of optical coherence tomography. In: Drexler W, Fujimoto JG, eds. Optical coherence tomography. Biological and medical physics, biomedical engineering. 47–82. Berlin, Heidelberg: Springer; 2008.

[17]   Myers CE, Klein BE, Meuer SM, Swift MK, Chandler CS, Huang Y, Gangaputra S, Pak JW, Danis RP, Klein R. Retinal thickness measured by spectral-domain optical coherence tomography in eyes without retinal abnormalities: The Beaver Dam Eye Study. Am J Ophthalmol. 2015; 159: 445–456.e1.

[18]   Aumann S, Donner S, Fischer J, et al. Optical coherence tomography (OCT): Principle and technical realization. In: Bille JF, ed. High resolution imaging in microscopy and ophthalmology: New frontiers in biomedical optics. Berlin, Heidelberg, New York: Springer; 2019, Chapter 3, 59–85.

[19]   Kim TS, Joo J, Shin I, Shin P, Kang WJ, Vakoc B, Oh WY. 9.4 MHz A-line rate optical coherence tomography at 1300 nm using a wavelength-swept laser based on stretched-pulse active mode-locking. Sci Rep. 2020; 10: 9328.

[20]   Drexler W, Liu M, Kumar A, Kamali T, Unterhuber A, Leitgeb RA. Optical coherence tomography today: Speed, contrast, and multimodality. J Biomed Opt. 2014; 19: 071412.

[21]   Tsai TH, Fujimoto JG, Mashimo H. Endoscopic optical coherence tomography for clinical gastroenterology. Diagnostics. 2014; 4: 57–93.

[22]   Gora MJ, Suter MJ, Tearney GJ, Li X. Endoscopic optical coherence tomography: Technologies and clinical applications. Biomed Opt Express. 2017; 8: 2405–2444.

[23]   Gora MJ, Sauk JS, Carruth RW, Gallagher KA, Suter MJ, Nishioka NS, Kava LE, Rosenberg M, Bouma BE, Tearney GJ. Tethered capsule endomicroscopy enables less invasive imaging of gastrointestinal tract microstructure. Nat Med. 2013; 19: 238–240.

[24] Seibel EJ, Carroll RE, Dominitz JA, Johnston RS, Melville CD, Lee CM, Seitz SM, Kimmey MB. Tethered capsule endoscopy, a low-cost and high-performance alternative technology for the screening of esophageal cancer and Barrett's esophagus. IEEE Trans Biomed Eng. 2008; 55: 1032–1042.

[25] Kaur M, Lane PM, Menon C. Endoscopic optical imaging technologies and devices for medical purposes: State of the art. Appl Sci. 2020; 10: 6865.

Further reading

Maitland KC, Wang TD. Endoscopy. In: Zouridakis G, Moore JE Jr, Maitland DJ, eds. Biomedical technology and devices. 2nd edition. Rota Bacon, Florida: CRC Press; 2013, Chapter 9, pp. 217–246.

Choi Y, Yoon C, Kim M, Yang TD, Fang-Yen C, Dasari RR, Lee KJ, Choi W. Scanner-free and wide-field endoscopic imaging by using a single multimode optical fiber. Phys Rev Lett. 2012; 109: 203901.

Bille JF, ed. High resolution imaging in microscopy and ophthalmology. Berlin, Heidelberg, New York: Springer Verlag, New Frontiers in Biomedical Optics; 2019. open access.

Useful website

http://www.endoatlas.com/atlas_1.html

3 Magnetic resonance imaging

Acronyms frequently used in this chapter:

BOLD	Blood oxygen-level dependence
CA	Contrast agent
CSF	Cerebrospinal fluid
DCE	Dynamic contrast enhancement
DNP	Dynamic nuclear polarization
DWI	Diffusion-weighted imaging
EPI	Echo-planar imaging
ETL	Echo train length
FEG	Frequency encoding gradient
FID	Free induction decay
fMRI	Functional MRI
FSE	Fast spin echo
hMRI	Hyperpolarization MRI
IR	Inversion recovery
mpMRI	Multiparameter MRI
MRI	Magnetic resonance imaging
MRSI	Magnetic resonance spectroscopic imaging
MRT	Magnetic resonance tomography
M_{xy}	Transverse magnetization
M_z	Longitudinal magnetization
PD	Proton density
PEG	Phase encoding gradient
PHIP	Para-hydrogen-induced polarization
rf	Radio frequency
SSG	Slice selection gradient
STIR	Short time of inversion recovery
$T1$	Longitudinal relaxation time, or spin–lattice relaxation time
$T2$	Transverse relaxation time, or spin–spin relaxation time
TE	Time to echo, also called spin-echo time
TR	Time of repetition

Selection of MRI physical parameters

Axial magnetic field	1.5–7 T
Proton Larmor frequency at 1 T	42.58 MHz
Axial slice resolution	~1 mm
Receiver frequency bandwidth	50–100 kHz
Frequency interval per voxel	Typically 80 Hz
Field gradients	Typically 20–30 mT/m
Acquisition time per slice	1–5 s
Frames/s in turbo-mode	10–20
Boil-off of liquid He	1.5–2 L/day

https://doi.org/10.1515/9783110757095-003

3.1 Introduction

It was an important scientific discovery when Rabi[1] first observed nuclear magnetic resonance (NMR) in atomic beams in 1938 [1]. However, unlike when Röntgen discovered x-rays, the implications for medical diagnostics were by no means obvious. The principle of NMR applied to condensed matter was demonstrated by Purcell[2] and Bloch[3] and was published in 1946 [2, 3]. The next breakthrough came in 1973 when Lauterbur[4] and Mansfield[5] independently demonstrated that local resonance conditions created by a magnetic field gradient can be used to image objects in space [4, 5]. Magnetic resonance imaging (MRI), also known as magnetic resonance tomography (MRT), has come a long way since those early demonstrations. MRI was introduced into clinical practice in 1984 to enable imaging of internal organs and to visualize tumor tissue. Since then, MRI has evolved into many different sub-branches, such as angio-MRI, multiparameter MRI (mpMRI), functional MRI (fMRI), gated MRI, diffusion-weighted MRI, or hyperpolarization MRI (hMRI).

The MRI technique has many complex facets, not all of which can be considered in this brief chapter. Rather, the basic principles should become clear. Almost all explanations refer to the so-called proton–MRI. A few more specialized procedures are discussed at the end of this chapter. References to more detailed literature on MRI theory and applications can be found under "Further reading".

Common to all MRI examinations is the generation of high-contrast cross-sectional images of body parts using radio wave proton resonance excitation and magnetic flux detection. Therefore, the terms "imaging" and "tomography" both correctly describe the procedures. However, the term "tomography" implies that three-dimensional (3D) images of body parts below the surface are produced, which is more than just "imaging" the surface. Therefore, the term "MRT" would more accurately describe this imaging modality than MRI. Unfortunately, the abbreviation "MRT" is already used in conjunction with many other biochemical and medical procedures. Therefore, following the convention in the international literature, we use the acronym "MRI" in this chapter.

3.2 Nuclear spin basics

NMR is an experimental method that determines resonance frequencies of nuclei with a finite *nuclear magnetic moment* in a magnetic field. Protons and neutrons are

---

1 Rabi, Isodor Isaac (1898–1988), American physicist and recipient of the Nobel Prize in Physics 1944.
2 Eduard Mills Purcell (1912–1997), American physicist, Nobel Prize in Physics 1952.
3 Felix Bloch (1905–1983), Swiss-Austrian physicist, Nobel Prize in Physics 1952.
4 Paul Christian Lauterbur (1929–2007), American chemist, Nobel Prize in Medicine 2003.
5 Peter Mansfield, British physicist, Nobel Prize in Medicine 2003.

Fermi[6] particles with an intrinsic nuclear spin $S = 1/2$. The magnetic moments associated with their spins are in terms of the nuclear magneton μ_N:

$$\text{Proton: } \mu_p = g_p S \mu_N = g_p \frac{1}{2} \mu_N = 2.793 \mu_N,$$

$$\text{Neutron: } \mu_n = g_n S \mu_N = g_n \frac{1}{2} \mu_N = -1.913 \mu_N, \tag{3.1}$$

where the nuclear magneton is

$$\mu_N = \frac{e\hbar}{2m_p} = 5.05783 \times 10^{-27} \text{ J/T}, \tag{3.2}$$

$g_{p,n}$ are the nuclear g-factors for protons and neutrons, e is the elementary charge, and m_p is the proton mass. The minus sign in eq. (3.1) indicates that in neutrons spin and magnetic moment are oriented antiparallel, whereas in protons spin and magnetic moment are parallel, as sketched in Fig. 3.1. The nuclear magneton μ_N is by a factor of 1836 smaller than the Bohr magneton of electrons because of the mass ratio $m_p/m_e = 1836.15$.

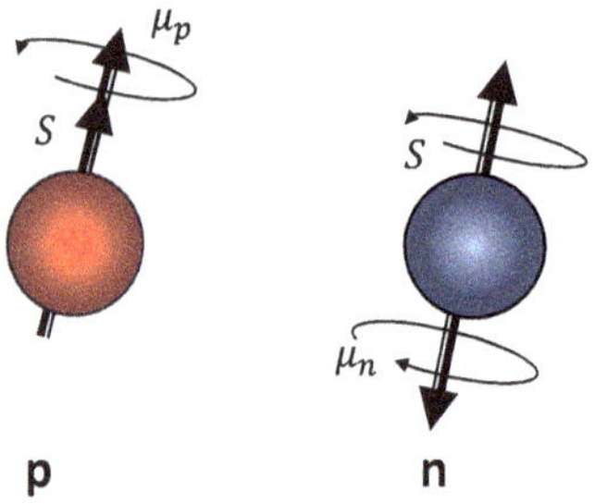

Fig. 3.1: Nuclear spin model. Protons have parallel spin and magnetic moment, and neutrons have antiparallel spin and magnetic moment.

According to Dirac,[7] Fermi particles with a spin $S = 1/2$ should have g-factor of 2 as for electrons. However, the fact that $\mu_p \neq 1\mu_N$ and $\mu_n \neq 0\mu_N$ shows that protons and neutrons are not simple Dirac particles. This remarkable effect indicates that neutrons contain charge currents without being charged. In fact, protons and neutrons are composed of quark particles, and their quark composition explains the unexpected magnetic moments.

When protons and neutrons combine in nuclei, spins of neutrons pair up antiparallel, and spins of protons also pair up. Even number isotopes, therefore, have zero *angular momentum* and zero magnetic moment. For instance, in ^{12}C there are six protons and six neutrons and these pair up like in atomic shells: two neutrons and two protons fit in the $1s_{1/2}$ nuclear shell, illustrated in Fig. 3.2. The next subshell $1p_{3/2}$ is occupied by four neutrons and four protons, resulting in a total nuclear angular

6 Enrico Fermi (1901–1954), Italian physicist, Nobel Prize in Physics 1938.
7 Paul Dirac (1902–1984), English theoretical physicist and Nobel Prize recipient 1933.

momentum $I = 0$. Nuclei with total angular momentum $I = 0$ cannot be used for NMR experiments. However, ^{13}C has one unpaired neutron spin, which goes into the sub-shell $1p_{1/2}$ with a total nuclear angular momentum $I = L + S = 1 - 1/2 = 1/2$. Here, L and S are the orbital and spin angular momentum of the nucleus in units of the Planck[8] constant $\hbar = h/2\pi$, respectively.

An interesting example is the case of oxygen. ^{16}O has a total nuclear angular momentum $I = 0$, as expected for an even–even nucleus. Adding one neutron to get the isotope ^{17}O, one would expect $I = 1/2$. But instead the angular momentum is $I = 5/2$. This is because the extra neutron occupies the nuclear subshell $1d_{5/2}$ with $L = 2$ and $S = 1/2$, yielding $I = L + S = 5/2$ (Fig. 3.2). ^{13}C and ^{17}O isotopes are used for studying carbon and oxygen perfusion in the body via MRI. Although the signal is rather weak and the spatial resolution is comparatively poor, it is one of the new trends that will be discussed in later sections.

Nuclei with total angular momentum $I > 1$ have a nonspherical magnetization distribution contributing a quadrupole moment to the resonance condition. In 98% of the cases, MRI is performed on single protons with spin $S = 1/2$ and nuclear magnetic moment $\mu_p = 2.793\mu_N$, featuring a spherical magnetization distribution. Therefore, in the following discussions we can safely neglect the quadrupole moment. More about nuclear properties is presented in Chapter 5 of this volume.

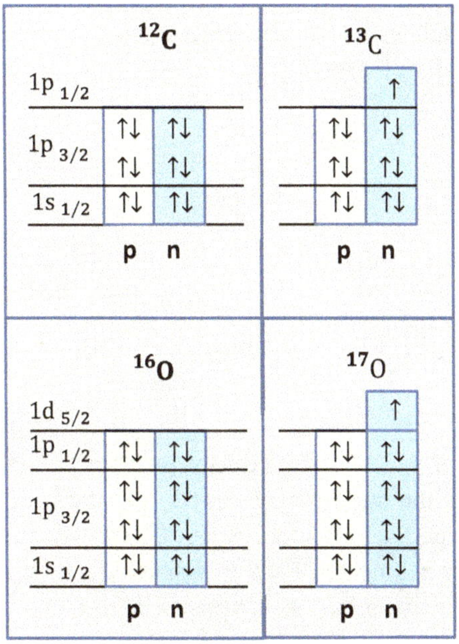

Fig. 3.2: Nuclear shell model and nuclear spin structure for isotopes of carbon and oxygen.

---

8 Max Planck (1858–1947), German theoretical physicist, Nobel prize in Physics 1919.

3.3 Nuclear magnetic resonance basics

The physics of NMR provides the basic concept for MRI. The following section gives a brief introduction into these concepts. Readers familiar with NMR may skip this section and go directly to Section 3.7, specific to MRI techniques. For those seeking a more fundamental introduction to NMR, the book listed under "Further reading" by Slichter[9] on the *Principles of Magnetic Resonance* is highly recommended.

3.3.1 Zeeman splitting

The magnetic moment of a nucleus with a total angular momentum $\vec{I}$ is

$$\vec{\mu}_I = g_N \frac{e}{2m_p} \vec{I} = \frac{g_N \mu_N}{\hbar} \vec{I} = \gamma_N \vec{I}. \tag{3.3}$$

$\vec{\mu}_I$ has the units: $[\vec{\mu}_I] = J/T$. Here $\mu_N = e\hbar/2m_p$ is the *nuclear magneton* (see eq. (3.2)), m_p is the proton mass, g_N is the nuclear g-factor, and γ_N is the *gyromagnetic ratio* that relates the angular momentum $\vec{I}$ to the nuclear magnetic moment, given by

$$\gamma_N = \frac{g_N \mu_N}{\hbar}. \tag{3.4}$$

For isolated protons with spin quantum number $S = 1/2\hbar$, the g-factor has the value $g_p = 5.585$. Therefore, the *gyromagnetic ratio for protons is*

$$\gamma_p = \frac{g_p \mu_N}{\hbar} = 2.675 \times 10^8 \text{ T}^{-1}\text{rad s}^{-1},$$

and

$$\frac{\gamma_p}{2\pi} = \frac{g_p \mu_N}{h} = 42.57 \frac{\text{MHz}}{\text{T}}.$$

The energy of a magnetic moment $\vec{\mu}_I$ in a *magnetic induction* field $\vec{B}$ is

$$E = -\vec{\mu}_I \cdot \vec{B} = -\gamma_N \vec{I} \cdot \vec{B}. \tag{3.5}$$

This expression is known as the *Zeeman*[10] *energy*. The unit of the magnetic induction B is Tesla (T).[11] The energy is lowered when the magnetic moment $\vec{\mu}_I$ and the magnetic induction $\vec{B}$ are parallel instead of antiparallel. Assuming that the magnetic induction is oriented parallel to the laboratory z-direction: $\vec{B} = B_z \vec{e}_z$, where $\vec{e}_z$ is a unit vector in the z-direction, then the Zeeman energy for the z-component is accordingly

9 Charles P. Slichter, (1924–2018), American Condensed Matter Physicist.
10 Pieter Zeeman (1865–1943), Dutch physicist, Nobel Prize laureate in Physics 1902.
11 Nikola Tesla (1856–1943), Serbian-Amercan phsicist, electrical engineer, and inventor.

$$E_z = -\mu_{I,z}B_z = -\gamma_N \hbar m_z B_z, \tag{3.6}$$

where m_z is the z-component of the angular momentum $\vec{I}$ with $(2I+1)$ degenerate energy eigenstates. In a magnetic field, these $(2I+1)$ eigenstates have different energies.

For protons, $I = S = \hbar/2$ and, therefore, we have two eigenstates for $m_z = \pm 1/2$. The energy splitting between these two eigenstates is (Fig. 3.3)

$$\Delta E = \gamma_N \hbar \frac{1}{2} B_z - \gamma_N \hbar \left(-\frac{1}{2}\right) B_z = \gamma_N \hbar B_z. \tag{3.7}$$

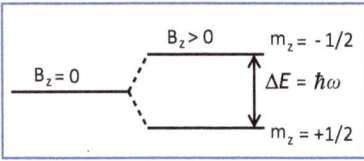

Fig. 3.3: Zeeman energy splitting of $S=1/2$ nuclear eigenstates in an external magnetic induction field B_z.

Expressing the Zeeman energy splitting in terms of frequencies:

$$\Delta E = \hbar \omega_L = \gamma_N \hbar B_z. \tag{3.8}$$

We find for the frequency:

$$\omega_L = \gamma_N B_z \tag{3.9}$$

ω_L is called the *Larmor frequency* of the proton. For the Larmor[12] frequency of protons, we find the following relation, where the units of B_z are in Tesla:

Tab. 3.1: Gyromagnetic ratios and Larmor frequencies of isotopes used for MRI.

Isotope	Nuclear spin	Gyromagnetic ratio γ_N (10^6 rad/s T)	Larmor frequency f (MHz/T)
^{1}H	1/2	267.513	42.576
^{3}He	1/2	−203.789	−32.434
^{13}C	1/2	67.262	10.705
^{15}N	1/2	−27.116	−4.316
^{17}O	5/2	−36.264	−5.772
^{19}F	1/2	251.662	40.052
^{31}P	1/2	108.291	17.235

Positive signs indicate that nuclear magnetic moment and angular momentum are parallel (proton-like), negative signs indicate antiparallel orientation (neutron-like). The precession is clockwise for positive sign and anticlockwise for negative sign.

---

12 Joseph Larmor (1857–1942), Irish physicist and mathematician.

$$\omega/2\pi = f(\text{MHz}) = 42.58 \times B_z(\text{MHz}). \tag{3.10}$$

This frequency is in the range of radio frequencies (rf) and sets the stage for MRI experiments, which usually operate in fields of 1–7 T, corresponding to frequencies from 40 to 300 MHz. The resonance frequency of 42.58 MHz equals 0.2 μeV on the energy scale or 2 mK on the temperature scale. The gyromagnetic ratio and the Larmor frequencies of some isotopes used for MRI are listed in Tab. 3.1.

ℹ️ Infobox I: Magnetic induction and magnetic field

These two terms are often mixed in the literature. However, they can easily be distinguished. Magnetic fields are created by electric currents such as in wires or solenoids. The Danish physicist Hans Christian Oersted (1777–1851) demonstrated for the first time that electric currents generate magnetic fields. Therefore, these magnetic fields H are called "Oersted fields," and their unit is $[H] = \text{A/m}$.

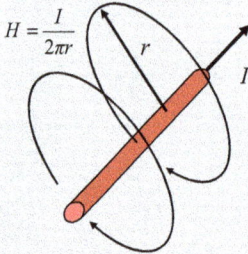

$$H = \frac{I}{2\pi r}$$

Oersted field produced by electric current in a wire.

There is a second source of magnetic fields: permanent magnets, like iron. Without any visible electric current, ferromagnets also produce magnetic fields, which we denote by M for magnetization. M has the same units as H: $[M] = \text{A/m}$. The sum of Oersted field and magnetization, multiplied with the magnetic field constant of the vacuum μ_0 yields the magnetic induction $\vec{B}$, now properly written as vectors:

$$\vec{B} = \mu_0\left(\vec{H} + \vec{M}\right).$$

μ_0 has the units: $[\mu_0] = \text{V s/A m}$ and the value $\mu_0 = 4\pi \times 10^{-7}\,\text{V s/A m}$. Therefore, the units of B is $[B] = \text{V s/A m} \times \text{A/m} = \text{V s/m}^2 = \text{T}$. Here, T stands for Tesla. The other symbols have their usual meaning: A, ampere; V, volt; s, second; m, meter.

⚠️ Larmor precessional frequency of protons in a 1 T magnetic induction field is: $f = 42.58$ MHz.

3.3.2 Equation of motion

Because of *angular momentum* conservation, changes of angular momentum $\vec{I}$ in space and time are only possible by application of a torque $\vec{T}$:

$$\frac{d\vec{I}}{dt} = \vec{T}. \tag{3.11}$$

Any magnetic induction which is not exactly parallel to the magnetic moment will cause a torque on the nuclear magnetic moment:

$$\vec{T} = \vec{\mu}_I \times \vec{B}, \tag{3.12}$$

such that the magnetic moment will precess about the external field. The *equation of motion* for the angular momentum $\vec{I}$ is then

$$\frac{d\vec{I}}{dt} = \vec{\mu}_I \times \vec{B}, \tag{3.13}$$

or

$$\frac{d\vec{\mu}_I}{dt} = \gamma_N \vec{\mu}_I \times \vec{B}. \tag{3.14}$$

The *precession* is clockwise for a positive particle like the proton (see Fig. 3.4).

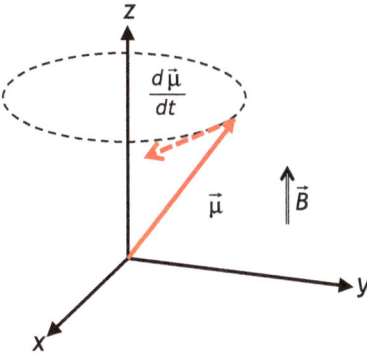

Fig. 3.4: Clockwise precessional motion of a nuclear spin about the external magnetic induction B.

3.3.3 Magnetization of a two-level system

In the following, we consider an ensemble of N nuclear moments $\vec{\mu}_I$ in a volume V, which together yield a *magnetization*:

$$\vec{M} = \frac{1}{V} \sum_{i=1}^{N} \vec{\mu}_{I, i}. \tag{3.15}$$

Therefore, the magnetization is defined as the sum vector of all nuclear moments in a unit volume. As these nuclear moments $\vec{\mu}_I$ are in random motion in space and time, the time average magnetization is $\vec{M} = 0$. Nevertheless, we take a statistical view and define the magnetization as the ensemble average over all magnetic moments in a defined volume:

$$\langle \vec{M} \rangle = \frac{1}{V} \left\langle \sum_{i=1}^{N} \vec{\mu}_{I,i} \right\rangle = \frac{N}{V} \langle \mu \rangle. \tag{3.16}$$

The brackets indicate an ensemble average over all configurations in equilibrium at temperature T. The temperature is defined by contact with a thermal bath of temperature T. Without field, $\langle \vec{M} \rangle = 0$. In a magnetic field, the moments partially line up and yield a finite magnetization $\langle \vec{M} \rangle > 0$. We neglect the brackets here and in the following since $\vec{M}$, in this context, is always understood as a statistical average over individual moments. Now we assume that $B = B_z$, i.e., parallel to the z-direction. Then in thermal equilibrium at temperature T, the magnetization components are

$$M_x = 0, \; M_y = 0, \; M_z = \chi B_z, \tag{3.17}$$

where χ is the *nuclear spin paramagnetic susceptibility*. In the simplest form, χ is a scalar and therefore M_z is linearly proportional to B_z.

For a two-energy level system of protons, the average magnetization in thermal equilibrium follows from the occupational number N_1 of magnetic moments being in the lower energy level, minus the number of moments N_2 being in the upper energy level (Fig. 3.5):

$$M_z = \frac{N_1 - N_2}{V} \mu_z, \tag{3.18}$$

where at finite temperatures $N_1 > N_2$.

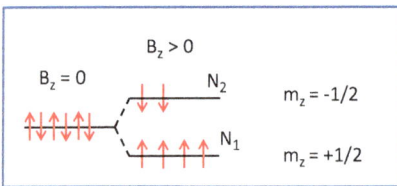

Fig. 3.5: Occupation of energy levels by 6 spins with $B_z = 0$ and $B_z > 0$ in thermal equilibrium.

In this *two-level system* with an energy difference of $\Delta E = \hbar \omega_L = \gamma_N \hbar B_z$ according to eq. (3.8), the ratio of the occupational numbers in thermal equilibrium is thus

$$\frac{N_2}{N_1} = \exp\left(-\frac{\gamma_N \hbar B_z}{k_B T}\right), \tag{3.19}$$

where k_B is the Boltzmann[13] constant. Inserting this ratio into the average magnetization of eq. (3.18) yields (for a derivation, see Exercise E3.1)

$$M_z(T) = \frac{N}{V}\langle\mu_z\rangle\tanh\left(\frac{\gamma_N\hbar B_z}{2k_B T}\right),\qquad(3.20)$$

where $M_z(T)$ is the equilibrium magnetization of all nuclear spins in a magnetic
field B_z at temperature T. Using realistic numbers for a 1 T field and a temperature of 300 K, the ratio $\Delta N/N$ is 6×10^{-6}. This means that out of 10^6 spins, the occupational difference between the upper and lower states will only be six spins, a
negligible amount! However, in 1 mol of protons, the occupational difference is
$6\times10^{-6}\times6\times10^{23}\approx4\times10^{18}$ protons, which becomes detectable, as we will see later.

In the case that the magnetization is not in equilibrium, it will change in time
proportional to the deviation from equilibrium $\Delta M = M_0 - M_z$ until it reaches equilibrium again:

$$\frac{dM_z(t)}{dt} = \frac{M_0 - M_z}{T_1},\qquad(3.21)$$

where we have set the equilibrium magnetization at temperature T, $M_z(T) = M_0$.
Time integration of this equation yields an exponential approach to equilibrium according to

$$M_z(t) = M_0\left[1 - \exp\left(-\frac{t}{T_1}\right)\right].\qquad(3.22)$$

T_1 is called the *longitudinal relaxation time,* and M_z is the *longitudinal magnetization.*
In MRI literature, T_1 is often written as $T1$. We will switch to this notation later. Note
that T_1 is not a temperature but a characteristic time, while t is the actual laboratory
time. Longitudinal means that the modulus of $M_z(t)$, i.e., $|M_z(t)|$ changes with time
during relaxation along the z-direction, parallel to the applied magnetic induction
B_z. For instance, $M_z(t)$ may grow from zero to a finite value, corresponding to thermal

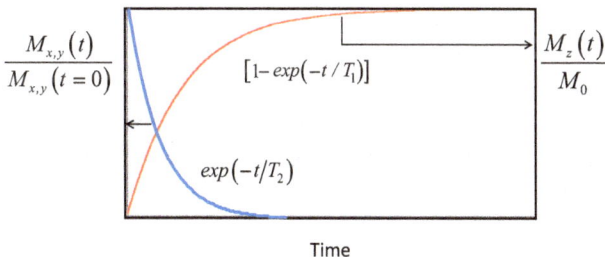

Fig. 3.6: Longitudinal and transverse spin relaxation.

---

13 Ludwig E. Boltzmann (1844–1906), Austrian theoretical physicist.

equilibrium after switching on a magnetic field. The time dependence is shown in Fig. 3.6.

3.3.4 Toy model of magnetization relaxation

A simple "toy" model for magnetization relaxation is shown in Fig. 3.7. Panel (a) shows the magnetic moment components (m_z) parallel to the z-axis and perpendicular to it (m_{xy}). The parallel component is also called longitudinal moment (or magnetization); the perpendicular component is referred to as the transverse component. The protons are omitted for clarity in panels (b) and (c), and only their moments are sketched. At the beginning (panel (b)), there are six magnetic moments equally distributed over two (almost degenerate) energy levels; the energy splitting is taken to be very small but sufficient to define the z-axis in space. The magnetization is zero, as all components of the magnetic moments cancel. As soon as a magnetic induction B_z is turned on, the magnetic moments have to redistribute over these two energy levels (panel (c)). Ultimately, one magnetic moment will flip from the higher energy level (moment antiparallel to the field) to the lower level (parallel to the field). In our simple example, the magnetization M_z versus time after switching on the field would be a step function. But assuming billions of magnetic moments in a field, the average magnetization will change continuously until saturation is reached, as shown in Fig. 3.6 and characterized by the relaxation time T_1. Each moment that flips from the upper to the lower energy level releases energy to the environment. $1/T_1$ is the *energy transfer* rate to the environment.

The *transverse magnetization* M_{xy} is not affected by the change of $M_z(t)$. M_{xy} remains zero during the entire relaxation process since the in-plane components of the magnetic moments are randomly distributed and not in phase.

Taking together the time dependence of the magnetization due to precession and longitudinal relaxation, the equation of motion with relaxation term is now:

$$\frac{dM_z(t)}{dt} = \gamma_N \left(\vec{M} \times \vec{B}\right)_z + \frac{M_0 - M_z}{T_1}. \tag{3.23}$$

The first term on the right side describes the precessional motion of the total magnetization in a magnetic field; the second term is the longitudinal relaxation term of $M_z(t)$ back to equilibrium. During longitudinal relaxation of the z-component, the transverse components of the magnetization $M_x(t)$ and $M_y(t)$ may temporarily be different from zero. In this case, the transverse x, y-components will relax to zero during a *transverse* relaxation time T_2:

$$\frac{dM_x(t)}{dt} = \gamma_N \left(\vec{M} \times \vec{B}\right)_x - \frac{M_x}{T_2}, \tag{3.24}$$

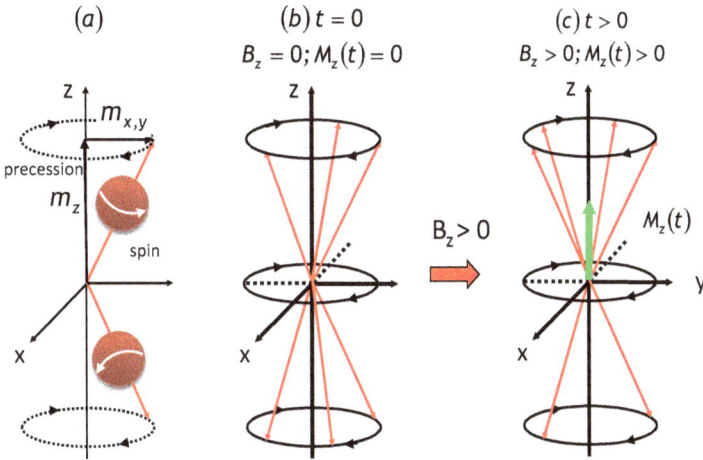

Fig. 3.7: Magnetic moment arrangement before and after turning on a magnetic induction field. (a) Definition of spin and precession of individual magnetic moments with projections parallel and perpendicular to the z-axis; (b) distribution of six magnetic moments over the two possible orientations before turning on a magnetic field; (c) redistribution of magnetic moments in a magnetic field. One moment flips from the higher energy level (antiparallel) to the lower energy level (parallel). The green arrow represents the magnetization M_z. The more moments flip to the lower energy level, the larger is the magnetization M_z.

$$\frac{dM_y(t)}{dt} = \gamma_N \left(\vec{M} \times \vec{B} \right)_y - \frac{M_y}{T_2}. \tag{3.25}$$

These three equations of motion (3.23)–(3.25) are known as the *Bloch equations*. The transverse relaxation time T_2 and the longitudinal relaxation time T_1 are unrelated and may have very different values. Usually $T_2 \ll T_1$. The longitudinal relaxation concerns the modulus of $|M_z(t)|$ and any change indicates flipping of proton spins. T_1 may therefore be considered as the *lifetime of antiparallel spins*. It is mainly controlled by *spin–lattice interaction*, i.e., the interaction of local proton spins with the thermal motion of their environment. The transverse relaxation time T_2 relates to a desynchronization of spin components in the x, y –plane due to stochastic processes and random interactions, also referred to as *spin–spin relaxation*. T_2 cannot directly be measured. But with special techniques described in the following, a field pulse can create some phase coherence of spins in the transverse plane. The dephasing time T_2, in MRI literature often referred to as *T2*, can then be detected by *spin-echo techniques*. Both relaxation times are illustrated with our toy model in Fig. 3.8. In contrast to the previous case, flipping of the spins has caused a finite in-plane magnetization M_{xy}, which vanishes quickly by dephasing, and only M_z remains. The corresponding relaxation times are plotted in Fig. 3.8.

T_1 is the spin-lattice relaxation time, T_2 is the spin-spin relaxation time.

$(a) t = 0$
$B_z = 0$
$M_z(t) = M_{x,y}(t) = 0$

$(b) t > 0$
$B_z > 0$
$M_z(t), M_{x,y}(t) > 0$

$(c) t \gg T_1, T_2$
$B_z > 0$
$M_z(t) = M_{equi}(z), M_{x,y}(t) = 0$

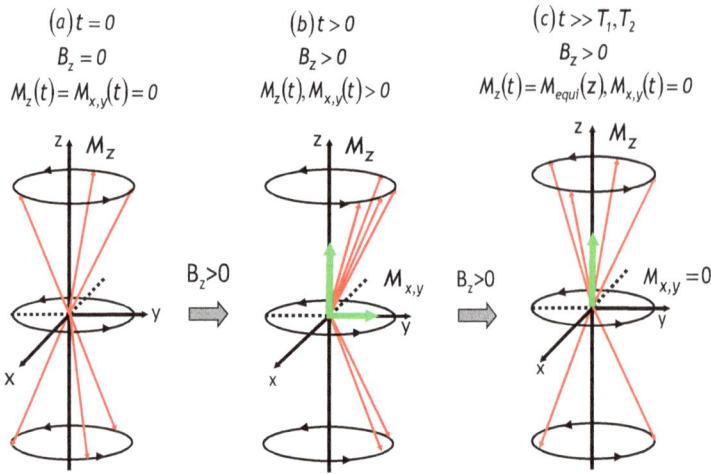

Fig. 3.8: Comparison of T_1 and T_2 relaxation times. (a) The starting situation at $t = 0$ has all moments equally distributed, yielding zero magnetization in the x, y-plane and no magnetization in the z-direction. (b) A pulse field B_z turns one magnetic moment from down to up and bunches up all moments along the y-direction. This causes a finite magnetization $M_z > 0$ and a transverse magnetization $M_{xy} > 0$. (c) While the magnetic field B_z is still turned on, the transverse components quickly dephase within the transverse relaxation time T_2, yielding $M_{xy} = 0$, whereas $M_z > 0$.

3.3.5 Resonance absorption

We are now prepared to design a simple NMR experiment, as illustrated in Fig. 3.9. An electromagnet generates a magnetic field in the z – direction, and the energy splitting of protons in the sample increases with increasing field strength. A rf coil provides an oscillating magnetic induction in the x-direction with a frequency ω_0 referred to as rf – *field*:

$$B_x(t) = B_{x0} \cos(\omega_0 t). \tag{3.26}$$

The x-component exerts a torque that flips the spins from the lower to the upper energy level. This transfer is most efficient if the energy of the oscillating magnetic induction matches the energy splitting:

$$\Delta E = \hbar \omega_0 = \hbar \omega_L. \tag{3.27}$$

Then the rf-power is resonantly absorbed from the coil to the proton spin system. The absorbed energy is detectable by a device sensitive to the field amplitude in the rf – coil. This device is usually a *"pick-up coil"* (see further). The NMR experiment can be performed in two alternative ways: 1. keeping the rf-field frequency constant and sweeping the magnetic induction in the z-direction or 2. keeping the B_z field

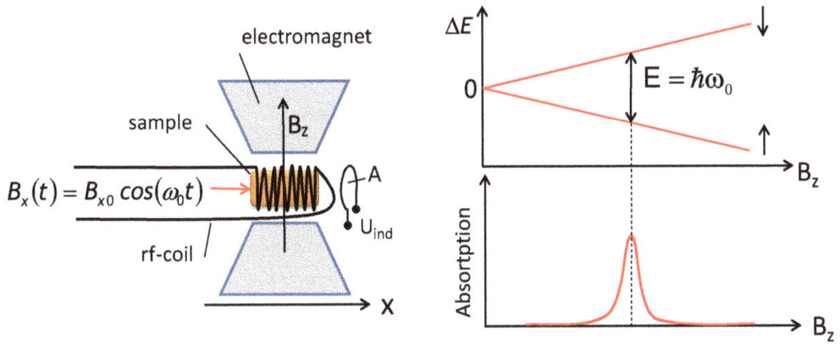

Fig. 3.9: Nuclear magnetic resonance experiment. Resonance absorption occurs at the energy of the rf-field that corresponds to the Zeeman energy splitting. Left panel shows the schematic setup with electromagnet for the B_z–field, rf-coil for the resonance field, and pick-up coil for detecting the oscillating magnetic flux.

constant and sweeping the frequency. Because of technical reasons, the first option is usually realized and schematically shown in Fig. 3.9.

Often in NMR experiments, the resonance field $B_x(t)$ is applied in the form of a short pulse. The *field pulse* is chosen such that the magnetization M_z dips into the xy-plane. This dip corresponds to a 90° rotation about the x-axis. M_z is then reduced or may even be zero during the pulse time. The spins are dragged into synchronization within the x,y plane and therefore create a M_{xy} component that rotates with the Larmor frequency (see Fig. 3.8). While rotating, the moments produce stray magnetic fields that can be measured as a magnetic flux

$$\phi_{xy} = B_{xy}A = \mu_0 M_{xy}A,$$ (3.28)

penetrating through pick-up coils; see Fig. 3.11(b). A is the area of the coil. As the spins precess, the magnetic flux generates an induced voltage according to Faraday's[14] induction law:

$$U_{\text{ind}} = -\frac{d\phi_{xy}}{dt} = -\mu_0 \frac{d(M_{xy}A)}{dt}.$$ (3.29)

The induced voltage is proportional to the damped oscillation of the transverse magnetization M_{xy}, known as *free induction decay* (FID). The corresponding transverse magnetization relaxation is described by

$$M_{xy}(t) = M_0 \sin(\omega_0 t) \cdot e^{-t/T_2}.$$ (3.30)

---

14 Michael Faraday (1791–1867), English physicist.

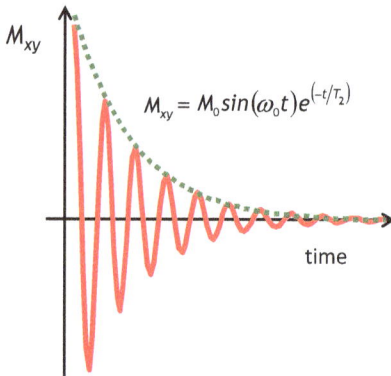

$$M_{xy} = M_0 sin(\omega_0 t)e^{(-t/T_2)}$$

Fig. 3.10: Free induction decay (FID) of transverse magnetization after pulse excitation.

The exponential decay is plotted in Fig. 3.6, and the damped M_{xy} – oscillation is illustrated in Fig. 3.10. Usually, many rf-pulses are needed to collect a sufficient signal amplitude.

The M_{xy} – component decays quickly according to the T_2^* relaxation time, and the spins will switch back into the z–direction until M_z is fully recovered according to the T_1 relaxation time. The magnetization makes a spiral motion about the z-axis, as illustrated in Fig. 3.11. While M_z increases continuously, M_{xy} decreases to zero. Both have their own and independent relaxation times, T_1 and T_2^*. The relaxation time T_2^* is not intrinsic but affected by many external factors, such as field inhomogeneities. Therefore, T_2^* is, physically speaking, not of interest, but very important for recording magnetic resonance images (MRI), as we will see later. T_2 and T_2^* are related through the equation

$$\frac{2\pi}{T_2^*} = \frac{2\pi}{T_2} + \gamma_N B_{inh}, \tag{3.31}$$

where $\gamma_N B_{inh}$ collects all contributions due to field inhomogeneities, referred to as susceptibility effect. The pulse technique allows determining T_1 *indirectly* but is not suitable for measuring T_2. Therefore, another method is necessary that is sensitive to the intrinsic T_2 relaxation time. This will be discussed in the next section on spin-echo procedures.

It is important to note that T_1 cannot be determined directly from the M_z relaxation as this relaxation does not produce a magnetic resonance (MR) signal. The only MR signal that can be recorded with an induction coil is the FID (Fig. 3.11(a)). Therefore, for measuring T_1, the 90° pulse has to be repeated several times with increasing delay time. The measurement scheme is shown in Fig. 3.12. After applying a 90° pulse, M_z relaxes up to the delay time t_1. Then another 90° readout pulse is exerted, and the subsequent FID from the M_{xy} decay is proportional to the amplitude $M_z(t_1)$ at time t_1. This procedure is repeated several times with increasing delay times t_n, always starting after the *time of repetition* (TR). Thus, after TR_1, we start again with

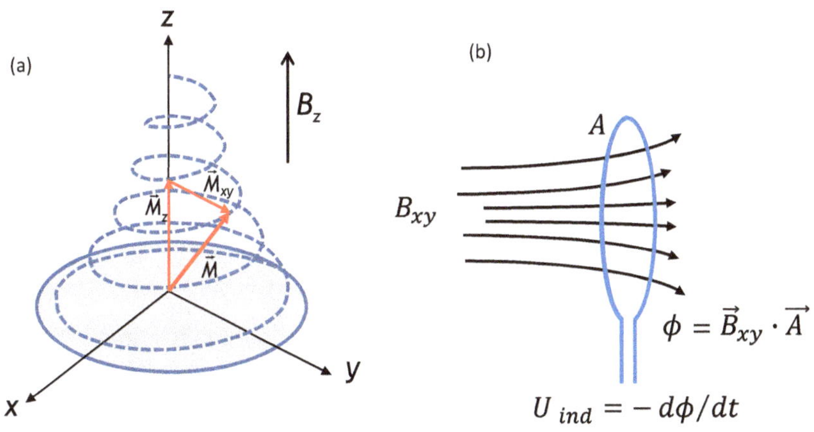

Fig. 3.11: (a) Relaxation of M_z and M_{xy} after a 90° pulse. The relaxation is spiral-like about the z-axis. M_{xy} vanishes completely once M_z is fully recovered. (b) Detection of the stray magnetic field during the M_{xy} relaxation by an induction coil.

an initial 90° pulse followed by a readout 90° pulse after delay time $t_2 > t_1$. The recorded FID is then representative for $M_z(t_2)$ at time t_2, etc. After a couple of more delay times, the relaxation time T_1 can be extracted.

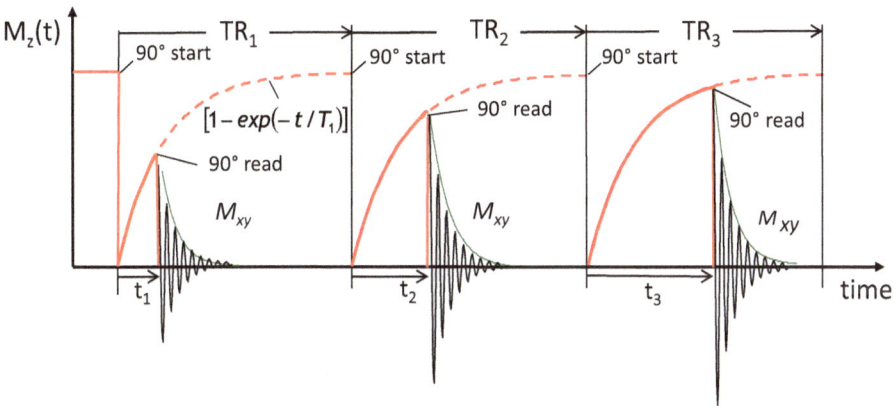

Fig. 3.12: Measurement procedure for determining T_1.

> **!** The longitudinal proton magnetization M_z is proportional to the magnetic induction B_z. An rf-pulse reduces the magnetization, which relaxes back within the characteristic relaxation time T_1.
>
> The transverse magnetization M_{xy}, which occurs after a 90° pulse, is unstable and relaxes back with the characteristic relaxation time T_2. The magnetic flux emanating from the spinning and relaxing M_{xy} magnetization can be recorded as an induced voltage.

3.4 Spin-echo techniques

For the following discussion, it is useful to change the coordinate system. So far, we have watched the precessional dynamics in a Cartesian coordinate system (x, y, z) as observers from the outside, referred to as a laboratory frame. Now we take our position in the rotating frame of the spins (x', y', z'), assuming that we rotate with the same Larmor frequency ω_L as the spins do. Then, in the rotating frame $\omega_L = 0$, and therefore the magnetic induction $B_z = \omega_L/\gamma_N$ in the z' – direction vanishes. The z – and z' – axes are parallel in the rotating frame, and the x, y-plane is parallel to the rotating x', y'-plane. From the perspective of the rotating (x', y', z') frame, the magnetization component M_z is at rest and parallel to the z' – axis (Fig. 3.13(a)).

Now we apply an rf-field B_1 parallel to the x'-axis in the rotating frame. This field exerts a torque on the magnetic moments and rotates the magnetization M_1 in the $y'z'$-plane according to the equation of motion:

$$\frac{dM_1(t)}{dt} = \gamma_N \left(\vec{M}_1 \times \vec{B}_1 \right) \tag{3.32}$$

with the frequency $\omega_1 = \gamma_N B_1$. The different steps are illustrated in Fig. 3.13. At time $t = 0$, $M_1 = M_z$ (panel (a)). In order to turn the magnetization by a finite angle α, a short pulse is applied of duration (panel (b)):

$$\Delta t_{\text{pulse}} = \frac{\alpha}{\gamma_N B_1}. \tag{3.33}$$

We assume that Δt_{pulse} is long enough to cause a 90° *flip* (panel (c)). Following the pulse, an excess magnetization $M_{x'y'}$ appears in the $x'y'$-plane parallel to the y'-axis. Now, $M_{x'y'}$ precesses with an angular frequency $\omega_L = \gamma_N B_z$, as already discussed in Section 3.3.3. However, since we rotate with the same frequency, the magnetization $M_{y'}$ will remain constant along the y'-axis (neglecting some relaxation).

Another pulse for a duration

$$\Delta t_{\text{pulse}}^{180°} = \frac{\pi}{\gamma_N B_1} \tag{3.34}$$

causes a 180° flip, turning the magnetization to the negative $(-y')$ axis (panel (d)). This outcome shows that successive field pulses can flip the transverse magnetization.

We recall that physically only magnetic moments can be flipped and the magnetization is the sum of all moments. The moments can only be flipped between the upper and lower states, but the resultant magnetization may have components parallel to the z'- or y'-direction. Here and in the following, we will only focus on the resultant effective magnetization.

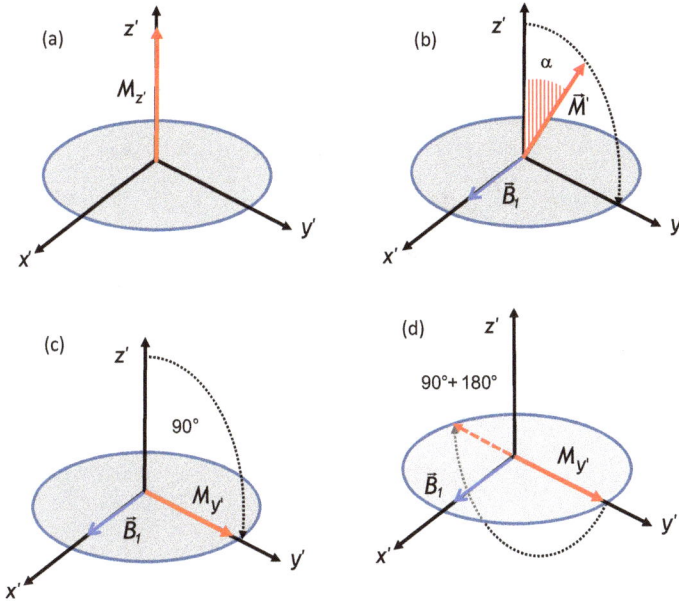

Fig. 3.13: Nuclear magnetic moments in the rotating frame x', y', z'. (a) Starting situation with $M' = M_z$ parallel to the z'-direction; (b) a 90° magnetic field pulse is applied parallel to the x'-direction, turning the magnetization M' into the y'-direction; (c) the magnetization has now only a y'-component; (d) another field pulse is applied parallel to the x'-direction, turning M'_y by 180° to the $-y'$-direction.

Now we go back once more and reconsider the transverse magnetization M'_y after applying the first 90° flip, referring to Fig. 3.14. The first 90° pulse sets the clock to $t = 0$. The transverse magnetization is composed of individual magnetic moments $m_{x'y'}$ all residing in different environments and all precessing at slightly different frequencies. The ones precessing exactly at the Larmor frequency $\omega_L = \gamma_N B_z$ appear not to move in the rotating frame, the faster ones move forward and the slower ones appear to move backward.

The time during which the magnetic moments spread out is called the *dephasing time*. As discussed earlier, the dephasing time is characterized by an FID with a relaxation time T_2^*. At time $t = t_0$ after the first 90° pulse, we apply a second pulse turning the spins by 180°. Since the rotation axis is parallel to the $x'-axis$, the moments at position a switch to position a', and moments at b switch to b' (Fig. 3.14(b,c)).

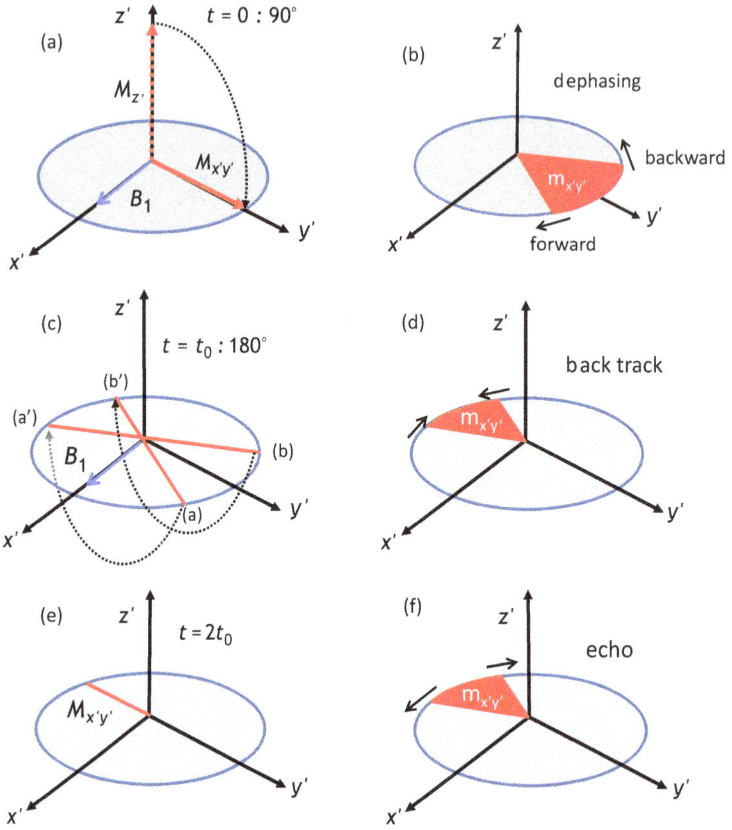

Fig. 3.14: Spin-echo procedure in the rotating frame. (a) Starting at $t = 0$ with a 90° flip of the magnetization M'_z into the $x'y'$ – plane; (b) dephasing of the magnetic moments $m_{x'y'}$ and fanning out. In the laboratory frame, the magnetization M_{xy} rotates with the Larmor frequency and is strongly damped with a decay time T_2^*; (c) at $t = t_0$, a 180° pulse turns the magnetic moments about the x'–axis; (d) the fanned-out magnetic moments $m_{x'y'}$ move back together; (e) the magnetic moments pile up in phase at $t = 2t_0$ and yield again a large $M_{x'y'}$, called echo signal; (f) dephasing starts again for $t > 2t_0$.

The same switching applies to all moments in the fanned-out red area in panels (b) and (c). During the time from t_0 to $2t_0$, we watch the flipped magnetic moments moving: the faster ones move clockwise, the slower ones move counterclockwise, i.e., the fanned out triangle closes at $t = 2t_0$, as indicated in Fig. 3.14(d,e). All strayed out moments are reassembled (neglecting loss due to other relaxation processes) and give a large $M_{x'y'}$ echo signal. For this reason, the 180° pulse is also referred to as *refocusing pulse*. At times $t > 2t_0$, the moments again dephase (Fig. 3.14(f)). This ingenious spin-

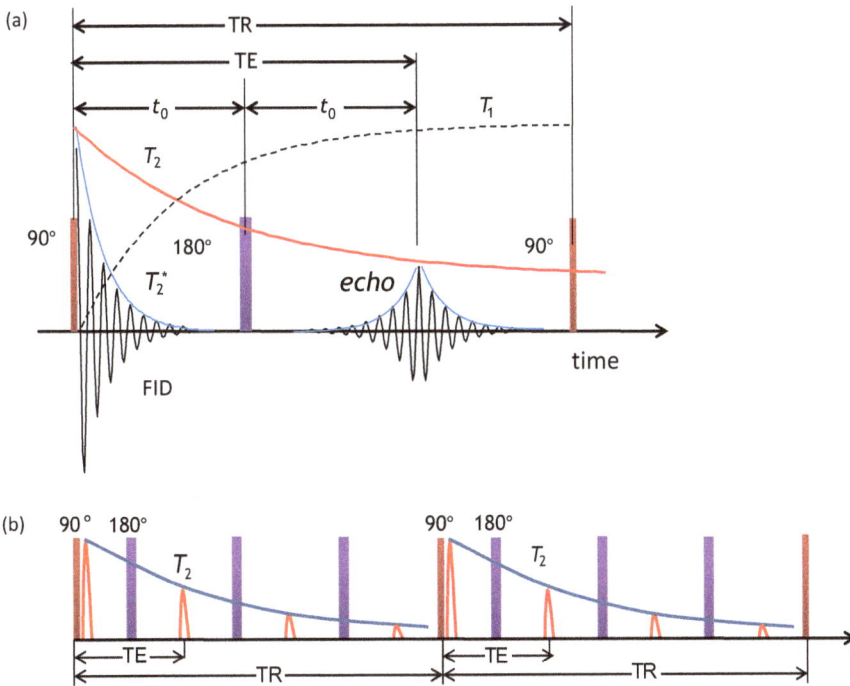

Fig. 3.15: Procedures for T_2 determination. (a) The spin-echo time $TE = 2t_0$ can be successively increased for determining the exponential decay of T_2. The dashed line is the M_z recovery curve. (b) After the first 90° pulse, the rephasing 180° pulse is repeated several times within one TR, and can be repeated in the next TR time span.

echo method was invented by Hahn[15] [6] and opened an entirely new area of pulsed NMR spectroscopy.[16] Nowadays, spin echo is the fundamental principle of MRI.

Figure 3.15 shows the scheme of a T_2 measurement. The first 90° pulse sets the clock. The $M_{x'y'}$-component rapidly decays from a maximum value $M_{x'y'}(0)$ via FID with the relaxation time T_2^*. After a waiting time t_0, a refocusing 180° pulse is applied. Then an echo signal appears in a pick-up coil after the *spin-echo time* $TE = 2t_0$ with the strength proportional to $M_{x'y'}(2t_0)$. The signal strength of this first echo is a measure of the *proton density* (PD) in the sample. From the ratio $M_{x'y'}(2t_0)/M_{x'y'}(0)$ one could determine the intrinsic relaxation time T_2. However, the initial amplitude $M_{x'y'}(0)$ is not well defined. Therefore, T_2 is generally determined by two different methods.

---

15 Erwin Louis Hahn (1921–2016), American physicist.
16 Spin-echo techniques for thermal neutron scattering was invented and described by Ferenc Mezei [7]. This technique is heavily used in thermal neutron spectroscopy, mainly for studying diffusion processes in soft matter.

(1) Referring to the top panel of Fig. 3.15: the spin-echo experiment is repeated with increasing waiting times t_0 and accordingly echo time $2t_0$. Then the echo signal will become smaller and smaller with increasing TE, from which T_2 can be determined. The procedure is similar to the one discussed for determining T_1.
(2) Referring to the bottom panel of Fig. 3.15: after the first spin echo at time TE, the 180° refocusing pulse is repeated several times, always resulting in a spin-echo signal with decreasing amplitude:

$$M_{x'y'}(n \times \mathrm{TE}) = M_0 \exp\left(-n \times \frac{\mathrm{TE}}{T_2}\right),$$

where n is the number of spin-echo repeats. From the decaying amplitudes, T_2 can again be determined.

In both cases, there should be sufficient time for recovering M_z with relaxation time T_1 before the spin-echo experiment is repeated after TR. The MR *signal strength* S produced by M_{xy} at the maximum of the spin echo at t = TE is

$$S \sim \rho_\mathrm{p} \cdot \left(1 - \exp\left(-\frac{\mathrm{TR}}{T_1}\right)\right) \cdot \exp\left(-\frac{\mathrm{TE}}{T_2}\right), \tag{3.36}$$

where ρ_p is the PD in a particular medium (tissue). The first bracket describes the M_z recovery during one SE time. The second exponential decay in eq. (3.36) allows for determining the transverse and intrinsic relaxation T_2. The signal strength S can be an induced voltage in a pick-up coil or any other suitable sensor such as a giant magnetic resistance sensor. The pick-up coil is usually identical with the RF coil used as receiver in between two pulses.

> **!** The spin-echo procedure inverts the transverse M_{xy} magnetization and prompts a refocusing echo pulse. During dephasing of the first pulse and refocusing after the second pulse, FID is observable.
> Spin echo is the single most important concept of MRI.

3.5 Autocorrelation and spectral density (for experts)

Protons in their environment, such as in water, have their characteristic wobbling frequency ω_p, which may be a combination of rotational, vibrational, or diffusional motion. Their characteristic relaxation rate $1/\tau_\mathrm{p} \sim \omega_\mathrm{p}$ depends on whether they are bound in molecules or freely diffusing. This frequency may not be well defined and may span a wide frequency range. In any case, energy transfer to the environment is most effective if the Larmor frequency is close to the characteristic frequency of the system: $\omega_\mathrm{L} \cong \omega_\mathrm{p}$. If the frequencies match, T_1 is short. If there is a large mismatch, energy transfer is inefficient and T_1 increases.

In order to understand the different relaxation times of protons in various environments, we need to know their frequency distribution. This information is obtained by looking at the *spectral density* $J(\omega)$, i.e., the weighted intensity of a particular movement with an associated frequency ω. $J(\omega)$ is of central importance for understanding the formation of contrast in MRI. Therefore, we will discuss this concept in more detail.

As an example, we consider the random diffusional motion of particles in liquids or gases referred to as *Brownian*[17] *motion*. Figure 3.16 illustrates a random walk of a particle in a liquid. At time t, the particle is at $x(t)$, and at a later time $t+\tau$, it is at another position $x'(t+\tau)$. According to Einstein[18] and Smolukowski,[19] the *mean square displacement* of a randomly diffusing particle in 3D during the time increment τ is expressed by [8]

$$\langle(\Delta x^2)\rangle = \langle(x'(t+\tau) - x(t))^2\rangle = 6D\tau, \tag{3.37}$$

where D is the diffusion constant, which can be expressed as follows:

$$D = \frac{k_B T}{6\pi\eta r}. \tag{3.38}$$

Here k_B is the Boltzmann constant, T is the absolute temperature, η is the viscosity of the solvent, and r is the radius of a suspended particle in a solvent. This relation allows to determine r, for instance of a cell nucleus, which was historically very important.

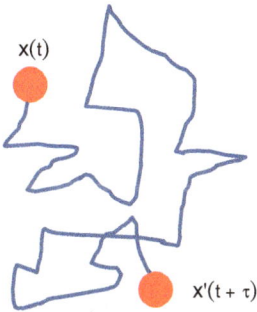

Fig. 3.16: Brownian motion of a particle in a solvent in equilibrium at temperature T.

---

17 Robert Brown (1773–1858), Scottish botanist.
18 Albert Einstein (1879–1955), German-Swiss-American theoretical physicist, Nobel Prize in Physics 1921.
19 Marian von Smoluchowski (1872–1917), Polish physicist.

For the diffusive motion, we define an *autocorrelation function* (also known as self-correlation function) $R(\tau)$ that relates the spatial coordinates of one and the same particle at times t and $t + \tau$:

$$R(\tau) = \lim_{t\to\infty} \frac{1}{2t} \int_{-t}^{+t} x'(t+\tau)x(t)dt. \tag{3.39}$$

Here the integral is taken over all times and is appropriately normalized with the result:

$$R(\tau) = R_0 \exp\left(-\frac{\tau}{\tau_c}\right). \tag{3.40}$$

The autocorrelation function for particles in liquids shows an exponential decay. The time constant is the *diffusional correlation time τ_c*. For times $\tau < \tau_c$, partial correlation exists; for $\tau > \tau_c$, correlation is lost.

Next, we take the Fourier transform of the autocorrelation function $R(\tau)$ by integrating over all possible relaxation times τ. The Fourier transform carries over the relaxation process from the time-space to the frequency space. This transformation yields the so-called *spectral density* of the autocorrelation function, defined by

$$J(\omega) = \int_{-\infty}^{+\infty} R(\tau)\exp(i\omega t)d\tau. \tag{3.41}$$

Inserting the result for $R(\tau)$ from eq. (3.40), and carrying out the integral, we find for the spectral density distribution:

$$J(\omega) = R_0 \int_{-\infty}^{+\infty} \exp\left(-\frac{\tau}{\tau_c}\right)\exp(i\omega t)d\tau \cong R_0 \frac{(1/\tau_c)}{\omega^2 + (1/\tau_c)^2}. \tag{3.42}$$

Thus, $J(\omega)$ has the shape of a Lorentzian curve. The maximum at $\omega = 0$ has the value $J(0) \sim R_0\tau_c$ and the half-height value at $\omega\tau_c = 1$ is $J(1/\tau_c) \sim R_0\tau_c/2$. This shows that the spectral density is high and narrow for long correlation times τ_c but low and broad for short correlation times. Diffusion in solids is comparatively slow, and accordingly, τ_c is large, in contrast to gases or liquids, where we will find much shorter relaxation times.

Let us assume that the gaseous, liquid, and solid materials considered here do not feature single correlation times τ_c but a broad distribution. Then the spectral density is stretched out, characterized by a broad plateau as a function of frequency and a cut-off beyond $\omega\tau_c = 1$.

The spectral densities for our three representative systems are shown qualitatively in Fig. 3.17: solids, viscous liquids, and watery liquids. Note that the area under each curve is identical, as imposed by the condition of thermal equilibrium.

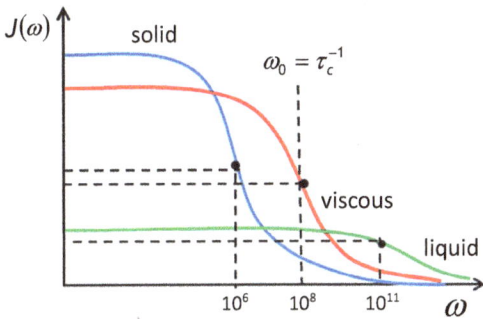

Fig. 3.17: Spectral density of different systems as a function of frequency.

Each system has the same total thermal energy distributed over all possible excitation frequencies at the same temperature.

In solids (blue line), the diffusion is very slow. Therefore, the spectral density is high and narrow at low frequencies. In liquids (green line), the diffusion is distributed over a wide range up to high frequencies. Viscose liquids (red curve) have a spectral density between solids and liquids.

Why do we have to consider the spectral density $J(\omega)$ of different systems? This is because the protons in the NMR experiment (see Fig. 3.9) are exposed to an rf-field and absorb energy from the electromagnetic field. During relaxation, the protons return this energy to the environment in which they sit. This energy exchange is most efficient when the frequencies match, i.e., the Larmor frequency ω_L of the protons and the characteristic frequencies prevailing in the environment. With a reasonable match, the transfer time for energy exchange is fast, and consequently, the T_1 relaxation time is short. Conversely, the T_1 relaxation time increases if the frequencies do not match well for $\omega\tau \gg 1$ or $\omega\tau \ll 1$. The spectral density distribution $J(\omega)$ tells us how good the agreement is. The matching frequencies, for which $\omega\tau = 1$ in different materials, are marked by a solid black dot in Fig. 3.17.

As an example, we consider the frequency $\omega_0 = 1/\tau_c$ in Fig. 3.17. If the Larmor frequency ω_L is close to ω_0, the energy transfer is most efficient in viscous liquids and less so in the other materials. Accordingly, the T_1 relaxation time for the viscous liquid is shorter than for the other materials. In a mixed system, the different materials are distinguishable by their T_1 relaxation times. In fact, these different relaxation times open up the possibility of creating contrasts, as we will see later.

In Fig. 3.18, the relaxation times for T_1 and T_2 are plotted as a function of hydrogen (proton) diffusivity in different environments. The dashed lines correspond to the optimal frequencies indicated by solid dots in Fig. 3.17. In the blue shaded area, the relaxation times are those which prevail in soft tissues of the body. Typical relaxation times for different tissues are listed in Tab. 3.2. Those are important for generating contrast in MRI. In general, $T_1 > T_2$ and $T_2 > T_2^*$ for all tissues. T_2^* is not listed in Tab. 3.2 as it is an extrinsic parameter and may vary depending on machine settings.

According to Tab. 3.2, bulk water has the longest relaxation time because of a mismatch between ω_L and ω_0. However, when protons are attached to proteins, their mobility and diffusivity are much reduced, relaxation becomes more effective, resulting in shorter relaxation times T_1.

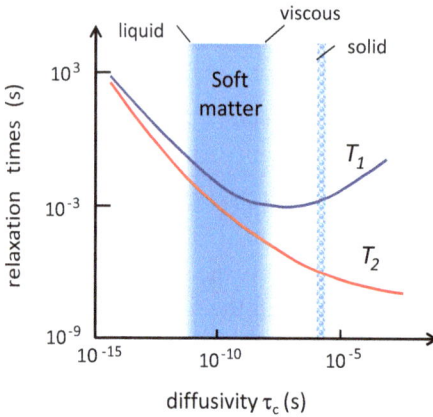

Fig. 3.18: Proton relaxation times T_1 and T_2 are plotted as a function of hydrogen diffusional relaxation time τ_c in different materials. The relaxation times may span over many orders of magnitude.

Tab. 3.2: Relaxation times for different tissues listed for a magnetic induction field of 1 T.

Tissue	T_1 (ms)	T_2 (ms)
Fat	210	80
Liver	400	40
Kidney	550	60
Muscle	750	45
bone marrow	800	100
White matter	650	90
Gray matter	800	100
CSF	2000	150
Bulk water	3000	3000

CSF, cerebrospinal fluid [9].

! The T_1 relaxation time is a matter of frequency match between the Larmor frequency and the frequency distribution in the tissue. The better the match, the shorter the T_1 relaxation time is.

3.6 NMR and MRI procedures

In NMR experiments, two more issues need attention.

3.6.1 Saturation

The amplitude of the rf-field should not be too high. Otherwise, the absorption saturates as soon as the number of spins in the upper and lower levels are equal. Contrast is lost in a saturated state (red line in Fig. 3.19(a)), whereas a signal strength of 50–60% still has variability, as indicated by the blue line.

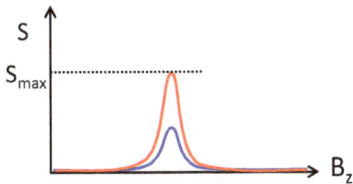

Fig. 3.19: Saturation effect in case of too high rf-power. S refers to signal strength due to absorption of the microwave.

3.6.2 Chemical shift

Protons sit in chemical environments with different *diamagnetic screening* effects. Diamagnetic screening lowers the effective magnetic induction at the location of protons compared to bare protons in a vacuum, described by

$$B_{\text{eff}} = \mu_r B_0 \leq B_0,$$

(3.43)

where μ_r is the *relative magnetic permeability* with values $0 \leq \mu_r \leq 1$. Due to diamagnetic screening, protons at different molecular sites have different resonance fields and can be distinguished according to their chemical environment (Fig. 3.20). For instance, protons in CH_3, CH_2, and OH environments are clearly distinguishable by their different characteristic resonance frequencies. This effect is known as *chemical shift*. The chemical shift occurs always in the direction to lower the resonance frequency compared to protons in vacuum. Moreover, an enhanced magnetic field is required to compensate for the screening effect. The origin of diamagnetism is explained in the Infobox II.

The chemical shift is expressed in terms of a relative frequency shift Δf per applied field and quoted with the dimensionless ratio in parts per million (ppm):

$$\delta = \frac{f_{\text{sample}} - f_{\text{ref}}}{f} = \frac{\Delta f}{f} \, (\text{ppm}).$$

(3.44)

The chemical shift of tetramethylsilane (TMS = $(CH_3)_4Si$) serves as a standard refer-ence. Compared to bare protons, TMS shows a particularly strong negative chemical shift, stronger than most other molecules. Therefore, if Δf of TMS is taken as refer-ence, smaller shifts Δf by other molecules are generally positive. An example can help to understand the notation.

Example: In a 2 T field, the resonance frequency without diamagnetic screening is expected to be at 2×42.6 MHz = 85.2 MHz. The chemical shift of fat is about $(\delta_{fat} = \Delta f/f)_{fat} = 1.1$ ppm/T. Therefore, In a 2 T field, fatty tissue shows a shift of the resonance frequency by $\delta_{fat} = 85.2$ MHz $\times 1.1$ ppm = 93.72 Hz. For water, the shift is 4.65 ppm/T. Water has a resonance frequency shift of $\delta_{water} = 85.2$ MHz $\times 4.65$ ppm = 396 Hz. Therefore, protons in water spin with a slightly higher frequency than in fat, and the difference between water and fat resonance frequency is 302 Hz at 2 T. These frequency shifts are not very large but noticeable. Figure 3.20 shows the chemical shifts for fat and water and the difference between both.

The chemical shift difference between water and fat is usually quoted as follows:

$$\Delta\delta = \left(\frac{\Delta f}{f}\right)_{water} - \left(\frac{\Delta f}{f}\right)_{fat} = 3.55 \,(\text{ppm}). \tag{3.45}$$

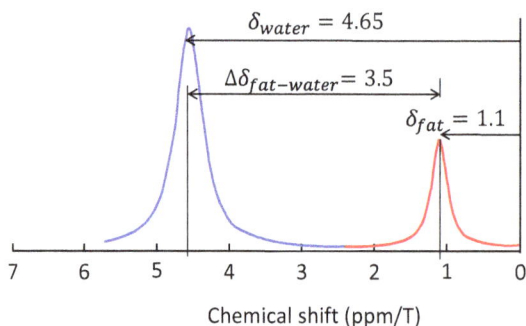

Fig. 3.20: Chemical shift of protons in fatty and watery environments. Note that zero shift $\delta = 0$ refers to the standard sample (TMS) with the highest chemical shift. The chemical shift of fat and water is quoted in reference to TMS.

When operating at higher B_z fields, the chemical shift becomes more pronounced and can cause artifacts and blurring of MR images. For example, protons in water and fat have significantly different resonance frequencies (see Exercise E3.7). When adding field gradients to spatially encode protons, chemical shifts can perturb the frequency encoding gradients (FEG). If fatty tissues and aqueous tissues are located one behind the other, they can appear laterally shifted. The typical voxel separation is 80–100 Hz. Therefore, a chemical shift of 300 Hz can shift an image by three vox-els, a fairly significant shift. Conversely, chemical shifts can be used for spectroscopic purposes, which is in fact used in spectroscopic MRI. Further information on the chemical shift can be found in [10, 11].

http://xrayphysics.com/chem_sh.html

Infobox II: Diamagnetism

When materials are exposed to a magnetic field, they react either by attracting or expelling magnetic fields. In the first case, the density of magnetic field lines increases (paramagnetic response); in the second case, the density decreases (diamagnetic response). The change of magnetic induction in the material is described by $B_{mat} = \mu_r B_0$, where B_0 is the magnetic induction in vacuum and μ_r is the relative magnetic permeability. μ_r is a dimensionless number and should not be confused with the magnetic field constant of the vacuum μ_0. The magnetic permeability $\mu_r > 1$ for paramagnets, and $0 \leq \mu_r \leq 1$ for diamagnets.

Alternately, we may consider the change of the magnetization M in an external field H: $M = \chi H$. The proportionality constant χ is called the magnetic spin susceptibility. Then the magnetic induction is $B = \mu_0 (H + M) = \mu_0 H (1 + \chi) = \mu_0 \mu_r H$. For diamagnets, χ is negative, and for paramagnets, it is positive. Superconductors are ideal diamagnets, for which holds: $M = -H$, and $\chi = -1$.

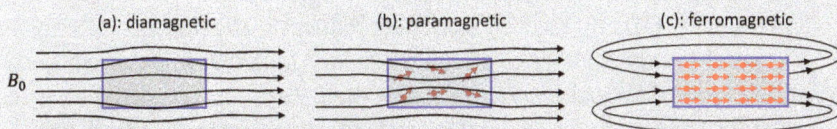

(a): diamagnetic   (b): paramagnetic   (c): ferromagnetic

All atoms and molecules with full electronic shells show a diamagnetic response in magnetic fields (a). This is the consequence of induced electron currents in a magnetic field that oppose the applied magnetic field, known as Lenz's law. In diamagnetism, μ_r is proportional to the total number of electrons in a molecule and its physical size. Diamagnetism prevails in organic matter with light atoms, noble gases, and noble metal atoms.

However, atoms with partially filled inner electronic shells often exhibit permanent magnetic moments $\vec{m}_e = \left(\vec{L} + 2\vec{S} \right) \mu_B$, where $\vec{L}, \vec{S}$ represent the orbital and spin angular momentum, and μ_B is the Bohr magneton. When local magnetic moments are exposed to magnetic fields, they partially turn in the direction of the field and enhance the field strength. This is the underlying mechanism of the paramagnetic response (b). The paramagnetic response is usually much stronger than the diamagnetic response. With increasing magnetic moment density or decreasing temperatures, paramagnets will become ferromagnets. This is because interaction energy between magnetic moments dominates over thermal energy. In ferromagnets (c), the magnetic moments are ordered, in contrast to paramagnets. The order is spontaneous below a critical temperature, called Curie temperature, and increases with decreasing temperature. The ferromagnetic order does not require the application of an external field for aligning the spins.

3.6.3 Standard nomenclature

MRI procedures use a certain amount of jargon; therefore, a summary of the main terms introduced in Sections 3.3–3.5 is given here before we continue with discussions of the contrast generation. Moreover, as we approach the core of MRI methods, we will switch over to the nomenclature that is used in standard MRI literature: $T_1 \to T1$ and $T_2 \to T2$.

TR: Data acquisition (DAQ) requires the repetition of a defined sequence of events within a time interval to repeat (TR). TR starts with an initializing 90° B_1 field pulse, followed by additional 90° and/or 180° pulses. During TR, $T2$ decay and full or partial $T1$ recovery takes place. Therefore, TR is longer than pulse times but can be shorter than relaxation times.

SE: Spin–echo technique is a nuclear resonance procedure, which uses a sequence of 90° and 180° B_1 field pulses. The first pulse at time $t = 0$ converts M_z into M_{xy}. The second pulse after time $t = t_0$ inverts all moments m_{xy} in the transverse plane such that dephasing of magnetic moments is reversed. Full M_{xy} amplitude is recovered at the SE time $t = 2t_0$.[20]

TE: The time to echo (TE) is the time t_0 between the first initializing B_1 90° field pulse and the 180° refocusing pulse, plus the time t_0 from the inversion pulse to the echo. In total, TE $= 2t_0$.

MR: The MR signal detected in a pick-up coil is solely produced by M_{xy} dephasing and decay by FID. M_z recovery does not produce an MR signal. In MRI maps, MR signal strength of a voxel is usually encoded in terms of pixel brightness.

M_z: M_z is the longitudinal magnetization in the z-direction proportional to the magnetic induction (applied magnetic field). In a two-level system like the one for protons, the magnetization M_z follows from the difference of the occupational numbers of the lower and upper energy states. In the lower energy state, the magnetic moments are parallel to the external field and antiparallel in the upper energy state. The energy splitting is linearly proportional to the applied magnetic field. The magnetization increases with an increasing field at a constant temperature because more moments go into the lower energy state.

M_{xy}: M_{xy} is the transverse magnetization obtained by rotating M_z via a 90° pulse into the transverse plane. M_{xy} is indifferent and decays quickly by relaxing back to M_z through a precessing spiral motion. But even if M_{xy} could be kept in the transverse plane, it would decay by dephasing of the magnetic moments m_{xy}, which sum up to the magnetization M_{xy}. The in-plane relaxation from a maximum value immediately after the 90° pulse to zero is characterized by the relaxation time $T2^*$.

FID: This is the oscillatory signal recorded by a pick-up coil that originates from precessing M_{xy} in the transverse plane. The precession is damped by dephasing with a relaxation time $T2^*$.

PD: PD is the proton density, i.e., the number of protons (or hydrogen atoms) per unit volume. The density depends much on the local tissue of interest.

---

20 Here and in the following, we neglect the prime referring to the rotating frame.

T1: The relaxation time $T1$ is the $(1-1/e)$ recovering time of M_z, i.e., at $t = T1$, 63.2% of the M_z magnetization has recovered. $T1$ values range from 200 ms (fatty tissue) up to 2000 ms (CSF) and are an indication of how effective energy can be dissipated from protons to the surrounding tissue at the resonance frequency. $T1$-weighted images emphasize differences of $T1$ in different tissues and de-emphasize differences due to $T2$. $T1$-weighted images are best for illustrating anatomic details.

T2: The relaxation time $T2$ is the $1/e$ decay time of the transverse magnetization M_{xy}, i.e., after $t = T2$, 36.7% of the original M_{xy} remains, after $2T2$, 13.5% remain, etc. $T2$ is the intrinsic transverse relaxation time due to spin–spin interaction, in contrast to $T2^*$, which is related to extrinsic dephasing sources. In general, $T2 > T2^*$. $T2$-weighted images emphasize differences of $T2$ in different tissues and de-emphasize differences due to $T1$. $T2$-weighted images highlight alterations in water content, which appears black in $T1$-weighted images.

3.7 Contrast generation

Medical imaging requires contrast, which is the ability to distinguish between different tissues and organs. In ultrasound imaging, different echo times of sound waves provide the necessary contrast, and endoscopy uses reflected light intensity for this purpose. Here we learn about several procedures to generate contrast in MRI. The procedures are all based on differences in the $T1$ and $T2$ relaxation times. In addition, there is also the possibility to generate contrast according to differences in the PD of dissimilar tissues. In general, the tissue with the higher $M_z(t_0)$ value at time t_0 just before turning on a 90° pulse has a higher brightness on MR images. Higher brightness implies a higher FID and consequentially a higher induced voltage in a pick-up coil.

Contrast requires two pieces of information: Intensity and location. In this section, we discuss various methods for contrast generation regardless of location, and in the next section, we will add the location information.

Before continuing, it is advisable to review the pulse sequences in Figs. 3.13 and 3.14 and the signal strength given by eq. (3.36).

3.7.1 $T1$ contrast

Using SE techniques, $T1$-weighted images are obtained by de-emphasizing $T2$ to maximize contrast between tissues of different $T1$. Figure 3.21 shows an example. The top panel displays the M_z relaxation to almost saturation after about 2000 ms for two different tissues (1) and (2). In the relaxed state, there is no contrast between different tissues. The highest $T1$ contrast occurs in the graph after about 500 ms. Therefore, we can achieve a high contrast by using a fairly short M_z repeat time TR.

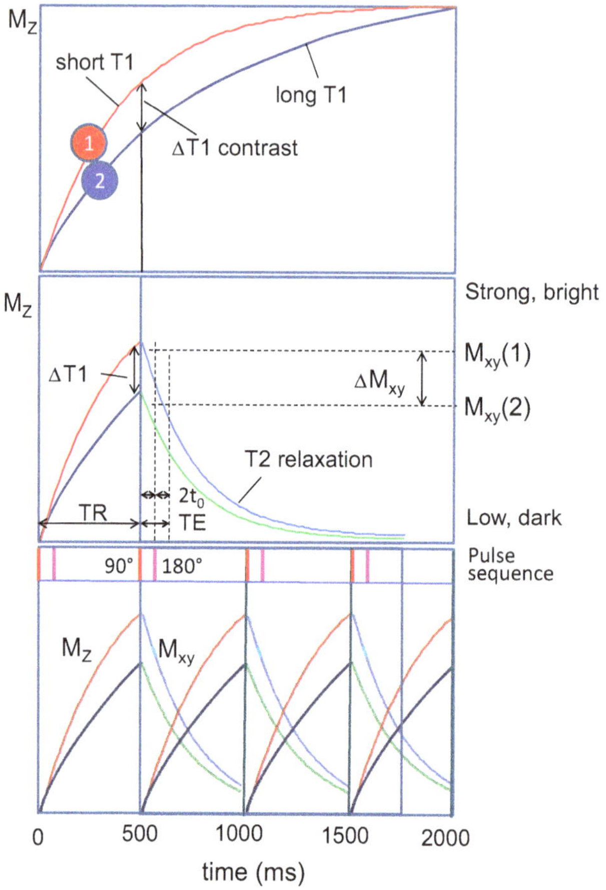

Fig. 3.21: T1 weighting. Top panel: Different T1–relaxation times for different tissues. Middle panel: Short TE for eliminating T2 contrast but maintaining T1 contrast. Bottom panel: Pulse sequence and sequence of M_z recovery with simultaneous M_{xy} decay.

A good choice is to use a TR on the order of the average $\langle T1 \rangle$ relaxation time, or roughly 300–800 ms. In Fig. 3.21, TR is set to 500 ms.

Now we have to solve two problems: maintaining the T1 contrast and generating a large FID signal. The middle panel shows the pulse sequence for maintaining the contrast: after TR = 500 ms, a 90° pulse at TR warrants a maximum $M_{x,y}$ contrast. Then after a very short dephasing time $t_0 = 15$ ms, a 180° refocusing pulse is applied, resulting in a TE of about 30 ms. During such a short TE, the FID signal is strongest and the M_{xy} contrast for tissues (1) and (2) is largest. Ideally, the M_{xy} contrast preserves the T1 contrast, if the the TE is chosen much shorter than TR. The MR signal detected is then proportional to the value of M_{xy} at the time of repeat:

$$S(t = \text{TR}) \sim \rho_\text{H}(1 - \exp(-\text{TR}/T1)). \tag{3.46}$$

Hence, at t=TR the tissue with the shorter T1 has the brighter signal S(TR). During the time TE, relaxation of M_z has already started and therefore the recovering time of M_z is simultaneously the repeat time TR. The bottom panel of Fig. 3.21 shows a sequence of M_z recovering curves following each 90° pulse and M_{xy} relaxation.

3.7.2 T2 contrast

Assuming that there is little difference in $T1$ for different tissues but sufficient difference in $T2$, $T2$ contrast can be achieved by using a long TR and a TE that is optimized for $T2$ contrast, usually 90–140 ms (Fig. 3.22). After a long TR (~1000–2000 ms), differences between $T1$ become less pronounced. Then the signal strength approaches the value:

$$S(t = \text{TE}) \sim \rho_\text{H}\exp(-\text{TE}/T2). \tag{3.47}$$

Differences in $T2$-weighted images are mainly due to PD and $T2$ differences.

3.7.3 PD contrast

To achieve PD contrast, $T1$ and $T2$ contrast should be down-weighted. $T1$ contrast is reduced by choosing a long TR on the order of 2–3 times the M_z relaxation time $T1$. Similarly, $T2$ contrast is lost by selecting a very short TE, shorter than typical M_{xy} decay time $T2$. Then the final contrast is just due to differences in PD (Fig. 3.23), as one can realize from eq. (3.36). For TR ≫ $T1$, $\exp(-\text{TR}/T1) \approx 0$; and for TE ≪ $T2$, $\exp(-\text{TE}/T2) \approx 1$. Thus, the signal strength is proportional to PD:

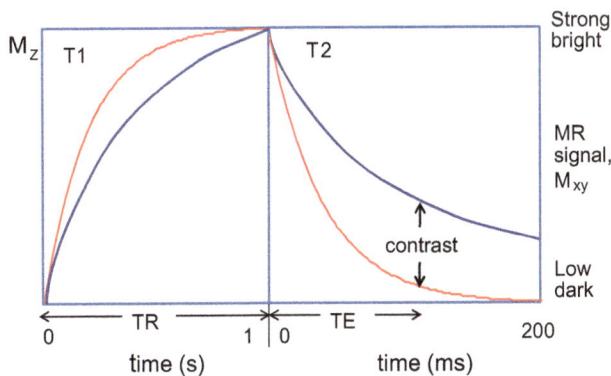

Fig. 3.22: T2 contrast is achieved by using a long TR and TE optimized for T2 contrast. Note the different timescales for T1 and T2.

$$S(t = \text{TR}) \approx \rho_{\text{H}}. \tag{3.48}$$

Under these conditions, fatty tissues, CSF, and lipids with a high proton content appear bright.

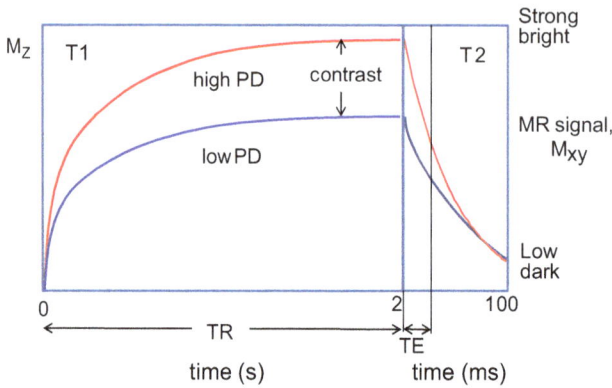

Fig. 3.23: Proton density weighting by using long TR and short TE.

The three different weighting schemes discussed above are summarized in Tab. 3.3 and their pulse sequences are compared in Fig. 3.24. Data acquisition of induced voltage in the pick-up coils always takes place during the spin echo time TE. Figure 3.25 shows MRI axial cross sections of the brain using the three different weighing schemes discussed earlier. Changes in the contrast of white and gray matter and in particular of the CSF is very pronounced. The CSF appears black in $T1$ weighting since it has a longer $T1$ than white or gray matter and therefore a lower M_z within the TR of 500 ms. In $T2$ weighting, the tissue contrast is high. CSF appears bright because it has a longer $T2$ than the surrounding white and gray matter

Tab. 3.3: Summary of TR and TE times for $T1$-, $T2$-, and PD-weighted contrast.

weighting	TE (ms)	TR (ms)
$T1$ contrast	Short	Short
Shorter $T1$ appears brighter	5–30	400–600
$T2$ contrast	Long	Long
Longer $T2$ appears brighter	60–150	2000–6000
PD contrast	Short	Long
Higher PD appears brighter	5–30	2000–6000

Data from Westbrook C., Handbook of MRI technique.
3rd edition. Wiley-Blackwell; 2008.

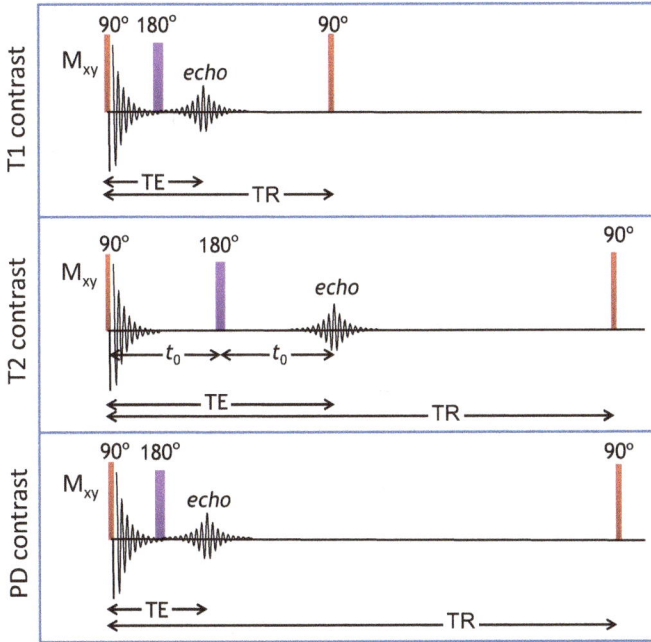

Fig. 3.24: Comparison of pulse sequences for generating $T1$, $T2$, and PD contrast.

Fig. 3.25: MRI cross sections of the brain with different weighting factors. $T1$ weighting: TR = 500 ms, TE = 20 ms, shorter $T1$ areas appear brighter. $T2$ weighting: TR= 6000 ms; TE = 70 ms, longer $T2$ areas appear brighter. PD weighting: TR = 2600 ms; TE = 20 ms, higher PD areas appear brighter (reproduced from https://openi.nlm.nih.gov/).

and therefore higher M_{xy}. The PD signal is high, but the contrast is moderate and therefore also the SNR is low. This is due to the fact that the CSF, white, and gray matter have similar proton densities.

> ! Short TE cancels $T2$ contrast, and long TR cancels $T1$ contrast. PD contrast requires different proton densities in adjacent tissues. Then the PD contrast can be enhanced by long TR and short TE.

Aside from the standard weighting procedures for $T1$, $T2$, and PD contrast, many more imaging modalities with special pulse sequences are available, but they are less frequently applied. Two of those, labeled IR and STIR, will be briefly described.

3.7.4 Inversion recovery (IR)

Inversion recovery (IR) is a pulse sequence that separates two different $T1$ relaxation times if the standard $T1$–weighting is insufficient. The pulse sequence is illustrated in Fig. 3.26. First, a 180° pulse is applied at time $t = 0$, turning $+ M_z$ to $- M_z$. Both systems relax back to $+ M_z$ passing through $M_z = 0$ at different time spans t_i. The relaxation of $M_z(t)$ has the form:

$$M_z(t) = M_{z,\text{sat}}[1 - 2\exp(- t_i/T1)]. \tag{3.49}$$

Immediately after applying the 180° pulse, $M_z(t = 0) = - M_{z,\text{sat}}$. For $M_z(t = t_i) = 0$, we find the condition: $t_i = T1 \ln 2$. The time t_i for the slower system with the longer $T1$ relaxation time is called the *time of inversion* TI: TI $= T1 \ln 2$. Therefore, if at TI a 90° pulse is applied, only those spins will return from recovered $+ M_z$ to the xy plane, which have already passed through the $M_z(t_0) = 0$ point and which have attained a

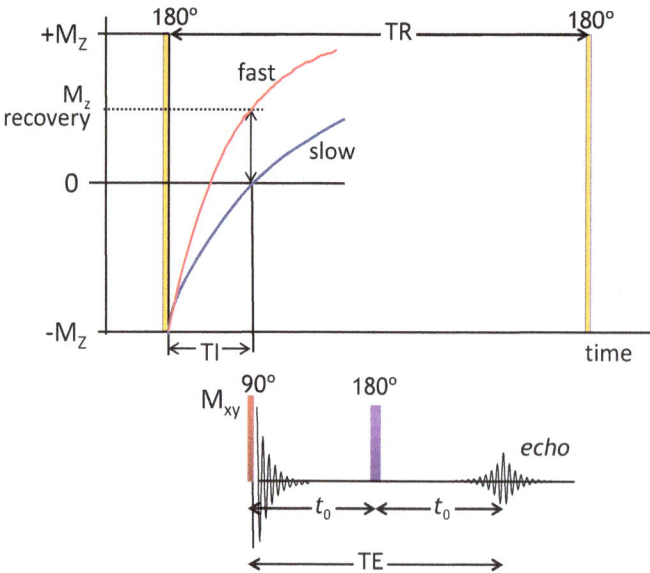

Fig. 3.26: Time dependence and pulse sequence for inversion recovery technique. TI, time of inversion.

positive $+M_z > 0$ value. This is the faster relaxing system with the shorter $T1$, which can be flipped back from recovered $+M_z$ into the transverse plane, generating an M_{xy} signal. Utilizing the M_{xy} component of the faster system, the usual SE procedure is followed, i.e., a 180° pulse is applied at the time t_0 after TI, inverting M_{xy} to $-M_{xy}$, followed by spin echo at TE = $2t_0$, which is detected in the usual way. The repeat time is TR > TI + TE. The IR procedure is also referred to as 180° – 90° – 180° sequence. Note that we apply here two different 180° pulses: one for inverting M_z (inverting pulse) and another one for inverting M_{xy} (refocusing pulse).

The IR method can nullify the signal of any tissue that has a slower recovery than another one. For instance, in brain imaging, the signal from the slow recovery of CSF, in comparison to white and gray matter, can be set to zero at TI, which is known as *fluid-attenuated inversion recovery* (FLAIR).

3.7.5 Short time inversion recovery (STIR)

Short time of inversion recovery (STIR) is a variation of the IR sequence. IR emphasizes the faster relaxing system. In contrast, STIR emphasizes the slower relaxing systems. This is demonstrated in Fig. 3.27. At t_i for the faster system, $M_z(t_i) = 0$ and the slower system has still some negative magnetization. This partially recovered negative magnetization is then turned by a $-90°$ pulse into the xy-plane, and usual SE procedures start again, however, now on the slower system. The STIR procedure

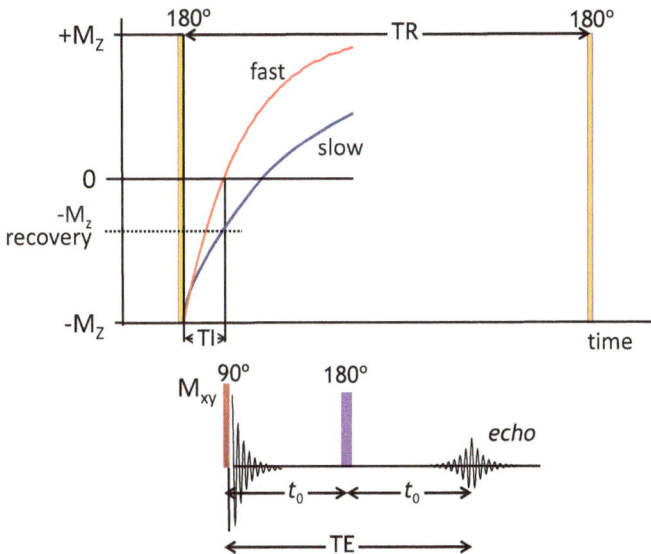

Fig. 3.27: Time dependence and pulse sequence for the short time inversion recovery (STIR) pulse structure.

is applied to cancel, for instance, fast relaxing fatty tissue so that slower relaxing parts nearby become better visible.

> MRI is based on three physical parameters:
> – T1 – relaxation
> – T2 – relaxation
> – PD – proton density or spin density
>
> Three adjusting screws exist to optimize image contrast
> – TE – spin-echo time
> – TR – repeat time
> – RF – radio frequency pulse intensity

3.8 MR signal localization

So far, we have discussed various signal generation procedures, but we have neglected the location from where the signals originate in the body. For tomographic imaging, 3D localization of individual volume elements (voxels) is essential. This can be realized in three steps, as illustrated in Fig. 3.28:
1. Slice selection in the Z-direction (usually from head to toe)
2. Column selection within the slice in the X-direction (usually from left to right)
3. Point selection within the column in the Y-direction (usually from back-to-front)

For the coordinate system of slices, columns, and voxels, we use capital letters in order to distinguish the spatial coordinates of the MRI machine and the patient's body from the coordinate system of the nuclear spins. However, the main axial field direction in MRI machines coincides with the z-direction (defined earlier), and the X,Y-plane of a slice coincides with the x,y and x',y'–planes in previous sections.

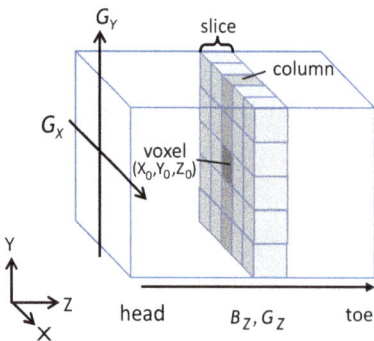

Fig. 3.28: Selection of a voxel within a column of a slice. G_X, G_Y, G_Z refer to field gradients in the respective directions.

3.8.1 Slice encoding gradient

The slice selection is performed as follows. First, a constant magnetic induction B_Z is applied in the horizontal Z-direction by a large DC solenoid that fits a patient (see Fig. 3.34). The technical realization will be presented in the next section. Superimposed on the magnetic induction B_Z is a linear gradient field $G_Z = dB_Z/dZ$. With the gradient field, each position along the Z-direction has a slightly different resonance frequency according to

$$\omega_L(Z) = \gamma_N(B_Z + G_Z \cdot Z). \tag{3.50}$$

Thus, the Larmor frequency depends on the position in the Z-direction, like the tones generated by a piano's keyboard. A narrowband rf-transmitter is tuned to generate rf-pulses (90°, 180°, etc.) at the local resonance frequency $\omega_L(Z_0)$ for a specific location Z_0 along the field gradient for just a small frequency range Δf, such that only in a thin slice ΔZ the protons are excited, as illustrated in Fig. 3.29.

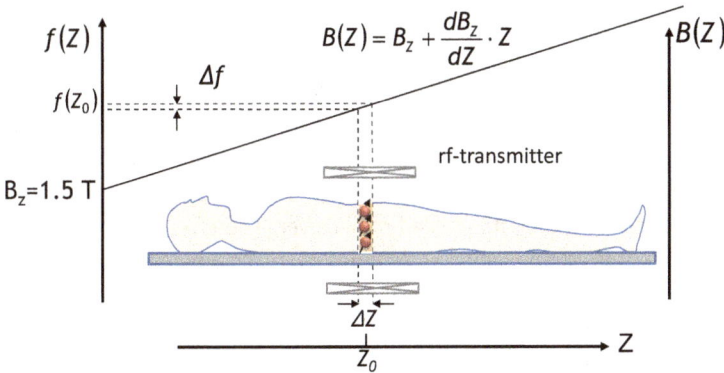

Fig. 3.29: Application of a gradient field in the Z-direction.

Fig. 3.29 indicates that the thickness of the slice depends on the field gradient $G_Z Z$ and the bandwidth Δf of the rf-transmitter. Both can be tuned to the desired spatial resolution, which typically ranges from 1 to 3 mm. The *slice selection gradient* (SSG) in the Z – direction is switched on only during the application of an rf-pulse, while B_Z is permanently turned on. The constant B_Z-field is provided by a superconducting solenoid, a normal conducting solenoid generates the field gradient $G_Z Z$.

The slice can now be subdivided into a matrix of 256×256 voxels. The MRI signals from these voxels are used to generate an image of the slice containing the same number of pixels as voxels. If the *field of view* (FOV) has the size 250×250 mm^2, then the pixel represents an area of about 1×1 mm^2 in the slice.

To identify the signals from individual voxels in the slice, additional field gradients in the X- and Y-directions are required, as indicated in Fig. 3.28. By superposition of all gradients, the voxels will fulfill the local resonance condition at (X, Y, Z):

$$\omega_L(X, Y, Z) = \gamma_N(G_X X + G_Y Y + (B_Z + G_Z Z)). \tag{3.51}$$

However, this scheme does not work. Within a slice, the sum of the local fields $G_X X + G_Y Y$ is not unique. Points on both sides of the diagonal $X = Y$ experience the same gradient field and, therefore, the same frequency. For encoding columns in slices and voxels in columns, different MRI procedures are needed and are applied. We distinguish between a *frequency encoding gradient* (FEG) in the X-direction and a *phase encoding gradient* (PEG) in the Y-direction, in addition to the field encoding gradient in the Z-direction.

3.8.2 Frequency encoding gradient (FEG)

First, we discuss the selection of columns by applying a field gradient in the X-direction. We turn on the Z-gradient and the rf-transmitter is prepared to generate a 90° M_z flip and a 180° M_{xy} refocusing pulse. Now during detection of the MR echo signal at time $2t_0$, the X-gradient is turned on for a few milliseconds. Protons in different columns will emit MR signals at slightly different frequencies and the rf-coil, now used as receiver instead of transmitter, will record all of them simultaneously. As this rf-coil has a double purpose, it is called *transceiver*. Therefore, the MR signal received by the transceiver is composed of a spectrum of rf-frequencies rather than a single frequency, as indicated in Fig. 3.30. The lower frequencies are, for instance, from columns on the left side of the slice and the higher frequencies are from columns on the right side of the slice. Furthermore, the amplitude of the received signals depends on the local PD. By Fourier analysis of the spectrum, these different frequencies and their amplitudes are identified and assigned to the 256 columns in the slice. Figure 3.30 shows an example for only two columns and four MR waves originating from four indicated voxels, labeled a, b, c, d. Voxels a and b in panel (a) belong to the same column and therefore have the same frequency f_1. The voxels c and d in the next column are exposed to a higher field and therefore precess at the higher frequency f_2. The signal amplitudes received from voxels a,b and c, d are different because of different proton densities or different $T1, T2$ relaxation times at their specific location (panel (b)). Superposition of all waves yields the sum signal shown in panel (c). Fourier transformation of the sum signal reveals their frequencies f_1 and f_2 and respective amplitudes marked in blue and yellow (panel (d)).

The FOV of the receiver coil depends on the bandwidth of the receiver and the field gradient. Since

$$dw = \gamma_N G_X dX, \qquad\qquad (3.52)$$

and setting $dX = $ FOV, we find

$$FOV = \frac{2\pi df}{\gamma_N G_X}. \qquad\qquad (3.53)$$

For example, a bandwidth of $df = 20$ kHz and a field gradient of $G_X = 20$ mT/m produce an FOV of 147.5 mm. The pixel width equals the FOV divided by the number of frequency components by which the frequency bandwidth is sampled. In our example, this is 147.5 mm/256 = 0.57 mm across. The frequency separation of each pixel is about 78 Hz.

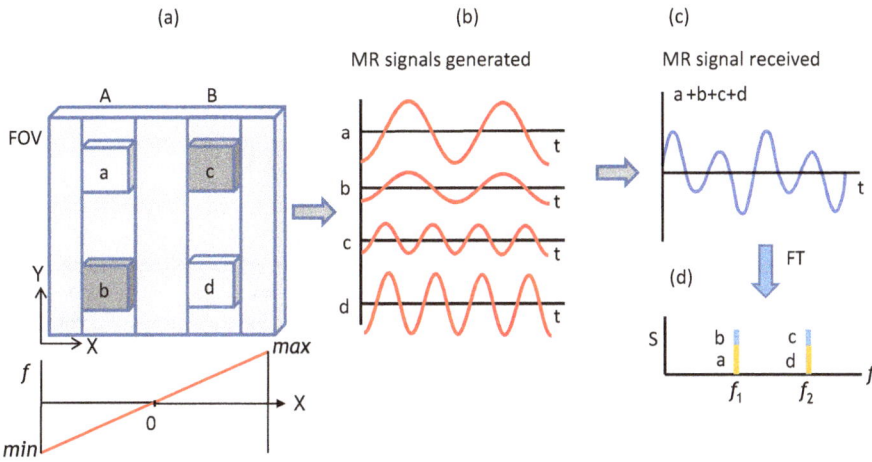

Fig. 3.30: Columns A and B generate MR signals with different frequencies and different amplitudes. The receiving coil senses a superposition of all frequencies from different columns. Fourier transform of the received signal yields the contributing frequencies and amplitudes marked in yellow and blue. Pixels a, b and c, d can only be separated by an additional gradient in the Y-direction. Note that the darker voxels b and c have lower amplitude.

3.8.3 Phase encoding gradient (PEG)

Finally we need to identify the voxels within a column. The standard procedure is the use of a PEG; see Fig. 3.31. After inversion from M_z to M_{xy} and before time t_0 for the 180° inversion pulse, a field gradient in the vertical Y-direction is applied for a few milliseconds. Because of the gradient, the protons in each column will precess with slightly different frequencies from top to bottom for a short time. After turning off the Y-gradient, the protons precess again with the same frequency as before, according to their specific location. But they will be out of phase. The ones which were precessing faster are still ahead in phase, and the ones which were precessing slower are still

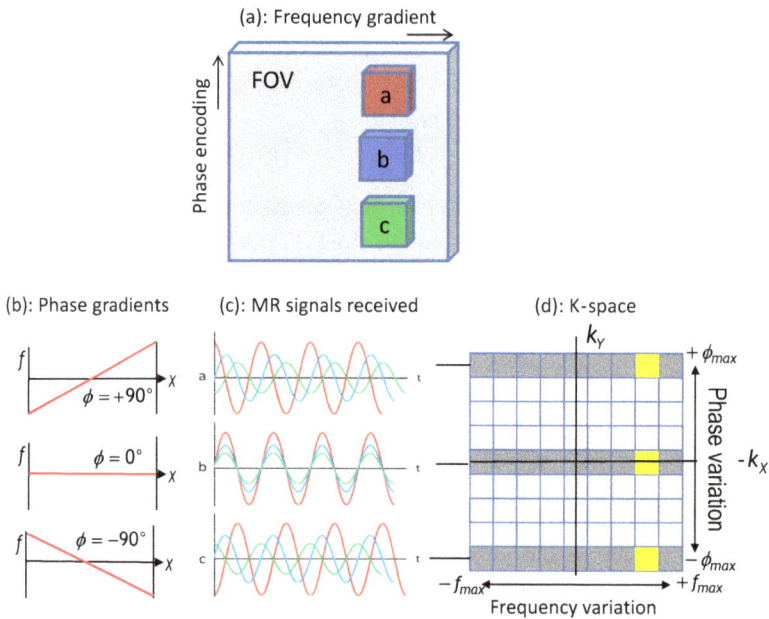

Fig. 3.31: Procedure of phase encoding. The top panel (a) shows three voxels within one column. They all have the same precessional frequency. Excitation with different G_Y-gradients provides phase information, shown in lower left panel (b). The MR signal received is a superposition of all waves, including frequency and phase (lower middle panel (c)). Fourier analysis reveals their frequency value (k_X) and phase (k_Y), plotted in the K-map (panel (d)).

lagging behind. Thus, the short Y-gradient imposes a phase gradient across all columns within the FOV, which can be detected. How this is done is the topic of the next section.

3.8.4 K-map

We consider a slice at Z and width ΔZ. The MR signal from each slice contains a frequency spectrum of bandwidth dw due to the G_X-gradient and a phase spectrum due to the G_Y-gradient. There are two possibilities to gain information on both:

(1) Frequency and phase are detected by two coils arranged at 90° to each other as to receive two independent projections of the precessing protons, k_X and k_Y.

(2) Only one coil is used for the MR frequency detection and the step-by-step phase angle variation is preadjusted via the G_Y-coil.

In both cases, the (k_X, k_Y)-points fill a 256 × 256 pixel K-map as illustrated in Fig. 3.31. k_X represents the frequency in a column and k_Y represents the phase. k_Y goes from maximum negative values to maximum positive values, and zero is at the center.

Frequencies cannot be negative. What is measured here is actually the frequency difference with respect to the precessional frequency of the slice at position Z and the center $X = 0$:

$$\Delta\omega_{\mathrm{L}} = \gamma_{\mathrm{N}}(G_X X + (B_Z + G_Z Z)). \tag{3.54}$$

Same arguments hold for the k_Y values.

Figure 3.31 shows an example of how the K-map is filled with data points. First, we concentrate on one particular column, which represents one particular frequency according to the G_X-gradient. This implies that protons in voxels a, b, and c precess with the same frequency but with different amplitudes due to differences in PD or $T1 - T2$ relaxation times, as discussed earlier. The precessional frequency, indicated by yellow color in the K-map, is not the highest possible one because the column chosen is not the last one in the FOV. Now we turn on a G_Y-gradient causing a maximum phase shift of $+90°$ between the wave trains received from voxels a, b, and c. Frequency and phase shift go into the top row of the K-map marked in yellow. In the next round, after completing one cycle within time TR, the phase shift is set to zero, providing one point in the middle row; in the third cycle, the phase is set to $-90°$, filling one yellow point in the last row.

This procedure encodes the location of voxels within a column. The wave with zero phase shift independent of G_Y must stem from the middle voxel; waves with positive phase shift during $+G_Y$ must originate from the top part, and waves with positive phase shift during $-G$ must come from the bottom part of the slice. The assignment of k-points to image points is done during the Fourier transform of the K-map to a real-space image.

In reality, the K-map consists of 256 rows, requiring 256 TR cycles for scanning all intermediate phase angles. Furthermore, not one column frequency is recorded in each cycle, but all 256 column frequencies simultaneously, spanning the entire FOV. During one cycle, the received MR signal is a superposition of all 256 different frequencies, including their phase shift. Fourier analysis allows separating and assigning them to the different k_X points within one row. The next row is filled in during the next cycle with a different phase angle, until the complete K-map is completed after 256 repeats.

Three gradients determine the voxel location in the body:
- Field gradient G_Z for an axial slice in the body
- Frequency gradient G_X for columns within a slice
- Phase gradient G_Y for rows within the column

!

The K-map is a Fourier representation of the real space from scanning one slice. The same argument holds for all slices, yielding a K-map in 3D. Back transformation should provide real-space images, which is indeed the case as shown in Fig. 3.32.

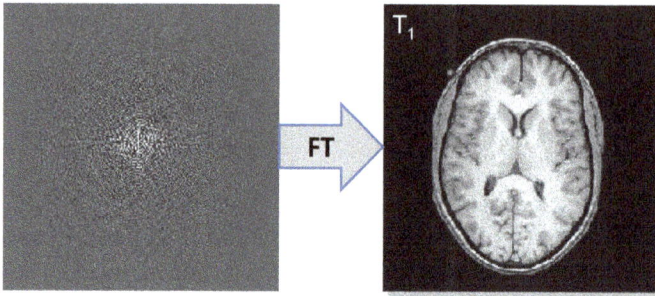

Fig. 3.32: Fourier transform of a K-space map into a real-space image.

3.8.5 Fourier transform

Mathematically, the Fourier transform can be shown as follows. First, we realize that any voxel within a slice at position (X, Y) with local PD $\rho_p(X, Y)$ has a specific resonance frequency:

$$\omega(X, Y) = \omega_0(Z) + \gamma_N(G_X X + G_Y Y), \tag{3.55}$$

where $\omega_0(Z)$ is a constant frequency that depends on the slice location Z. The frequency shift in the X, Y plane at constant Z can be associated with a phase change $\phi = \omega \int dt$, integrated over the pulse time:

$$\phi(X, Y) = \gamma_N \left(\int G_X X dt + \int G_Y Y dt \right). \tag{3.56}$$

Assuming that the signal strength is dominated by the PD $\rho_p(X, Y)$ at constant Z, we find

$$dS(t) = \rho_p(X, Y, t) \exp(i2\pi\phi(X, Y, t)) dX dY. \tag{3.57}$$

Now we convert the phases into wave numbers and rephrase the wave numbers in vectorial notation:

$$\vec{K} = k_X \vec{e}_x + k_Y \vec{e}_y;$$

With

$$k_X = 2\pi\gamma_N \int G_X dt; \quad k_Y = 2\pi\gamma_N \int G_Y dt;$$

$$\vec{K} \cdot \vec{r} = k_X X + k_Y Y, \tag{3.58}$$

where $\vec{r}$ is a spatial vector in the (X, Y) plane of the slice and $\vec{e}_x, \vec{e}_y$ are unit vectors. Integration over the slice we find for the signal strength in K-space:

$$S(k_X, k_Y, t) = \iint \rho_p(X, Y, Z, t) \exp\left(i\vec{K} \cdot \vec{r}\right) dX dY. \tag{3.59}$$

This is a standard 2D Fourier transform of an object with density distribution $\rho_p(X, Y)$ in real space (Z is fixed). From this expression we conclude that MRI measurements in K-space represent 2D Fourier transforms of the PD distribution $\rho_p(X, Y)$ in real space. The Fourier transform provides all the necessary phase and frequency information within a slice. The inverse Fourier transformation into real space provides the image we are looking for

$$\rho_p(X, Y) = \iint S(k_X, k_Y) \exp\left(-i(k_X X + k_Y Y)\right) dk_X dk_Y$$

$$= \iint S(k_X, k_Y) \exp\left(-i\vec{K} \cdot \vec{r}\right) dk_X dk_Y. \tag{3.60}$$

3.8.6 Data acquisition

Summarizing the spatial encoding for MRI, the Z-gradient defines the slice thickness, which is usually a transverse cut through the body. In the case of a full-body scan, the slices start from the head and continue to the toe. The Z-gradient is applied as soon as the RF-pulse is turned on. The X-gradient produces a frequency change and is applied during receiving of the MR echo signal. The Y-gradient produces a phase shift and is applied just before the 180° M_{xy} pulse reversal. The time relations of the different gradients applied in a standard scan are shown in Fig. 3.33. DAQ takes place during the x-scan.

The gradients can be used for controlling the slice thickness, FOV, and pixel size. The steeper the Z-gradient, the thinner is the transverse slice. The steeper the frequency gradient and the phase gradient, the smaller is the FOV. The bandwidth of the transceiver also controls the FOV. The bigger the bandwidth, the larger is the FOV.

Frequency and phase encoding operate on very different timescales. While frequency encoding is a matter of 10–20 ms during recording the echo signal, phase encoding takes much longer. Because for each column, the gradient has to be changed in a sequence of TR scans. This can take as many as a few seconds up to minutes for one slice. During this time, the examined person has to be immobilized. Otherwise, motion artifacts show up on the image.

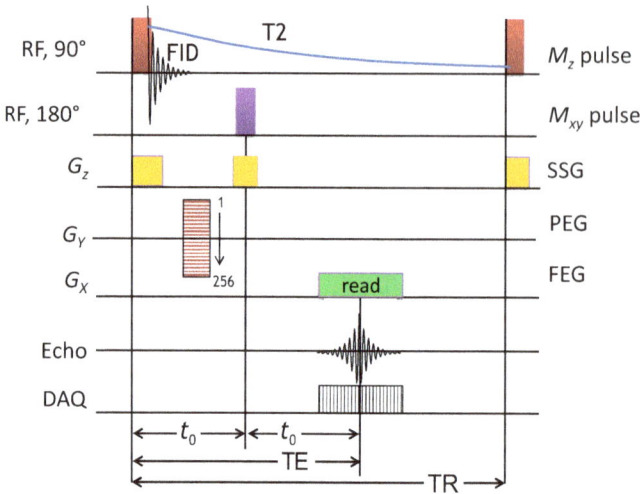

Fig. 3.33: Time sequence of different field gradients applied. SSG, slice selection gradient; PEG, phase encoding gradient; FEG, frequency encoding gradient; DAQ, data acquisition.

3.9 Magnets and coils

Figure 3.34 shows a cutaway view of an MRI scanner, displaying the essential components: main coil for the constant field B_Z, gradient coils for the X-, Y-, and Z-gradients, and an rf-transceiver. The patient lies on a coach, which can be slid into a borehole of about 70 cm diameter. All parts are cylindrically arranged about the Z-axis.

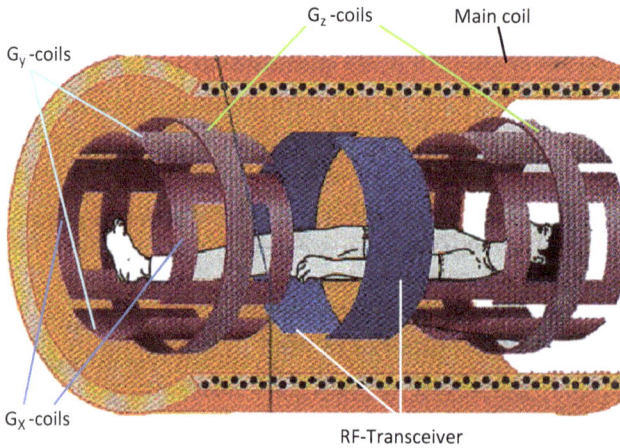

Fig. 3.34: Constant field and gradient field coils for magnetic resonance imaging.

3.9.1 Main coil

The most expensive component of an MRI system is the constant B_Z field magnet. Since the required field is above 1 T, conventional electromagnets cannot be used. The combination of a high magnetic field and a large borehole to fit a whole body is a challenging task requiring the use of *superconducting solenoids*. They are constructed like a long Helmholtz[21] coil but with the difference that the wires are made of metals that become superconducting below a critical temperature T_c. Usually, NbTi alloy wires are used for this purpose, which have a transition temperature of about 9 K. This implies that the wires have to be cooled to temperatures below 9 K, which is done by a combination of liquid N_2 (LN_2) and liquid He cooling. The wires are bathed in liquid He at 4 K, surrounded by a heat radiation shield cooled by LN_2 to 77 K. Depending on specifications, *superconducting magnets* between 1 and 7 T are used; the standard ones for clinical applications have B_Z fields in the range of 1.5–3 T; for research, often higher fields are applied. With larger fields, larger M_z and consequentially larger MR can be accomplished. However, larger fields also imply a longer recovery time $T1$ and, therefore a longer repeat time TR, which contributes to an increased total time for taking images. Also artifacts due to chemical shifts increase with increasing field.

Field homogeneity along the Z-direction and over the FOV with tolerance as low as 1 ppm is one of the most important specifications of MRI scanners. Permanent ferromagnetic sheets are used for fine-tuning (shimming) the main magnet.

As soon as current flows in the coil, it will continue to flow as long as the coil is kept at liquid helium temperatures. The danger occurs when the superconducting coils quench by unplanned warming above the critical temperature. Then the stored energy of some 20 kWh is transferred to the surrounding liquid gases, which rapidly vaporize. The scanner room must be prepared with sufficient ventilation for such a circumstantial accident. In modern MRI machines, the boil-off of liquid He during standard operation can be kept to a minimum 0.1–0.2% daily of the 1500 L reservoir, reducing the operating cost of a scanner.

It would certainly be more cost-effective if high-temperature superconducting (HTS) wires with T_c above LN_2 temperature (>77 K) were used instead of conventional low-temperature superconducting wires. This would abolish expensive liquid He consumption and simplify the cooling system of MRI machines considerably. So far, the brittle ceramic high T_c materials have hindered the fabrication of wires needed for winding coils. However, this obstacle has been overcome recently by using second-generation HTS wires. The major manufacturers of MRI machines are now implementing this new technology, which likely will be soon on the market [12].

---

21 Hermann von Helmholtz (1821–1894), German physicist and physiologist.

3.9.2 Gradient coils

Resistive wires are used to generate field gradients, operating at room temperature. A Z-gradient field G_Z is achieved with an *anti-Helmholtz-type coil* sketched in Fig. 3.35 (a). The current in both coils flows in opposite directions, creating a magnetic field gradient between them. At the center between both coils, the B_z field goes through zero. Anti–Helmholtz coils are also known as magnetic quadrupole coils. With such gradient coils, field gradients in the order of 30 mT/m can be produced. Fast switching of gradient coils causes a loud knocking noise in MRI scanners because of the exertion or release of Lorentz[22] forces on the wires.

The X- and Y-field gradients are generated by pairs of so-called *butterfly coils*, also known as Golay[23] coils. Because of their half-circle shape and current flow direction, the field is oriented parallel to the Z-axis for each pair of coils on one side and in opposite direction on the other side. The immediately opposing loops (marked in red in Fig. 3.35(b)) can be considered as half Helmholtz coils, and the rest of the loop is simply returning the current. Therefore, a field gradient develops in between, similar to the one in the Z-direction. It is important to realize that all gradient coils produce magnetic fields in the Z-direction. However, the *field gradients* have varying components in the X-, Y-, and Z-directions. More information on gradient coils and design features can be found in [13].

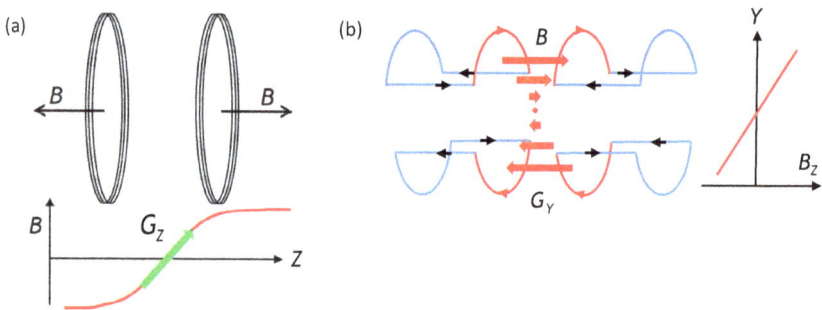

Fig. 3.35: Gradient magnetic field coils. (a) For producing a Z-gradient G_Z, anti-Helmholtz coils are used. (b) For the G_X and G_Y gradients, pairs of half-circle Helmholtz coils generate field gradients in the X- and Y-directions. Shown is only the G_Y gradient coil; the G_X gradient coil works according to the same principle.

---

22 Hendrik Antoon Lorentz (1853 – 1928), Dutch physicist, Nobel Prize in Physics 1902.
23 Marcel J. E. Golay (1902–1989), Swiss-American electrical engineer and information scientist.

3.9.3 rf-coils

rf-coils can be tuned to the resonant frequency like a radio. In return, the coil acts as a Faraday coil for receiving electromagnetic waves via a measurable induction. Three different types of transceivers are distinguished:
1. *Standard body coils* for transmitting rf-pulses and for picking up MR signals when imaging large parts of the body such as the chest or the abdomen.
2. *Head receiver coils* included in a helmet specifically used for brain imaging; see Fig. 3.37.
3. *Surface coils*, designed to be used locally for small area scans such as lumbar spine and knee.

These additional coils are fixed before the patient is slid into the MRT machine. They provide smaller voxels and give better resolution over a smaller FOV.

3.9.4 MRI machine specifications

Most common commercial MRI systems are specified with main magnetic fields of either 1.5 or 3 T. Most applications can be performed with high quality using 1.5 T systems. However, the most advanced applications such as functional imaging, diffusion-weighted imaging (DWI), and time-resolved imaging requires a 3 T model. Typical machine specifications are listed in Tab. 3.4. A 1.5 T scanner is shown in Fig. 3.36. An eight-channel head scanner for high-resolution and high-speed brain imaging is shown in Fig. 3.37.

Tab. 3.4: Some parameters of standard 1.5 and 3 T MRI systems.

	1.5 T	3 T
Field homogeneity 40 cm DSV ppm	0.2–0.4	0.1–0.5
Max FOV isotropic (mm)	500	500
Min FOV isotropic (mm)	5	5
Bore diameter (cm)	70	70
Bore length (cm)	150	150
Field gradient (mT/m)	33	50
He refill (years)	3	1
He capacity (L)	1800	2000
He boil-off per day	0.1–0.2%	0.1–0.2%
Weight (kg)	5000	8000

DSV, defined spherical volume.

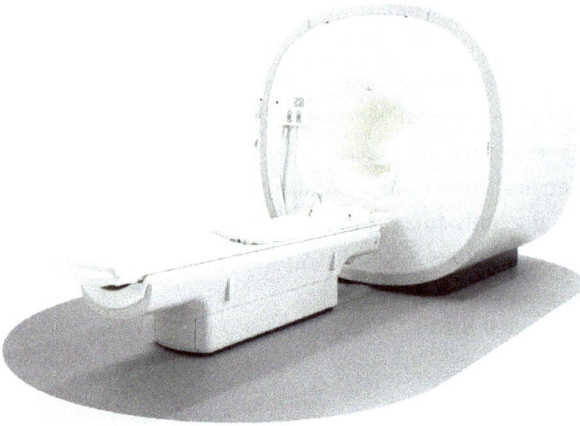

Fig. 3.36: Clinical 1.5 T MRI scanner with an RF power of 1 kW, field gradient of 33 mT/m, FOV of 53 cm, and a power consumption during scanning of 22 kW. Vendor Philips, The Netherlands.

> ! 1.5 and 3 T full-body scanners are currently the most frequently used MRI devices. In the future, more compact, lighter, lower (or higher) fields, lower maintenance costs, cryogenic-free, and more specialized scanners will come onto the market.

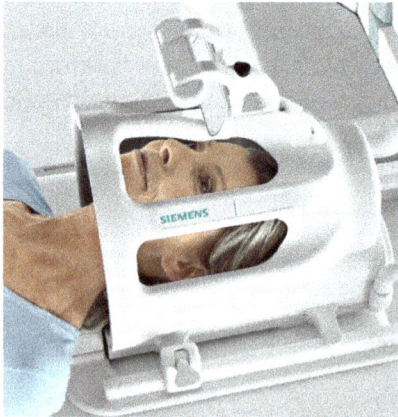

Fig. 3.37: Head scanner for brain imaging in a MRI system. Vendor: Siemens, Germany.

One of the main applications of MRI is the inspection of diseases and injuries in the area of joints at the extremities, such as knees, elbows, hands, wrist, and ankle. For such screenings, full-body MRI scanners are not needed, but much smaller scanners in length and borehole size are sufficient. This has a dramatic impact on the design of such systems and finally on the operational cost. With a smaller borehole, main-tenance free permanent magnets can be used, which provide a B_z field of about 0.3 T, field gradients of 20 mT/m, and the total power consumption may be as low as 1 kW in contrast to 20 kW for a standard 1.5 T machine. A small unit extremity scanner is shown in Fig. 3.38.

Another approach has recently been reported for overcoming limitations in whole-body MRI scanners [14]. A dedicated 3 T MRI system was designed with an inner diameter of 42 cm. The diameter is sufficient for scanning the head and extremities while providing a high homogeneous B_Z field and G_z gradient of 80 mT/m. The conventional superconducting coil is fabricated of NbTi wires and cooled with a cryo-cooler that consumes no liquid He. The advantage of this design is a much lighter and smaller scanner for specific applications with superior imaging quality due to the high field gradient.

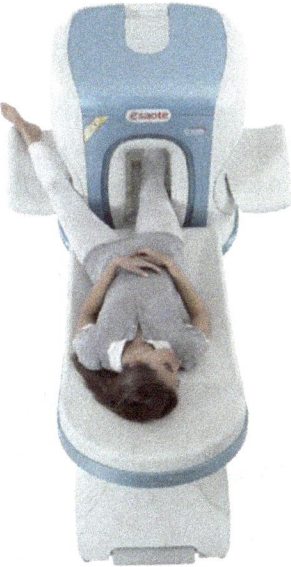

Fig. 3.38: Extremity scanner using a permanent magnet of 0.3 T, maximum RF power of 1.5 kW, field gradient of ±20 mT/m, and FOV of 14 cm (vendor: Esaote, Spain).

A serious obstacle of standard MRI scanners is their weight and size, which often requires special building measures. Moreover, these machines must be installed in large rooms for safety reasons regarding magnetic stray fields and quenching of the superconducting coils. Therefore, conventional MRI scanners cannot be installed in hospitals close to the ambulance or operating theater where they would be needed. For all these reasons, one manufacturer has now designed a novel full-body, low-field 0.55 T MRI scanner with a comfortable 80 cm inner diameter and a dry cryo-cooler [15]. Despite the larger inner diameter, the scanner is smaller and lighter than the standard machines. As the stray magnetic fields are much reduced, and danger due to quenching is eliminated, the safety measures can also be softened, enabling MRI screening in closer proximity to emergency and intensive care rooms. The smaller field gradient G_Z is compensated for by using software based on neural network analysis [16] that guarantees high sensitivity and excellent SNR.

3.10 Applications of MRI

Most frequent applications of MRI scanners are for imaging joints, brain and the entire head, mammary, and prostate. To image moving organs like the heart was challenging in the past. But advanced techniques have solved this problem. Many specialized MRI techniques were introduced during recent years, including diffusion-weighted MRI, angio-MRI, multiparameter MRI, functional MRI, gated MRI, or hyperpolarization MRI. MRI is also of great benefit for detecting brain impairments via tumors, dementia, stroke, etc. For localizing brain activity in response to outside stimulus, fMRI has been developed, which is also of great interest for brain research, cognitive sciences, and psychiatry. A few MRI applications are discussed as follows.

3.10.1 Joints

There is a high MRI contrast between bones and tissue, which allows inspection of injuries in the area of joints. The contrast is mainly due to differences in $T1$ and PD of bones and surrounding tissues. A distinction of muscles, tendons, cartilage, fat, ligaments, and fluids is clearly possible. Therefore, MRI is of benefit to disciplines such as sports medicine and orthopedics. In Fig. 3.39, some examples are shown. Recording of these images is done with standard $T1$- and $T2$-weighting procedures.

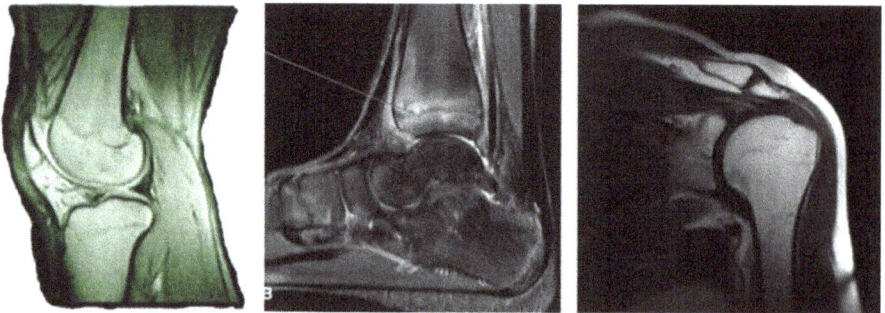

Fig. 3.39: MRI scans of joints. From left: sagittal scan of knee, foot, and coronal scan of shoulder, $T1$-weighted images (reproduced from https://openi.nlm.nih.gov/).

3.10.2 Dynamic contrast enhancement (DCE) MRI

In case $T1$ and $T2$ weighting does not produce sufficient contrast in soft tissue, such as in the brain, there is the possibility of enhancing the contrast with the use of paramagnetic ions. Paramagnetic ions generate magnetic dipole fields, which are five orders of magnitude higher than magnetic fields from proton spins. The main effect of *contrast agents* (CA) is a shortening of $T1$ and/or $T2$ relaxation times.

When applying a CA, the time dependence of the CA distribution in the body and the washout time are important imaging parameters. Therefore, MRI with the use of CAs is called DCE-MRI analysis. Figure 3.40 shows three types of time dependencies, after administering a CA at time t_0:

Type I: The distribution of the agent in the body is very slow and does not level off

Type II: The distribution is fast at the beginning and very slow after reaching a plateau

Type III: The distribution is fast at the beginning and washout starts quickly after reaching a maximum

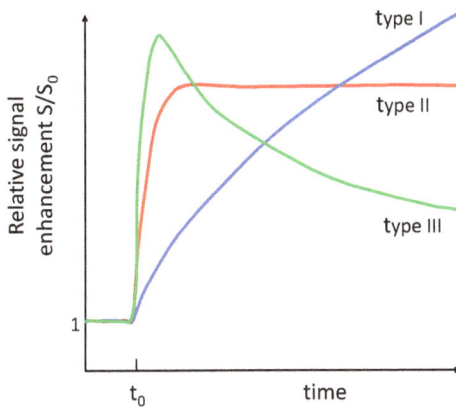

Fig. 3.40: There are three types of time dependencies for the contrast agent distribution and washout in the body after administering a CA at time t_0.

Type II is ideal from an MRI perspective. After reaching the plateau, images can be taken like under steady state conditions. From a biological and safety point of view, however, type III is more favorable. The shape of the curve can depend not only on the CA administered but also on the type of tumor examined. For example, it has been shown that a distinction between benign and malignant breast tumors is possible by means of a signal shape analysis since the residence time of CAs in malignant tumors is higher than in benign tumors [17].

CAs have the purpose of enhancing the *relaxivity* of the relaxation times $T1$ and $T2$. *Relaxivity* is defined as the relaxation rate $R = 1/T$, where T is the relaxation time (not temperature). For instance, $R_1 = 1/T1$ is the relaxivity for the longitudinal relaxation time $T1$. This may vary between different tissues. However, assuming constant magnetic induction B_z and constant temperature, it is generally assumed that the relaxivity is linearly proportional to the concentration x_{CA} of CA administered:

$$R_1 = r_1 x_{CA} + R_{10}, \tag{3.61}$$

where r_1 is the specific relaxivity (units: $Mol^{-1}\,s^{-1}$) and R_{10} is the baseline relaxivity without CA. Using dynamic acquisition with different rates R_1 before, during, and after CA administering allows for qualitative and quantitative characterization of specific tissues with typical differences between healthy and malignant behavior [18].

The rare earth metal ion gadolinium Gd^{3+} is most frequently used as a CA [19]. Gd^{3+} has a half-filled 4f shell with a spin state of $S=7/2$ and a total magnetic moment of nearly 8 Bohr magnetons. Only the outer $5d^1 6s^2$ electrons take part in chemical bonding, while the 4f shell and the full magnetic moment remain preserved. Figure 7.7 of Volume. 3 shows schematically the energy-level scheme of the Gd^{3+} 4f-shell. Gd^{3+} packed in a *chelate complex* such as *diethylenetriamine penta-acetic acid* (short *DTPA*) is biocompatible. The cytotoxic heavy metal ion is strongly bonded and hindered from leaching into the cellular environment. The chemical structure is shown in Fig. 3.41. It is one of a family of chelates that are used for contrast enhancement. Gd-chelates belong to type III CAs: after intravenous injection, it is distributed within 12 min and cleared out from the plasma within a short blood circulation half-life of about 100 min, sufficient for taking MRI images. Excretion in an unchanged form takes place via the kidneys. The atomic weight of Gd-DTPA lies between 1000 and 2000 and therefore passes the glomerular filter of the kidneys easily (see Tab. 10.1 in Vol. 1).

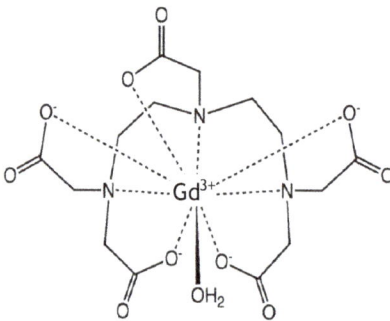

Fig. 3.41: Chemical structure of the Gd^{3+}-chelate diethylenetriamine penta-acetic acid (DTPA), used for $T1$ contrast enhancement of cranial and spinal MRI.

Gd-chelates preferentially shorten $T1$ values in tissues where it accumulates, rendering them bright on $T1$-weighted images. Gd^{3+}-chelates do not pass the intact *blood–brain barrier* (BBB) because of their hydrophilic properties. However, in case of BBB breakdown due to tumors or stroke, Gd-chelates can enter the brain tissue and support contrast enhancement. In Fig. 3.42, an example is shown of a $T1$-weighted coronal section through the brain taken with and without Gd-CA. The Gd-enhanced image shows clear changes in the region of the BBB breakdown [20].

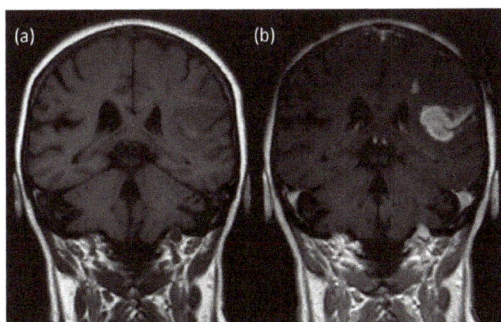

Fig. 3.42: $T1$-weighted MRI scans of the brain without (a) and with Gd-chelate enhancement (b). The $T1$-weighted coronal sections show in comparison defects of the blood–brain barrier that occurred after stroke. Without these defects, Gd-chelate contrast agent would not be able to penetrate the BBB (reproduced from Wikipedia © creative commons).

Some tumors have little contrast to their surrounding healthy tissue, making differentiation difficult. In those cases, Gd-CAs are often used for better identification and delineation of the tumor volume. A discussion on the identification of tumors in the abdomen, pelvis, brain, and spine using Gd-CAs is given in [21].

Gd-chelates are called positive image CAs, as the shortening of the $T1$ relaxation time lets those tissues, where the agent accumulates, appear brighter. This method plays a vital role in early tumor recognition.

Superparamagnetic iron oxide nanoparticles (see Chapter 7 of Volume 3 for details) are more effective for shortening the $T2$ relaxation time, producing negative image contrast, as shorter $T2$ produces darker images (see Tab. 3.3).

There are many more contrast CAs in use and still being tested for specific tasks and targets. Nevertheless, the majority of MRI CAs used in clinics are Gd^{3+}-chelates. This is due to their favorable properties in terms of contrast enhancement, high chemical stability, short biological half-life, and inertness in the body. However, there are also some shortcomings of Gd-CAs with respect to the very short circulation half-life and decreasing r_1 values in high fields that make $T2$ CAs attractive. More information on contrast enhancement agents can be found in [22, 23].

CAs are used to enhance the contrast between different tissues with similar $T1, T2$ relaxation times. Most CAs contain paramagnetic ions. All CAs shorten the $T1, T2$ relaxation time and increase the relaxivity. Biocompatibility and washout characteristics are important parameters for their applicability.

3.10.3 Angio-MRI

Venous blood consists of 70% paramagnetic deoxyhemoglobin. On the contrary, arterial blood is 95% nonmagnetic oxyhemoglobin. This provides a contrast mainly in $T2$-weighted images. However, since blood flows, images of blood vessels appear blurred. With the help of Gd-chelates, the contrast can be dramatically increased. However, the distinction between venous and arterial blood is lost [24]. An impressive image of the blood vessels taken by full-body MRI in several sections is shown in Fig. 3.43. The distinction between oxyhemoglobin and deoxyhemoglobin is essential for fMRI, which is discussed in Section 3.6.7.

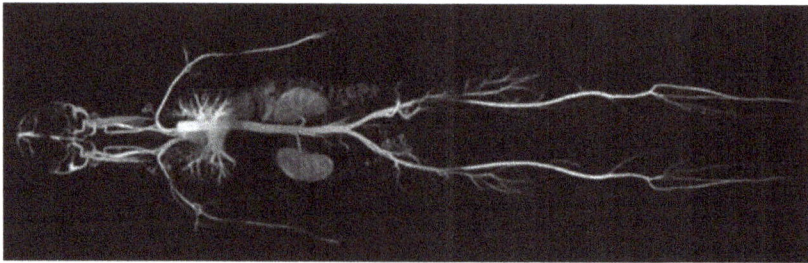

Fig. 3.43: Angio–MRI of the blood vessels. The contrast is enhanced by Gd-chelate (reproduced from https://www.healthcare.siemens.de/magnetic-resonance-imaging).

3.10.4 Diffusion-weighted imaging (DWI)

DWI can be applied to any body part for studying diffusion and perfusion of liquids and gases through tissues. DWI is particularly valuable for investigations of *neuroactivities*.

Randomly diffusing particles have a mean square displacement in three dimensions (see also eq. (3.37)):

$$\langle x^2 \rangle = 6D\tau, \tag{3.62}$$

where D is the diffusion constant and τ is the incremental time during which the displacement occurs. When protons diffuse, this will lead to an additional damping of the SE signal, i.e., the $T2$ time will be shortened. Diffusion can be tested by applying a gradient field pulse G_D just before the 180° refocusing pulse, and another one of the same magnitude but opposite in direction just after the 180° pulse. The first G_D pulse marked in yellow in Fig. 3.44 will dephase the proton spins in the $x, y-$ plane in addition to the already existing dephasing due to T_2^* relaxation, while the second opposing pulse will reverse the dephasing of the first pulse. If the system is static, these two G_D pulses cancel each other, and no additional change will occur. However, if the protons are in diffusional motion, the cancellation is incomplete and the MRI signal strength will experience additional damping. The attenuation due to diffusion is expressed as follows:

$$S_{DWI} = S_0 \exp\left[-(\gamma G_D \delta)^2 \left(\Delta - \frac{\delta}{3}\right)D\right] = S_0 \exp[-bD]. \tag{3.63}$$

S_0 contains the usual relaxation terms due to $T1$ recovery and $T2$ decay:

$$S_0 = K\rho_p \left(1 - \exp\left(-\frac{TR}{T_1}\right)\right) \exp\left(-\frac{TE}{T_2}\right). \tag{3.64}$$

Combined we have

$$S_{DWI} = K\rho_p \left(1 - \exp\left(-\frac{TR}{T_1}\right)\right) \exp\left(-\frac{TE}{T_2}\right) \exp[-bD]. \tag{3.65}$$

This equation is known as the Stejskal[24]–Tanner[25] equation [25]. γ is the gyromagnetic ratio, G_D is the pulse amplitude, δ is the pulse length, and Δ is the pulse separation. $K\rho_p$ is a constant proportional to PD. The symbols are also explained in Fig. 3.44. Usually the prefactors in the exponent are combined in a b-factor: $b = (\gamma G_D \delta)^2 (\Delta - \delta/3)$, which has the units of s/m^2, b is a machine parameter that the operator can control, and D is the intrinsic diffusion constant of the system. Because of the complexity of all contributing factors, D is also called an *apparent diffusional constant* (ADC). The gradient can be applied in any direction X, Y, or Z to analyze the main diffusional direction of fluids. Furthermore, diffusional MRI can be combined with other already discussed sequences for time-resolved and fast MRI methods. However, the Stejskal–Tanner equation is only valid for tissues where diffusion shows no spatial anisotropy, i.e., purely Brownian-type motion.

If only one measurement is performed with one b-value, it is difficult to distinguish whether the attenuation is due to diffusion or additional T_2^* relaxation. However, using two different b-values, the diffusional contrast can be separated from all other effects:

$$\ln \frac{S_{DWI,1}}{S_{DWI,2}} = -(b_1 - b_2)D. \tag{3.66}$$

D or better ADC can then be characterized for specific locations, for instance, in the white or gray matter of the brain, and differences can be identified for healthy tissues compared to those which have suffered damage or injuries.

DWI methods have been used successfully in clinical studies. However, it turned out that the diffusion behavior of bound and unbound water, especially in

---

24 Edward O. Stejskal (1932–2011), American chemist.
25 John E. Tanner (1930), American chemist.

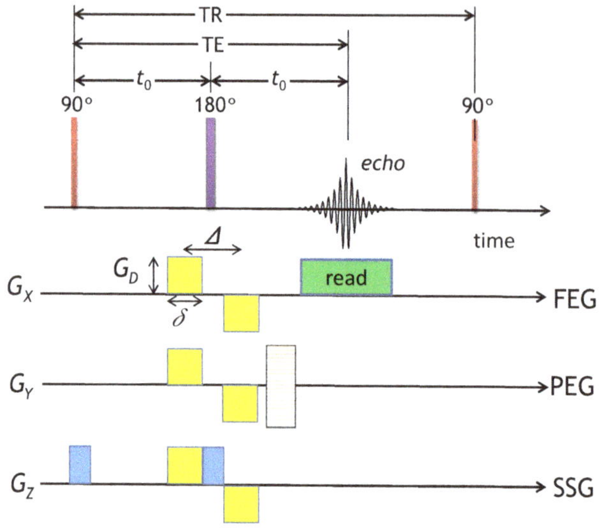

Fig. 3.44: Time sequence for diffusion-weighted imaging. An additional gradient field pulse G_D (marked in yellow) is applied shortly before and shortly after the 180° refocusing pulse. The G_D pulse can be applied in any direction in addition to the usual pulse sequence. FEG, frequency encoding gradient; PEG, phase encoding gradient; SSG, slice selection gradient.

Fig. 3.45: Wiring of the human brain visualized by tracking the movement of water molecules using diffusion tensor imaging (DTI). The nerve fibers run through the mid-sagittal plane. Particularly noticeable are fibers that connect the two hemispheres through the corpus callosum and those which descend toward the spine (blue, within the image plane) (reproduced from Wikipedia https://en.wikipedia.org/wiki/Diffusion_MRI, © creative commons).

the brain, is so complex that a full anisotropic diffusion model has to be used, in which D and b are treated as tensors. The ratio of the signal strength with and without activated G_D is then

$$\ln\frac{S_{\text{DWI}}}{S_0} = b_{ij}D_{ij}, \tag{3.67}$$

where sums are taken over double indices according to tensor notation. Imaging via this method is known as *diffusion tensor imaging*. The aim is to visualize the strongly anisotropic diffusion in nerve fibers; an example is shown in Fig. 3.45. For a review and further details, we refer to [25, 26].

3.10.5 Multiple parameter MRI (mpMRI)

mpMRI is becoming an increasingly important modality for early cancer detection with superior reliability compared to PET or US screening. The mpMRI is comparable to fMRI discussed in the next section, since it provides morphological and functional information about certain tissue volumes. In addition to $T2$-weighted contrast imaging, mpMRI combines DWI, DCE-MRI, and magnetic resonance spectroscopy imaging (MRSI). By combining these methods, the size and morphology of potential tumors can be determined and their physiological response, for example, in prostate carcinoma [27]. Therefore, it is possible to differentiate between clinically irrelevant tumors and malignant tumors with significantly increased sensitivity and specificity. With mpMRI, biopsies needed to confirm tumors can be placed much more precisely. Unnecessary surgery can be avoided. mpMRI is also important after radiation treatment of prostate carcinoma to evaluate tumor response without accumulating further radiation.

3.10.6 Functional MRI (fMRI)

fMRI is based on changes of $T1$ and $T2$ relaxation times ($R1$ and $R2$ relaxivities) of those protons close to hemoglobin in the brain. Neural activity enhances oxygen-rich blood flow. Without oxygen bonding (deoxyhemoglobin), Fe^{2+} is in a high-spin $S = 2$ paramagnetic state associated with a magnetic moment of 5.4 μ_B. In contrast, with O_2 bonding, Fe^{2+} is in a low-spin $S = 0$ state with zero magnetic moment. Further details on the high-spin–low-spin transition of Fe^{2+} in hemoglobin can be found in Chapter 8/Infobox III (Vol. 1). In the high-spin deoxy state, the spin–spin interaction is large, similar to the case of Gd^{3+} discussed in Section 3.10.2. The T_2^* relaxation time is accordingly short. In contrast, in the oxy state, T_2^* is normal. Thus, Fe^{2+} acts as an endogenous CA that can, in addition, differentiate between oxyhemoglobin and deoxyhemoglobin, which, in turn, is related to brain activity. Therefore, the T_2^* relaxation time can be utilized for mapping out the brain and correlating locations of brain activity with enhanced blood flow. The amount of oxygen delivered by the blood flow to neuronal activity centers exceeds the amount required by the surrounding tissue. Thus, an enhanced oxyhemoglobin to deoxyhemoglobin ratio signals local brain activity. The signal change is small but is reliably detectable by taking difference images collected at rest and during brain activity.

The previous discussion suggests that fMRI can be used to relate high metabolic activity of the brain with external stimuli for locating responsive centers in the brain. The stimuli may be visual, audible, physical (tipping a finger), or cognitive (associative, problem-solving, etc.). An example of an associative stimulus is shown in Fig. 3.46. Children were given the task of associating objects with physical activity, such as "ball" and "throw" [28]. A high correlation between different locations in the brain needed for this task is clearly visible in the $T2$–weighted MRI cross sections. This scanning technique is named *blood oxygen-level-dependent* (BOLD) fMRI acquisition.

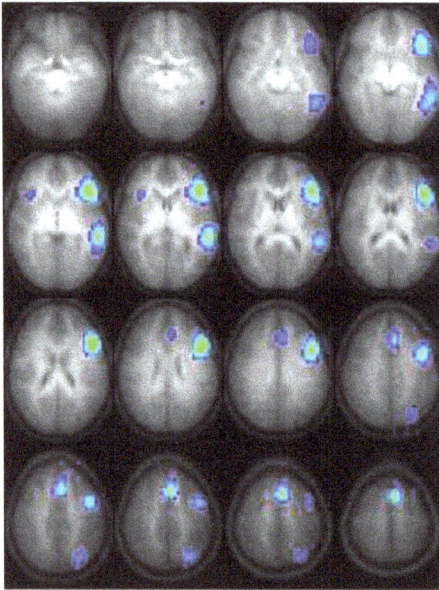

Fig. 3.46: Sequence of fMRI maps of the brain after stimulation through word association. Sixteen axial slices are shown ranging from below the intercommissural line (top left) to above (bottom right) (reproduced from [28] by permission of John Wiley and Sons Inc.).

BOLD fMRI has been used for many different tasks mapping out most of the brain and localizing the centers for language, music, different peripheral movements, etc. In fact, the spatial resolution has become so high that single words can be identified and "read" by fMRI mapping. Another active research branch of fMRI is studies of neural interconnectivity of the brain in the resting state, i.e., in a state without specific tasks. An overview on current brain research with fMRI is provided in [29] and fMRI study of language development in early age is presented in [30].

One of the most intriguing problems in brain research is the question about plasticity. Nerve cells in the brain develop at an early age, but neurons and neural networks are never refurbished after reaching adulthood. This, at least is the traditional view. In Fig. 3.48, we see fMRI images upon light response of a blind dog and a healthy dog for control [31]. The blind dog has a gene defect and was therefore blind since birth. After gene manipulation, retinal visibility could be reestablished. However, the question was whether the blind dog would ever develop optic nerves and a visual cortex for recognizing light. Can blindness be cured even if born blind? Figure 3.47 demonstrates that the blind dog regained his visual perception after only 2 months since gene therapy. fMRI BOLD scans shows visual sensitivity at the same

WT Canine [E946]

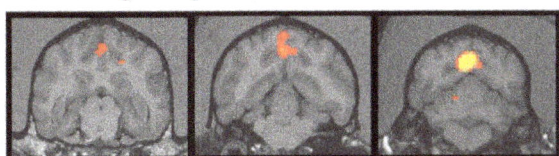

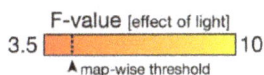

F-value [effect of light]
3.5 [⎯⎯⎯⎯⎯⎯] 10
▲ map-wise threshold

RPE65-mutant Canine [BR235]

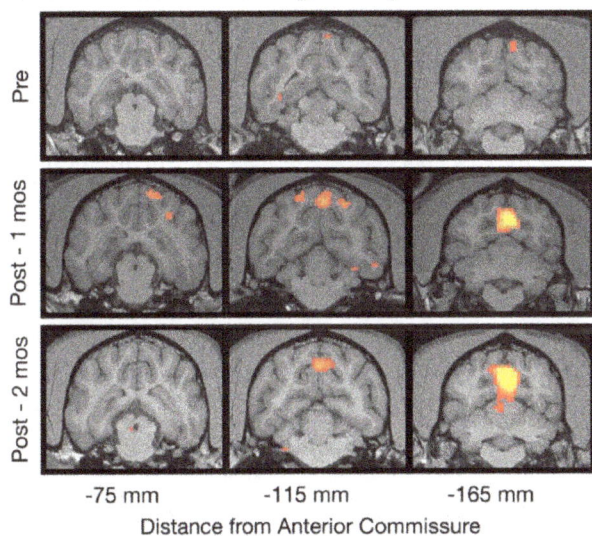

Pre

Post - 1 mos

Post - 2 mos

-75 mm   -115 mm   -165 mm
Distance from Anterior Commissure

Fig. 3.47: fMRI responses of a blind dog before and after gene therapy. Three coronal slices through the brain are shown at different distances from the anterior commissure. Red and yellow colors indicate the location of significant responses to light stimulation. Top row: visual responses of a healthy control dog labeled WT canine [E946]. Lower three rows: pretreatment and posttreatment data from a blind dog. Posttreatment data were taken during two separate sessions separated by 1 month. The MRI scans confirm normal response in both sessions (reproduced from open-access Ref. [31]).

location of the visual cortex as the healthy dog. The images are taken at three different coronal slices behind the anterior commissure, which is a fix point in the brain.

3.10.7 Real-time MRI

Standard MRI has a time resolution of 1–5 s for scanning a complete slice. Rhythmically moving organs such as the heart with a cycle time of about 700–900 ms at rest cannot be imaged without additional procedures. Two techniques are applied for those cases:

(a) Stroboscopic imaging

Stroboscopic imaging is performed by gating the transceiver with the ECG signal. This has the advantage of high spatial and temporal resolution. However, the method can only be used for periodic processes. The procedure fails in the case of cardiac arrhythmias, which are typical of heart diseases. Nevertheless, recent advances in imaging techniques have shown that self-gating techniques can be used to image arrhythmic heartbeats [32].

(b) Echo-planar imaging (EPI)

Echo-planar imaging (EPI), also known as *turbo-MRI*, is an imaging modality in real time with a time resolution of about ca. 10–20 frames/s. With such a high temporal resolution also aperiodic processes can be recorded, such as perfusion studies of heart, lung, kidneys, or brain [33, 34]. The disadvantage is a poor spatial resolution of the frames. The technique is based on the *fast spin-echo* (FSE) method. In FSE, the 180° refocusing pulse is applied several times before reaching TR. Each time the pulses will be refocused but consecutive amplitudes of the SE signal will decrease while still detectable, as shown in Fig. 3.15. Furthermore, for each 180° pulse the PEG is changed between echoes, which allows for filling up several lines in the K-map within one TR interval. The FSE pulse sequence shown in Fig. 3.48 is a combination of the MRI protocols displayed in Figs. 3.15 and 3.33. The number of echoes recorded in a given TR interval is labeled *echo train length* (ETL) or *turbo factor*. The ETL typically ranges from 4 to 32 for routine imaging; in Fig. 3.46 it is only 3. For an ETL of 32, only 8 TRs are required to scan an entire K-map, which is a remarkable time-saving.

EPI is a variant of FSE. A sequence of consecutive spin echoes are generated by 180° refocusing pulses similar to FSE (see Fig. 3.49). However, after each echo, the sign of the readout FEG G_X–gradient is alternated and the initial PEG is reduced sequentially by opposing smaller G_Y gradients. The consequence is a faster filling of the K-map, by running from left to right in the top most row, then right to left in the next row, etc. Finally, the K-map is completely filled within just one or a few TR intervals. Only one TR interval with an ETL of 128 is used for time-resolved measurements

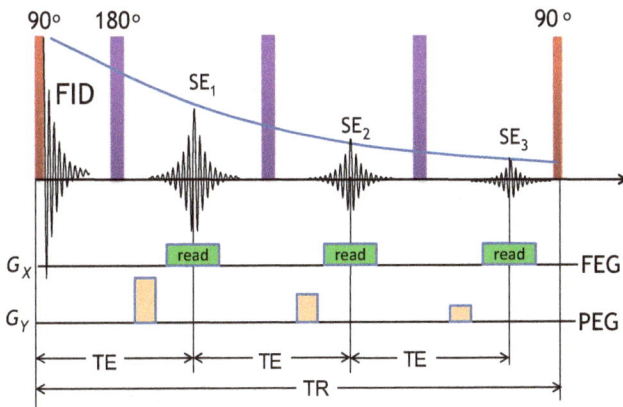

Fig. 3.48: Time sequence for fast spin-echo (FSE) scans. SE is repeated many times within one TR. After each TE, the phase gradient is changed for acquiring multiple lines of the K-map within one Y interval.

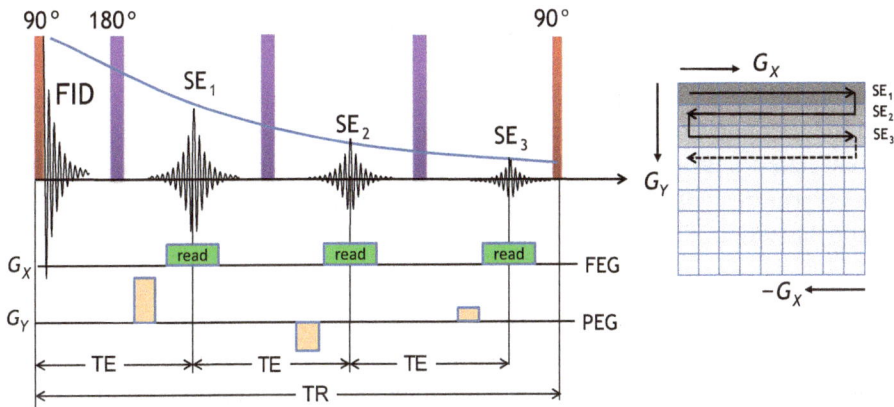

Fig. 3.49: Echo-planar imaging method is a fast spin-echo sequence for time-resolved imaging of moving organs such as the heart.

to fill a 128×128 K-map. Several other sequences are being used for time-resolved MRI. MRI movies of the beating heart are posted on the internet, for instance at [35].

3.11 Hyperpolarization MRI

All MRI procedures discussed so far were based on proton spin resonance, also known as ^{1}H-MRI or more simply pMRI. In the following, we present MRI methods based on isotopes different from protons. Recalling from the introduction in Section 3.2, isotopes

(I) with unpaired nuclear spins are NMR active. Examples are ^{3}He, ^{13}C, ^{15}N, ^{17}O, ^{19}F, and ^{31}P. These light nuclei are not as abundant in the body as ^{1}H. Their lower density ρ_I must be compensated for by an artificially induced higher polarization p_I in order to achieve a high signal strength

$$S \sim p_I \rho_I \mu_I \qquad (3.68)$$

and a good signal-to-noise ratio (SNR), where μ_I is the respective nuclear magnetic moment of the isotope I. Hyperpolarization refers to the fact that the artificial polarization of the nuclei is beyond thermal equilibrium and therefore unstable with lifetimes ranging from seconds to hours. The respective imaging modality is referred to as hMRI. The nuclear hyperpolarization is always performed ex situ. Three methods are commonly used for polarizing the nuclei:

a. Laser excitation

The isotope ^{3}He can be polarized by laser excitation, a process referred to as metastability-exchange optical pumping (MEOP) [36]. This method first performs optical pumping on the metastable 2^3S triplet state of ortho-^{3}He atoms. By optical pumping with a polarized laser, the 2^3S state becomes polarized. Due to an efficient hyperfine coupling between electrons and nuclei of 2^3S He atoms, this electronic polarization also induces nuclear polarization. Finally, collisions between metastable 2^3S and ground state 1^1S He atoms transfer the nuclear polarization to the ground state. This process requires only low magnetic fields and room temperature conditions. Nuclear polarization of more than 90% can be achieved.

b. Dynamic nuclear polarization (DNP)

DNP implies a transfer of high electron spin polarization of paramagnetic ions to nuclear spins via electron-nuclear spin–spin interaction. For this purpose, the target containing the isotope in question is doped by paramagnetic ions. Polarization transfer to the isotopes is achieved by microwave irradiation with frequencies close to the electron spin resonance in high fields (~3 T) and low temperatures (~1 K) [37].

c. Para-hydrogen-induced polarization (PHIP)

Similarly, PHIP uses the low-temperature properties of protons in H_2 molecules. At room temperature, H_2 molecules are either in a para- or in an ortho-state with respect to their nuclear spins. In para-H_2, the proton spins are oriented antiparallel, in ortho-H_2 parallel, with a ratio 1:3. The 1:3 occupancy is because para-hydrogen molecules are in a singlet state with the proton spins adding to a total nuclear spin $I = 0$, while ortho-hydrogen molecules are in a triplet state with $I = 1$. The energy difference is little: 1.45 kJ/mol ($\triangleq$15.1 meV or $\triangleq$175 K). Hence, the ortho- and para-states are equally populated at room temperature. However, the ground state of hydrogen molecules is para-H_2. Upon approaching low temperatures, an ortho–para conversion takes place. When in contact with other molecules, the para spin state of H_2 causes hyperpolarization of, for instance, CAs. By this process, the NMR signal strength becomes dramatically enhanced. In all cases, hyperpolarization takes place in vitro. The administering is done either by inhalation or intravenously by injection.

Unlike CAs for DCE-MRI, hyperpolarized isotopes do not contribute to changes in $T1$ or $T2$ in standard proton-based MRI (^{1}H-MRI). Instead, these isotopes offer the opportunity to perform their own MRI imaging. Therefore, from now on, we will distinguish proton-based MRI (^{1}H-MRI) and isotope-based MRI I-MRI, I = ^{3}He, ^{13}C, ^{15}N, ^{17}O, ^{19}F, ^{31}P). The use of specific resonance energies of selected isotopes like ^{23}Na-MRI and ^{31}P-MRSI, to scan tissues, is also known as *magnetic resonance spectroscopy imaging* (MRSI) [38]. When combined with hyperpolarized isotopes, MRSI provides specific metabolic information in addition to the morphological contrast. In this respect, the hMRI is similar to the fMRI presented in Section 3.10.6. In the following, we present a few examples for the application of hMRI and MRSI.

3.11.1 ^{3}He-hMRI

The respiratory system is difficult to image using ^{1}H-MRI as the PD is rather low. Nevertheless, MRI has advantages over x-ray projection radiography or CT in those cases when ionizing radiation is not tolerable, for instance, during pregnancy, for small children, or when multiple radiographs are wanted, while the total accumulated dose should be kept as low as possible. Because of the low contrast, contrast enhancement methods should be applied. One possibility is filling the respiratory volume with Gd-containing nanoparticles. But the risk of this procedure is quite high, as the nanoparticles, similar to natural fine dust, can impact the immune system and enhance toxicity. An interesting alternative has been proposed, using polarized helium for inhalation [39]. Most people have already experienced by themselves or by demonstrating the impact of He gas on the voice when inhaling and exhaling. ^{3}He is an isotope of standard ^{4}He with one missing neutron and therefore with an uncompensated neutron spin that can be polarized by laser pumping, as described earlier. The hyperpolarization is performed in a magnetic field external to the patient. But even if taken out of the field and inhaled from a flask, the polarization remains sufficiently high for acquiring MRI images. By using FSE techniques, a sequence of images can be taken to study the pulmonary ventilation during inhaling and exhaling, as shown in Fig. 3.50. One can easily recognize the flow of ^{3}He gas through the bronchial branches, distributing over the entire tidal volume, and retracting again during exhaling. From such images, damages to the lung tissue through emboli, tumors, smoking, or asthma can be recognized and analyzed [40].

Now we try to estimate the signal strength $S_{He} \sim p_{He}\rho_{He}\mu_{He}$ of ^{3}He-MRI. In the case of protons, the polarization in a field of 1 T is in the order of 10^{-6} at room temperature, for ^{3}He the polarization is nearly 1 upon hyperpolarization. The ratio of the densities is about $\rho_H/\rho_{He} \approx 2500$. The magnetic moments have different signs but otherwise are almost equal (see Tab. 3.1). Considering all these factors, the ratio of the signal strengths for ^{3}He imaging versus proton (^{1}H) –imaging is about $S_{He}/S_H > 10$. This is indeed a sizeable enhancement factor. The only disadvantage of ^{3}He-MRI is the extreme rarity of ^{3}He and, therefore, its high prize. Meanwhile, other hyperpolarized rare gases have been tested to overcome the cost barrier, such as ^{129}Xe and ^{83}Kr. A discussion of the present status of hyperpolarized gases for pulmonary MRI studies is provided in [41].

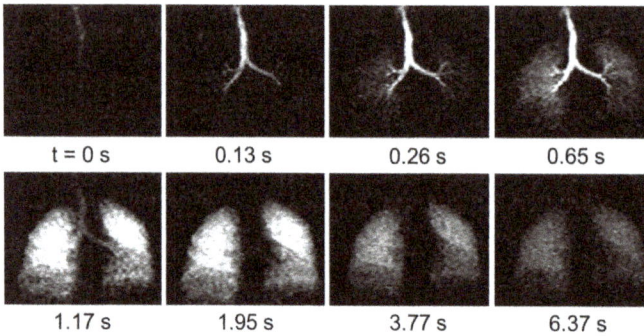

| t = 0 s | 0.13 s | 0.26 s | 0.65 s |

| 1.17 s | 1.95 s | 3.77 s | 6.37 s |

Fig. 3.50: MRI sequence of images taken during inhaling and exhaling polarized ^{3}He gas (courtesy Werner Heil, Johannes Gutenberg Universität Mainz, Germany).

3.11.2 ^{13}C-hMRI

For assessing prostate cancer, ^{13}C-hMRI has been used [42]. ^{13}C has a natural abundance of 1.1%. Because of its larger mass, the gyromagnetic ratio is smaller than for protons: $\gamma_{13C} = 0.673$ rad T^{-1} s^{-1}, compared to $\gamma_{1H} = 2.675 \times 10^8$ rad T^{-1} s^{-1}, and the resonance frequency is accordingly lower:

$$\omega_{13C}/2\pi = f(\text{MHz}) = 10.7\,B_z\,(\text{MHz}),$$

as compared to protons with:

$$\omega_{1H}/2\pi = f(\text{MHz}) = 42.58\,B_z\,(\text{MHz}).$$

Therefore, higher magnetic fields are required for inducing a Zeeman splitting similar to the one for protons. Furthermore, molecules used for studying metabolic pathways, such as pyruvate ($C_3H_4O_3$), can be enriched with ^{13}C to increase the isotopic density. In addition, DNP is applied for hyperpolarization. DNP increases the SNR by 4 to 5 orders of magnitude, rendering in vivo imaging of various metabolites and their enzymatic conversion into other species possible. The main limitation of this imaging technique is the short half-life of the polarization, which is only 30–40 s, meaning that the hyperpolarized signal is useful only for mere 2–3 min. Therefore, administering enriched and hyperpolarized ^{13}C-containing molecules to the target has to proceed very fast.

3.11.3 ^{17}O-hMRI

^{17}O-hMRI is predestinated for performing studies of oxygen metabolism in general and in particular of the brain [43]. Like ^{13}C, ^{17}O also has a small gyromagnetic ratio of $\gamma_{17O} = -0.362 \times 10^8$ rad T^{-1} s^{-1} and a correspondingly low resonance frequency:

$$\omega_{17O}/2\pi = f(\text{MHz}) = -5.77\, B_z\ (\text{MHz})$$

The minus sign indicates that angular momentum and nuclear magnetic moment are antiparallel, which does not impact MRI protocols. The natural abundance of ^{17}O is only 0.037%, which is a severe limitation and makes the method costly. Enrichment of the oxygen isotope is absolutely necessary as well as recycling, similar to ^{3}He. When ^{17}O is used in water (H$_2^{17}$O), hyperpolarization is possible via DNP, but presently not in the gaseous state.

3.11.4 ^{19}F-hMRI

Among the above-listed isotopes, ^{19}F features the most favorable NMR properties. The natural abundance of ^{19}F is 100%, the spin 1/2, gyromagnetic ratio $\gamma_{19F} = 2.51 \times 10^8$ rad T^{-1} s^{-1} and resonance frequency:

$$\omega_{19F}/2\pi = f(\text{MHz}) = 40.08\, B_z\ (\text{MHz}).$$

Those values are very close to ^{1}H-MRI. Only trace amounts of ^{19}F can be found in the body in solid form (bones and teeth). Therefore, the signal is naturally background-free, and MRI signal strength can be directly related to local ^{19}F concentrations. Hyperpolarization of ^{19}F with PHIP has been demonstrated [44]. The potential for ^{19}F MRSI studies is high.

Many drugs contain fluorine, and MRI can be used to study the pharmakinetics of drug delivery. In addition, many fluorine-containing molecules, polymers, and nanoparticles have been developed to target specific tissues. ^{19}F-MRSI also has the potential to replace positron emission tomography (PET) with fluoro-deoxyglucose (^{18}F-FDG).

^{18}FDG-PET is mainly used for the early detection of carcinomas because they have an enhanced glucose metabolism. Further details on ^{18}FDG-PET can be found in Chapter 10. In any case, the use of ^{18}FDG-PET requires an extensive infrastructure for producing and handling short-lived ^{18}F radioisotopes. Maintaining such an expensive infrastructure would become obsolete with the introduction and routine use of ^{19}F-MRSI. Despite all these favorable properties, especially low background, high sensitivity, and no radiation risk, the clinical applications of fluorinated CAs are still limited. This is likely to change shortly. A current overview on this topic can be found in [45].

3.12 Further remarks

3.12.1 New trends and comparisons

MRI methods and equipment have reached a very high technical standard. Many methods and protocols are available to amplify signals, increase time resolution, determine diffusion, perfusion, and brain activity. The 1.5 and 3 T MRI scanners have emerged as the most useful devices for clinical applications. Scanners up to 9 T are used for research, but the benefit of these high fields is not obvious. The high spatial resolution is offset by a number of artifacts due to susceptibility and chemical shift issues. Smaller devices dedicated to scanning extremities (hands, elbows, feet, and knees) are coming onto the market, with significant advantages over their larger sisters in terms of maintenance, room temperature operation, and capital and operating costs. Of course, the versatility of full-size scanners is not available, but high-contrast imaging of static and anatomical lesions is still possible. It is foreseeable that in the near future almost every doctor will have a small MRI scanner, as is already the case with US scanners. It is also foreseeable that full-size MRI scanners will take over x-ray radiography, CT, SPECT, and PET in the long term. MRI has the compelling advantage of posing no or negligible radiation hazards to patients and requiring no radioisotopes, including infrastructure to produce and dispose of short-lived isotopes.

MRI development would fare even better than has already been impressively demonstrated if certain paradigms were challenged. The two most important paradigms are: (1) high fields and (2) inductive detection. They go hand in hand. However, brain imaging was only detected at 130 μT with SQUID detectors [46]. In general, emerging ultra-low-field MRI techniques may even be applicable to imaging tumors without needing contrast-enhancing agents and imaging protocols requiring at least 1.5 T fields. The message of these promising developments is that as long as induction coils are used to acquire MR signals, high frequencies are required since the induced voltage increases with frequency. High frequencies require high fields. Using pickup sensors based on SQUIDs or atomic probe devices, MRI could be performed at much lower fields and become more affordable than today's units. There is still a lot of room for further developments.

3.12.2 Advantages–disadvantages and hazards

Due to the high magnetic field operation, extreme precautions must be taken to keep metal objects out of the MRI treatment room. The same applies to patients. All metal wristwatches, jewelry, pacemakers, glasses, hearing aids, metal implants, unless diamagnetic, must remain outside.

The rf-coils act like a microwave oven. Care must be taken to maintain performance at a level that does not raise body temperature by more than one degree.

Some people suffer from claustrophobia and find confinement in a closed tube frightening. For these cases, open C-frame machines have been developed. However, the magnetic field homogeneity, which is essential for the resolution, is naturally not as perfect as in a closed cylindrical system.

Due to the rapid switching of the coils, the MRI devices tend to generate very loud and continuous hammering noises during scanning. The rapid switching of gradient coils can also create eddy currents in electrolytes. However, no negative effects have been reported so far.

Some patients may be too large to fit in the tube. The current standard size is 70 cm in diameter. Open C-frame machines can also be a solution in these cases.

Some patients may be allergic to Gd-chelated CAs. In these cases, alternative CAs must be used.

MRI scans require patients to remain very still for long periods of time. But with rapid scanning techniques, this condition has been significantly alleviated.

MRI systems are expensive to purchase and operate. MRI examinations are correspondingly expensive. This will change in the future when smaller units with greatly reduced investment and operating costs will become available.

There is no radiation hazard for patients or staff. However, potential hazards arise from the high magnetic fields and if the superconducting coils quench.

Before concluding, we compare the three imaging modalities discussed in the first three chapters that provide images without radiation hazards: ultrasound imaging, endoscopy, and MRI. Table 3.5 lists the technical specifications of the three methods and Tab. 3.6 compares advantages and disadvantages. Fig. 3.51 shows the respective lateral and axial resolution in comparison.

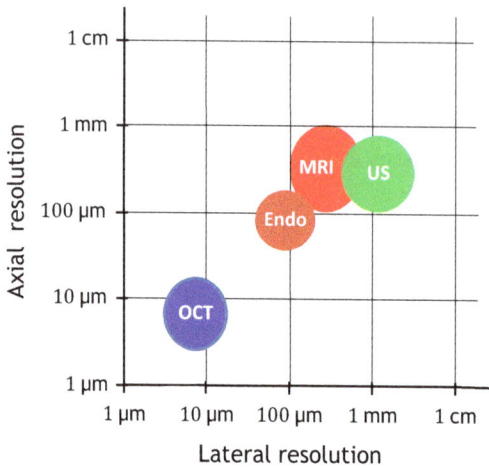

Fig. 3.51: Comparison of the lateral and axial resolution for three imaging modalities: ultrasound, endoscopy (including OCT option), and MRI.

Tab. 3.5: Technical specifications of the three imaging modalities discussed so far, which are free of radiation hazards.

	US	Endoscopy	MRI
FOV	Scan size	140°	5–500 mm
Penetration depth	3–25 cm	5 μm	No limit
Axial resolution	150–800 μm	Not applicable	0.5–1 mm
Lateral resolution	1–2 mm	100 μm	0.5–1 mm
Scanning lines per frame	100	Not applicable	256
Video frames per second	20–50	50	20
Pulse repetition frequency	3 kHz	No pulsing	0.05–5 s
Contrast	Boundary between different $z = \rho v$	Light scattering at surfaces	Tissues with different $T1$, $T2$, or PD
Contrast enhancement	Not applicable	Fluorophores, stains	Contrast agents, special isotopes
Extras	Doppler, duplex	NBI, chromo-E, CLE, OCT-E, capsule-E	fMRI, angio, hMRI, DWI, mpMRI, DCE, MRSI

Tab. 3.6: General comparison of the advantages and disadvantages of the three imaging methods discussed in Part A.

	Pros	Cons
Ultrasound	– Not invasive – High accuracy – Can be repeated any time – High acceptance by patients – "cheap" weith respect to investment and operational costs – Bed-side application – Real-time imaging – Interactive application	– Impedance mismatch of air and bones – Low lateral resolution – Informational value strongly dependent on examination qualification
Endoscopy	– Imaging of inner cavities – Important medical check-up – OCT option with very high spatial resolution – Support of minimally invasive surgery – Low cost but versatile applications	– Low probability of injury
MRI	– High-resolution 3D images – Very good contrast of connective tissues – High-contrast brain imaging – Functionality of the nervous system via fMRI with time resolution <1 s – No danger to the patient – Painless	– High investment cost – High maintenance cost – Scanning takes much time – Implants cause hazards – Possibility of claustrophobia – Loud noises during scanning – No support of invasive procedures – High qualification of the staff required

3.13 Summary

S3.1   Nuclei with odd atomic numbers exhibit weak nuclear paramagnetism.

S3.2   In an external field, nuclear moments precess with a characteristic frequency, i.e., the Larmor frequency.

S3.3   Applying an RF-field perpendicular to the main field, nuclear magnetic moments can be rotated by various angles, most prominently by 90° or 180°.

S3.4   Spin-echo refocusing is the single most crucial feature for image generation.

S3.5   Transverse slices are defined by field gradient in the z-direction.

S3.6   The columns in a slice are encoded by a frequency gradient; rows are encoded by a phase gradient.

S3.7   Frequency and phase are recorded in a K-map, which delivers a real-space picture upon Fourier transformation.

S3.8   The longitudinal relaxation time $T1$ into the equilibrium state depends on the spin–lattice interaction, i.e., the interaction of the nuclear spin with its surrounding.

S3.9   The transverse relaxation time $T2$ for dephasing of spins in the plane perpendicular to the main field depends on spin–spin interactions.

S3.10   Chemical shift is an artifact that can blur the MRI image. It is due to the diamagnetic environment, in which protons are embedded.

S3.11   MRI offers a high-resolution, noninvasive and radiation-free 3D imaging modality of tissues with contrast generated by $T1$, $T2$, or PD differences.

S3.12   Contrast enhancement can be achieved by specific $T1$-, $T2$-, and PD-weighting schemes or by using paramagnetic contrast-enhancing agents.

S3.13   DWI uses FSE techniques for imaging diffusion and perfusion in nerve fibers.

S3.14   MRI with protons (^{1}H-MRI) is supplemented by MR spectroscopic imaging using various stable isotopes: ^{3}He, ^{13}C, ^{15}N, ^{17}O, ^{19}F, ^{31}P, ^{83}Kr, and ^{129}Xe for specific tasks.

S3.15   fMRI of the brain is based on enhanced oxygen consumption in active centers and can be used for cognitive studies, brain development, and brain diseases.

S3.16   EPI is a high-speed MRI modality that is fast enough to capture videos of repetitive moving organs (heart) and perfusion processes.

? Questions

Q3.1   Which nuclei are suitable for MRI?

Q3.2   Which type of MRI is most prominent?

Q3.3   How is the Larmor frequency defined?

Q3.4   What is the typical precessional frequency range of protons in a magnetic field?

Q3.5   The nuclear spin precession is damped by which two independent processes?

Q3.6   What is the FID good for?

Q3.7   When does FID occur?

Q3.8   How can magnetization be turned from the z-direction to the in-plane xy-direction?

Q3.9   Which relaxation time controls the return of the in-plane component M_{xy}?

Q3.10   Which relaxation time controls the magnetization component M_z?

Q3.11   What is meant by spin echo?

Q3.12   What is the pulse sequence in a spin-echo detection?

Q3.13   At what time does a spin echo occur if the 180° pulse was applied at time t_0 after the 90° pulse?

Q3.14   What is the time sequence of RF pulses applied for $T1$ and $T2$ contrast?

Q3.15   What is the time sequence of RF pulses applied for PD contrast?

Q3.16   What is the idea of IR, and how is it applied?

Q3.17   Discuss why the inverse relaxation time $1/T1$ should be on the same order as the resonance frequency f_0 for a maximum signal in NMR experiments.

Q3.18   Why does the magnetic induction B_z vanish in the rotating frame?

Q3.19   How is a slice in the body selected by MRI methods?

Q3.20   How are the X- and the Y-directions selected to define a voxel ΔX, ΔY, ΔZ in the body by MRI methods?

Q3.21 How is the FOV determined?
Q3.22 What is a *K*-map?
Q3.23 How is a real-space image generated from the *K*-map?
Q3.24 When is the phase gradient applied during TR?
Q3.25 What are typical MRI field specifications?
Q3.26 How is the main magnetic field of 1–3 T generated?
Q3.27 How are field gradients generated?
Q3.28 What is the typical power consumption during MRI scanning with a conventional MRI system?
Q3.29 What is the effect of a contrast-enhancing agent?
Q3.30 What kind of ion do CAs always contain?
Q3.31 What is meant by hMRI?
Q3.32 What is the advantage of hMRI compared to conventional MRI? What is the main disadvantage?
Q3.33 What is meant by DWI, and how is it realized?
Q3.34 What is the main application of DWI?
Q3.35 What is the basic principle of fMRI?
Q3.36 How can moving organs like the lungs and heart be imaged with MRI?
Q3.37 How many frames per second can be achieved by the application of turbo MRI?
Q3.38 What are the new trends in MRI technology?
Q3.39 What are the three main hazards which need to be considered for the application of MRI scanning?

Attained competence checker   +   0   –

I know what the Larmor frequency is and how it depends on the magnetic induction.

I can distinguish between $T1$ and $T2$ relaxation.

I know which isotopes can be used for MRI.

I know what FID is and how it can be exploited.

I understand the spin-echo process and I know what it is good for.

I know how $T1$-, $T2$-, and PD-weighted contrast can be generated.

I know why M_z relaxation does not generate an FID.

I know how a slice in the body is selected by MRI methods.

I know how to encode a voxel in a slice.

K-map is a map in _____ space.

I know how many coils are in an MRI scanner.

I know why CAs are used and what effect they have.

I can describe the main principle of fMRI.

I know the spatial resolution of MRI and I can set this in perspective to other imaging modalities.

I can name the main advantages/disadvantages of MRI.

Exercises

E3.1   **Magnetization:** Show that eq. (3.20) is correct.

E3.2   **Pulse sequence:** Why is a pulse sequence with TR = short and TE = long not used in MRI scanning modes?

E3.3   **T2 contrast elimination:** How is the T2 contrast eliminated in Fig. 3.18?

E3.4   **Time of inversion:** Show that for contrast enhancement via IR, the inversion time is given by $t_i = T1 \ln 2$.

E3.5   **Liquid He consumption:** The daily boil-off of liquid He is 0.15% of the 1500 L liquid He tank. What is the consumption per year and when is refill necessary if the tank should not drop below 50%.

E3.6   **G_Z-gradient:** For the main coil, a slice thickness of 1 mm is specified. The rf-coil has a bandwidth of 100 Hz. What field gradient G_Z should be applied to meet these specifications?

E3.7   **T1 relaxation:** According to Tab. 3.2, protons in water and fat exhibit considerably different T1 relaxation times: 3000 s for water and 200 s for fat. Try to give a physical explanation for this large difference.

E3.8   **Water contrast:** In T1-weighted images water appears black. Why is this so?

E3.9   **Power consumption of solenoid:** A local field of 30 mT at the center of a circular solenoid is turned on for 5 s. The coil has a diameter of 80 cm and consists of $N = 100$ Cu windings. The Cu wire has a diameter of 5 mm and a resistivity of $\rho = 1.7 \times 10^{-8}$ Ohm·m. What is the power consumption of the coil and the energy required?

E3.10   **Frequency gradient:** If the field gradient for the B_Z field is 10 mT/m, what is the frequency gradient per mm?

References

[1]   Rabi II, Zacharias JR, Millman S, Kusch P. A new method of measuring nuclear magnetic moment. Physical Review. 1938; 53: 318.

[2]   Bloch F, Hansen W, Packard M. The nuclear induction experiment. Physical Review. 1946; 70: 474–485.

[3]   Purcell EM, Torrey HC, Pound RV. Resonance absorption by nuclear magnetic moments in a solid. Physical Review. 1946; 69: 37–38.

[4]   Lauterbur P. Image formation by induced local interactions: Examples employing nuclear magnetic resonance. Nature. 1973; 242: 190–191.

[5]   Mansfield P, Grannell PK. NMR 'diffraction' in solids?. J Phys C Solid State Phys. 1973; 6: L422–L426.

[6]   Hahn E. Spin echoes. Phys Rev. 1950; 80: 580.

[7]   Mezei F. Neutron spin-echo: A new concept in polarized thermal neutron techniques. Zeitschrift Für Physik A Hadrons and Nuclei. 1972; 255: 146–160.

[8]   Einstein A. Über die von der molekularkinetischen Theorie der Wärme gefordete Bewegung von in ruhenden Flüssigkeiten suspendieren Teilchen. Ann Phys. 1905; 17: 549–560.

[9]   Bottomley PA, Foster TH, Argersinger RE, Pfeiffer LM. A review of normal tissue hydrogen NMR relaxation times and relaxation mechanisms from 1–100 MHz: Dependence on tissue type, NMR frequency, temperature, species, excision, and age. Med Phys. 1984; 11: 425–448.

[10]   http://xrayphysics.com/chem_sh.html

[11] Slichter CP. Principles of magnetic resonance. 3rd. Springer Series in Solid-State Sciences; 1990.

[12] Parizh M, Lvovsky Y, Sumption M. Conductors for commercial MRI magnets beyond NbTi: Requirements and challenges. Supercond Sci Technol. 2016; 30: 014007.

[13] Hidalgo-Tobon SS. Theory of gradient coil design methods for magnetic resonance imaging. Concepts Magn Reson Part A. 2010; 36A: 223–242.

[14] Foo TKF, Laskaris E, Vermilyea M, Xu M, Thompson P, Conte G, Van Epps C, Immer C, Lee SK, Tan ET, Graziani D, Mathieu JB, Hardy CJ, Schenck JF, Fiveland E, Stautner W, Ricci J, Piel J, Park K, Hua Y, Bai Y, Kagan A, Stanley D, Weavers PT, Gray E, Shu Y, Frick MA, Campeau NG, Trzasko J, Huston J, Bernstein MA. Lightweight, compact, and high-performance 3T MR system for imaging the brain and extremities. Magn Reson Med. 2018; 80: 2232–2245.

[15] Biber S. MAGNETOM Free.Max: Access to MRI – How to Make it Big Inside and Small Outside. Technology MAGNETOM Flash. 78(1): 2021.

[16] https://www.siemens-healthineers.com/de/magnetic-resonance-imaging/technologies-and-innovations/deep-resolve

[17] Petralia G, Summers PE, Agostini A, Ambrosini R, Cianci R, Cristel G, Calistri L, Colagrande S. Dynamic contrast-enhanced MRI in oncology: How we do it. Radiol Med. 2020; 125: 1288–1300.

[18] Wahsner J, Gale EM, Rodríguez-Rodríguez A, Caravan P. Chemistry of MRI Contrast Agents: Current Challenges and New Frontiers. Chem Rev. 2019; 119: 957–1057.

[19] Laurent S, Elst LV, Muller RN. Comparative study of the physicochemical properties of six clinical low molecular weight gadolinium contrast agents. Contrast Media Mol Imaging. 2006; 1: 128–137.

[20] Zhou Z, Lu ZR. Gadolinium-based contrast agents for magnetic resonance cancer imaging. Wiley Interdiscip Rev Nanomed Nanobiotechnol. 2013; 5: 1–18.

[21] Pruvo J, Vilgrain V, Roy C, Peretti P, Halimi P, Ernst O, Valette P, Matos C, El-Khoury C. Characterisation of central nervous system, liver, and abdomino-pelvic tumours using meglumine gadoterate: Pooled phase III studies. Internet J Radiol. 2010; 13: 2.

[22] Xiao Y, Paudel R, Liu J, Ma C, Zhang Z, Zhou S. MRI contrast agents: Classification and application (Review). Int J Mol Med. 2016; 38: 1319–1326.

[23] Hu X, Tang Y, Hu Y, Lu F, Lu X, Wang Y, Li J, Li Y, Ji Y, Wang W, Ye D, Fan Q, Huang W. Gadolinium-chelated conjugated polymer-based nanotheranostics for photoacoustic/ magnetic resonance/NIR-II fluorescence imaging-guided cancer photothermal therapy. Theranostics. 2019; 9: 4168–4181.

[24] Fenchel M, Nael K, Seeger A, Kramer U, Saleh R, Miller S. Whole-body magnetic resonance angiography at 3.0 Tesla. Eur Radiol. 2008; 18: 1473–1483.

[25] Winston GP. The physical and biological basis of quantitative parameters derived from diffusion MRI. Quant Imaging Med Surg. 2012; 2: 254–265.

[26] Ranzenberger LR, Snyder T. Diffusion tensor imaging. In: StatPearls [Internet]. Treasure Island (FL): StatPearls Publishing; 2022.

[27] Bui TL, Glavis-Bloom J, Chahine C, Mehta R, Wolfe T, Bhatter P, Rupasinghe M, Carbone J, Haider MA, Giganti F, Giona S, Oto A, Lee G, Houshyar R. Prostate minimally invasive procedures: Complications and normal vs. abnormal findings on multiparametric magnetic resonance imaging (mpMRI). Abdom Radiol (NY). 2021; 46: 4388–4400.

[28] Szaflarski JP, Holland SK, Schmithorst VJ, Byars AW. An fMRI study of language lateralization in children and adults. Hum Brain Mapp. 2006; 27: 202–212.

[29] Papageorgiou TF, Christopoulos GI, Smirnakis SM, editors. Advanced brain neuroimaging topics in health and disease – Methods and applications. In Tech; 2014. Open access available at: www.intechopen.com/books/advanced-brain-neuroimaging-topics-in-health-and-disease-methods-and-applications

[30] Olulade OA, Seydell-Greenwald A, Chambers CE, Turkeltaub PE, Dromerick AW, Berl MM, Gaillard WD, Newport EL. The neural basis of language development: Changes in lateralization over age. Proc Natl Acad Sci. 2020; 117: 23477–23483.

[31] Aguirre GK, Komaromy AM, Cideciyan AV, Brainard DH, Aleman TS, Roman AJ, Avants BB, Gee JC, Korczykowski M, Hauswirth WW, Acland GM, Aguirre GD, Jacobson SG. Canine and human visual cortex intact and responsive despite early retinal blindness from RPE65 mutation. PLoS Med. 2007; 4: e230.

[32] Laubrock K, von Loesch T, Steinmetz M, Lotz J, Frahm J, Uecker M, Unterberg-Buchwald C. Imaging of arrhythmia: Real-time cardiac magnetic resonance imaging in atrial fibrillation. Eur J Radiol. 2022; 9: 100404, 1–8.

[33] Pedersen M, Klarhöfer M, Christensen S, Ouallet JC, Østergaard L, Dousset V, Moonen C. Quantitative Cerebral Perfusion Using the PRESTO Acquisition Scheme. J Mag Res Imaging. 2004; 20: 930–940.

[34] Deichmann R, Gottfried JA, Hutton C, Turner R. Optimized EPI for fMRI studies of the orbitofrontal cortex. NeuroImage. 2003; 19: 430–441.

[35] www.biomednmr.mpg.de/index.php

[36] Batz M, Nacher PJ, Tastevin G. Fundamentals of metastability exchange optical pumping in helium. J Phys Conf Ser. 2011; 294: 012002.

[37] Goertz S, Meyer W, Reicherz G. Polarized H, D and He-3 targets for particle physics experiments. Prog Nucl Phys. 2002; 49: 403.

[38] Posse S, Otazo R, Dager SR, Alger J. MR spectroscopic imaging: Principles and recent advances. J Magn Reson Imaging. 2013; 37: 1301–1325.

[39] Kauczor HU, Ebert M, Kreitner KF, Nilgens H, Surkau R, Heil W, Hofmann D, Otten EW, Thelen M. Imaging of the lungs using 3He MRI: Preliminary clinical experience in 18 patients with and without lung disease. J Magn Reson Imaging. 1997; 7: 538–543.

[40] Fain S, Schiebler ML, McCormack DG, Parraga G. Imaging of lung function using hyperpolarized helium-3 magnetic resonance imaging: Review of current and emerging translational methods and applications. J Magn Reson Imaging. 2010; 32: 1398–1408.

[41] Lilburn DML, Pavlovskaya GE, Meersmann T. Perspectives of hyperpolarized noble gas MRI beyond 3He. J Magn Resonance. 2013; 229: 173–186.

[42] Serrao EM, Brindle KM. Potential clinical roles for metabolic imaging with hyperpolarized [1-^{13}C]Pyruvate. Front Oncol. 2016; 6: 59.

[43] Hoffmann SH, Begovatz P, Nagel AM, Umathum R, Schommer K, Bachert P, Bock M. A measurement setup for direct 17O MRI at 7 T. Magn Reson Med. 2011; 66: 1109–1115.

[44] Kuhn LT, Bommerich U, Bargon J. Transfer of parahydrogen-induced hyperpolarization to 19F. J Phys Chem A. 2006; 110: 3521–3526.

[45] Tirotta I, Dichiarante V, Pigliacelli C, Cavallo G, Terraneo G, Bombelli FB, Metrangolo P, Resnati G. (19)F magnetic resonance imaging (MRI): From design of materials to clinical applications. Chem Rev. 2015; 115: 1106–1129.

[46] Inglisa B, Buckenmaier K, SanGiorgio P, Pedersen AF, Nichols MA, Clarke J. MRI of the human brain at 130 microtesla. PNAS. 2013; 110: 19194–19201.

Further reading

Slichter CP. Principles of magnetic resonance. 3rd. Springer Series in Solid-State Sciences. Berlin, Heidelberg 1990.

Ladd ME, et al.. Magnetresonaz tomographie und -spektroskopie. In: Schlegel W, Karger CP, Jäkel O. Medizinische Physik. Heidelberg, Berlin: Springer Spektrum; 2018.

Liang ZP, Lauterbur PC. Principles of magnetic resonance imaging: A signal processing perspective. New York: Wiley-IEEE Press; 1999.

Hornak JP The basics of NMR. Available online at: http://ww.cis.rit.edu/htbooks/nmr/bnmr.htm

Westbrook C. Handbook of MRI technique. 5th. edition New York, London, Sydney, Toronto: Wiley-Blackwell; 2021.

Bushberg JT, Seibert JA, Leidholdt EM, Jr, Boone JM. The essential physics of medical imaging. 3rd. Philadelphia, Baltimore, New York, London: Lippincott Williams & Wilkins, Wolters Kluwer; 2012.

Useful websites

Image gallery of various imaging modalities: https://openi.nlm.nih.gov/ (accessed on 26 January 2023)

Questions and answers on MRI: http://mriquestions.com/index.html (accessed on 26 January 2023)

http://xrayphysics.com/review/mri_b.html (accessed on 26 January 2023)

Part B: **X-ray and nuclear methods**

4 X-ray sources and generators

Physical parameters of x-ray sources	
Mo-$K_{\alpha 1}$ photon energy	17.48 keV
W $K_{\alpha 1}$ photon energy	59.3 keV
Typical peak accelerating potential for radiography	100–150 kVp
Typical accelerating potential for radiotherapy	5–20 MeV

4.1 Introduction

In 1895 Röntgen[1] discovered a new type of radiation called x-rays. He was uncertain about their properties but noticed that they were able to penetrate opaque matter. To demonstrate this, he took x-ray images from his spouse's hand, the first x-ray images ever, which made him and the new method instantly famous worldwide. Since then, x-rays have been known to the general public mainly for their medical use.

Medical x-ray imaging is only one of many other uses of x-rays. The others include x-ray scattering, x-ray spectroscopy, and x-ray microscopy in all fields of science and technology. X-rays from low to high energies are so ubiquitous that a world without x-rays is hard to imagine. Without x-rays, we probably would not know about the helical structure of the DNA, the complex folding of proteins such as myoglobin and hemoglobin, the rich structure and functionality of ribonucleic acid (RNA), or the spiky nature of the SARS-COVID 2 virus.

X-rays are electromagnetic (EM) waves with energies ranging from 50 eV up to several MeV, correspondingly to wavelengths λ from 25 nm (50 eV) down to 0.0012 nm (1 MeV). The conversion factor is derived from the equation for the photon energy:

$$E = hf = \frac{hc}{\lambda} \tag{4.1}$$

or

$$\lambda(\text{nm}) = \frac{1239 \, \text{eV} \cdot \text{nm}}{E(\text{eV})}, \tag{4.2}$$

where $h = 6.623 \times 10^{-34}$ Js is the Planck constant, $c = 299\,792\,458$ m/s $\cong 3 \times 10^8$ m/s is the vacuum velocity of EM waves, and f is the frequency. Electronvolt (eV) is the natural unit in the context of x-rays. Therefore, we make here an exemption from the standard use of SI units.

---

[1] Wilhelm Conrad Röntgen (1845–1923), German Physicist, first Nobel Prize in Physics 1901.

https://doi.org/10.1515/9783110757095-004

At the lower energy end, x-rays overlap with far-ultraviolet radiation. At the upper energy scale, they overlap with γ-radiation. It is not primarily the energy or the respective wavelength that characterizes x-rays; it is how x-rays are generated. Three kinds can be distinguished:

1. *Bremsstrahlung*, radiation produced by de-acceleration of high energy electrons;
2. *Characteristic radiation*, occurring after excitation of core electrons of atoms;
3. *Synchrotron radiation* emitted by the radial acceleration of electrons in high-energy storage rings.

One can simplify this categorization and reduce it to two main effects: acceleration and excitation. Bremsstrahlung and synchrotron radiation arise from the acceleration of electrons: deceleration (stopping) of high-energy electrons in a target in one case and radial acceleration of electrons on a circular orbit in the other case. Characteristic radiation arises from the excitation of atomic core-shell electrons. Although energetically overlapping with x-rays, the term "γ-radiation" is reserved for EM radiation emitted by radioisotopes (see Chapter 5).

For medical x-ray diagnostics (radiography) and x-ray cancer treatment (radiotherapy, XRT) solely bremsstrahlung is used. Their specifications are, however, very different. Hence not the same x-ray equipment can be employed for both applications. X-ray diagnostics requires bremsstrahlung up to about 150 keV, whereas XRT entails x-ray energies up to 25 MeV. This chapter will introduce basic concepts of x-ray production for both applications: radiography (Chapter 8) and radiotherapy (Chapter 2 in Volume 3).

4.2 General components of x-ray tubes

Standard *x-ray tubes*, which are frequently found in research laboratories, use a high voltage difference between cathode and anode to accelerate free electrons over a short distance in a vacuum. These x-ray glass tubes are permanently evacuated and sealed off like traditional light bulbs. The principal features of such a tube are shown in Fig. 4.1. The cathode is connected to a high negative voltage supply between –10 kV and –100 kV, whereas the anode is grounded. A tungsten filament in the cathode is heated to very high temperatures just below the wire's melting temperature to generate free electrons via thermionic emission. The current going through the filament of the cathode, $i_{cathode}$, controls the filament temperature and, consequently, the rate of electrons emitted into the vacuum tube.

Electrons entering at the high negative potential into the vacuum are immediately accelerated toward the anode following the potential difference eU_{acc}. A cap with a small aperture surrounding the filament, called the Wehnelt cylinder,[2] acts

---

2 Arthur Rudolph Berthold Wehnelt (1871–1944). German physicist and inventor.

as an electrostatic lens that keeps the electrons from straying. Obviously, there is a close connection between the cathode current heating the filament $i_{cathode}$ and anode current i_{anode} (unit $[i]$ = Ampere, A) hitting the target. With increasing $i_{cathode}$, the anode wire becomes hot and electrons start being emitted from the wire into the vacuum of the x-ray tube. This process is referred to as *thermionic emission* and is expressed by the *Richardson's law*:[3]

$$i_{cathode} = AT^2 \exp(-W/k_B T), \tag{4.3}$$

where A is a constant, T is the absolute temperature, k_B is the Boltzmann constant, and W is the work function, acting as a potential well for electrons escaping from the metal host into the vacuum. By applying an electrostatic potential difference eV_{acc} between anode and cathode, the work function W is effectively lowered and increases the anode current. Electron emission that is controlled by temperature T and electrostatic potential eU_{acc} is known as *Schottky emission*[4] and expressed by the modified Schottky–Richardson law [1]:

$$i_{cathode} = AT^2 \exp(-(W - eU_{acc})/k_B T), \tag{4.4}$$

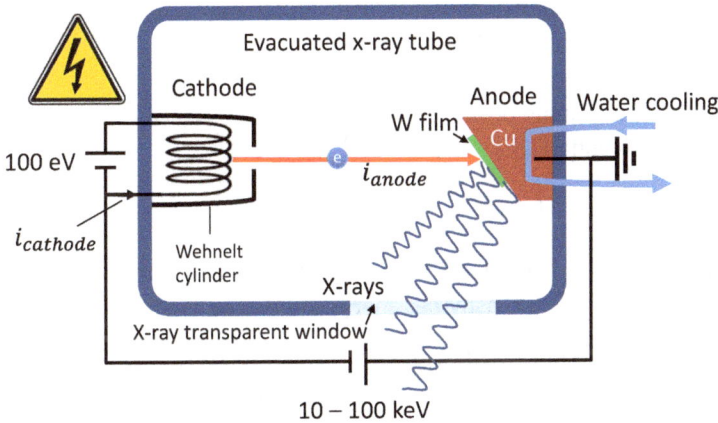

Fig. 4.1: Schematic outline of a sealed x-ray tube.

The anode (target) consists of a water-cooled Cu block coated with a metal film. For medical applications, the anode is usually coated with a tungsten film. In the anode, the free electrons are rapidly stopped by trapping in the Coulomb[5] potential of the

3 Owen Willans Richardson (1879–1959), English physicist, Nobel Prize in physics 1928.
4 Walter H. Schottky (1886–1976), German physicist.
5 Charles Augustin de Coulomb (1736–1806), French physicist.

target nuclei (see Fig. 4.2). While circulating about the target nuclei, the electrons emit EM dipole radiation. They may emit just one photon, carrying away the total kinetic energy of an electron, or they may split their energy into several photons.

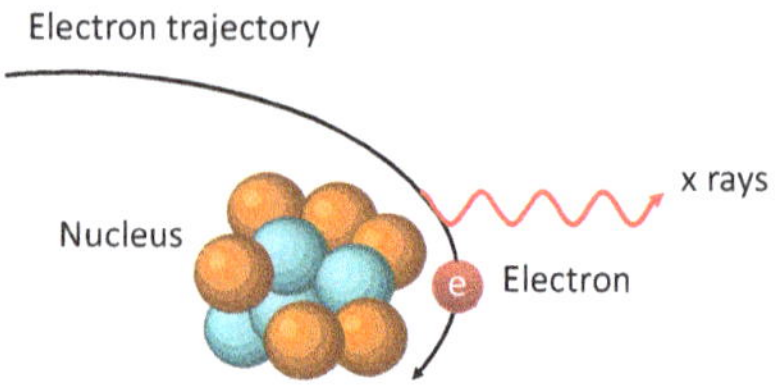

Electron trajectory

Fig. 4.2: Generation of bremsstrahlung in the Coulomb potential of target nuclei.

4.3 Bremsstrahlung radiation

The de-acceleration of high-energy electrons in the anode causes the emission of x-ray photons with a wide spectral distribution of photon energies (or wavelengths). This radiation is called *bremsstrahlung*. The German expression, common in the literature, can be translated as "de-acceleration radiation" or "stopping-radiation." Bremsstrahlung is polychromatic, lacks phase coherence, and spreads in all directions. Such a radiation source is known as an incoherent broadband photon source, sometimes also referred to as "white x-ray source" in analogy to traditional light bulbs. The total integrated intensity over all energies and angles is proportional to the product of the anode current i_{anode} and the square of the potential difference U_{acc} between cathode and anode (accelerating potential) [2]:

$$I_{x\text{-ray}} \sim Z i_{anode} U_{acc}^2. \tag{4.5}$$

where Z is the atomic number. The intensity $I_{x\text{-ray}}$ has the units of number of photons N per area A and time t. The intensity per energy interval ΔE is then:

$$\frac{\Delta I_{x\text{-ray}}}{\Delta E}. \tag{4.6}$$

The x-ray intensity per energy interval is plotted in Fig. 4.3 versus photon energy and for four different accelerating potentials. The intensity of the (unfiltered) radiation can be approximated by a linear equation, known as Kramers[6] equation (dashed lines in Fig. 4.3):

$$\frac{\Delta I_{x\text{-ray}}}{\Delta E} = kZ\left(E_{max} - E_{x\text{-ray}}\right). \tag{4.7}$$

---

6 Hendrik Anthony Kramers (1894–1952), Dutch physicist, mainly known for the Kramers-Kronig relation.

where k is a constant and E_{max} is the cutoff energy:

$$E_{max} = hf_{max} = eU_{acc} = \frac{hc}{\lambda_{min}} \qquad (4.8)$$

The cutoff energy is defined by the maximum energy (minimum wavelength) that a photon can attain when the complete kinetic energy of the accelerated electron is converted into EM wave energy. The SI unit of the accelerating potential is volts (V) or kilovolts (kV). The unit for the maximum potential applied across an x-ray tube is termed in the medical literature as *kilovoltage peak* (kVp). The accelerating potential U_{acc} determines the hardness of the radiation. The higher the potential difference between cathode and anode, the higher the energy of x-ray photons that are produced. High energy photons penetrate matter deeper than low energy photons.

For taking x-ray radiographs of the body, an aluminum (Al)-filter is always inserted that absorbs the low energy part of the x-ray bremsstrahlung spectrum. Otherwise, the low-energy x-ray photons would be preferentially absorbed in the skin and contribute to the dose without enhancing the image quality in x-ray radiographs. The modified bremsstrahlung spectrum with an Al-filter of 2 mm thickness inserted is shown in Fig. 4.3 by the bold lines for several kVp.

When operating an x-ray generator, the user has control over the kVp, the tube current, and the exposure time. Typical values for taking an x-ray radiograph are: voltage between 80 and 140 kVp, anode current of 500 mA, and exposure time of 50–100 ms. This yields a tube energy of 2–7 kWs. Production of x-ray photons by means of bremsstrahlung is not efficient. About 99% of the electron energy is lost in heat production, and only 1% is converted into photons. Therefore x-ray tubes must be cooled very well to protect them from overheating. Water cooling is obligatory in

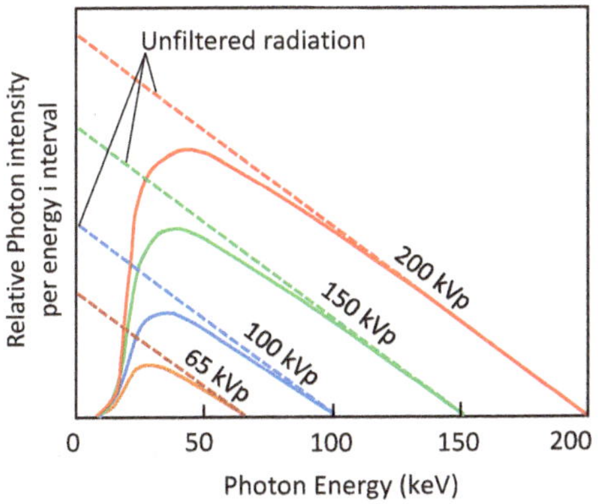

Fig. 4.3: Bremsstrahlung spectrum for four different accelerating potentials. Note the intensity increase with increasing kVp and the linear intensity drop terminating at the maximum photon energy. The dashed lines refer to unfiltered x-ray intensity, the bold lines to x-ray intensity with a 2 mm Al foil inserted. kVp = kilovoltage peak.

laboratory x-ray tubes that run for extended periods of time. Air cooling is preferred for medical x-ray tubes used only for short exposure times.

> **!** The Bremsstrahlung spectrum is continuous and featureless. At the high energy end of the spectrum, the Bremsstrahlung is cutoff at the maximum electron energy that can be converted into x-rays. The maximum electron energy corresponds to the kVp of the x-ray tube.

4.4 Characteristic radiation

4.4.1 Atomic transitions

When electrons strike a target, a second process of x-ray generation occurs: simultaneously with bremsstrahlung, *characteristic radiation* is emitted The incoming electron may collide with an electron of the inner atomic shell of the target material, such as with an electron of the K-shell. If the energy of the incoming electron E_{in} is higher than the binding energy of the core electron E_K, it may kick out the core electron leaving a hole in the K-shell as depicted in Fig. 4.4. This core hole is sequentially filled with an electron from the next higher shell, and so on, until after a cascade of transitions, the hit atom returns to the ground state. When an atom's electron transits from a higher to a lower shell, equivalently from a higher to a lower energy level, it will emit the difference in binding energy in terms of EM waves, i.e., dipole radiation. This radiation is called *characteristic x-ray radiation*, i.e., characteristic of the atom hit in the target material.

As an example, we assume that the kinetic energy of the accelerated cathode electron in the tube is sufficient to kick out one of the electrons from the inner molybdenum K-shell. During this process, sketched in Fig. 4.4, the cathode electron scatters, looses energy by energy transfer to the K-electron, and finally stops somewhere in the target. The former K-electron of the target atom will leave the Mo-atom and thermalize on the path through the solid target. The minimum (threshold) energy required for kicking out a K-shell electron of the target atom to the vacuum level is given by its binding energy:

$$E_{threshold} = E_K - E_{vacuum} = E_K - 0 \tag{4.9}$$

E_K is called the binding energy, the threshold energy, or the absorption energy, depending on the spectroscopic intention. Subsequentially, the missing electron ("hole") in the K-shell is filled with an electron from the L-shell, giving off its extra energy as a photon with energy:

$$E_{K\alpha}^{photon} = E_K - E_L \tag{4.10}$$

where E_K and E_L are the binding energies of electrons in the K and L-shells, respectively. The energy difference $E_K - E_L$ is very well-defined; the characteristic x-ray emission line is sharp on the energy scale, typically on the order of $\Delta E/E = 2 \times 10^{-4}$ [3, 4].

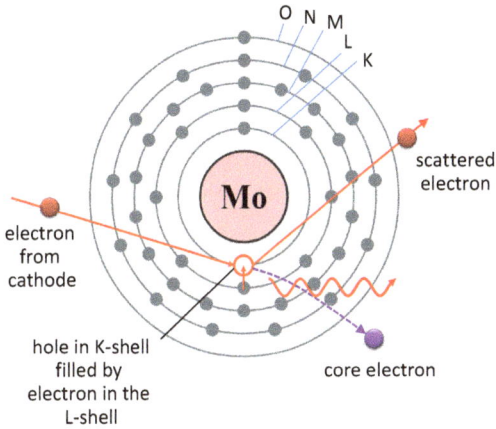

Fig. 4.4: A highly energetic incident electron arriving from the cathode kicks out a core electron in a target atom, here a Mo-anode, creating a hole in the K-shell. An electron from the L-shell fills the hole and simultaneously emits an x-ray photon.

Note that in the case of characteristic x-ray photons, the excitation energy is higher than the emission energy. The difference is:

$$\Delta E = E_K^{\text{excitation}} - E_K^{\text{emission}} = E_K - (E_K - E_L) = E_L \qquad (4.11)$$

Therefore, in x-ray science, we have to distinguish between excitation (or absorption) spectroscopy and emission spectroscopy. The respective energies are distinctively different.

The above explanations are based on a simplification of the actual atomic structure. To correct this we need to consider the dipole selection rules and the fine structure of the L-shell. Due to spin-orbit coupling, the L-shell splits up into three sublevels: $2s_{1/2}$, $2p_{1/2}$, and $2p_{3/2}$. The selection rule for dipole transitions requires that the change of the orbital angular momentum Δl must be ±1 and that the spin angular momentum change Δs must be 0. This excludes a dipole transition from $2s_{1/2}$ to $1s$. However, the other two transitions $2p_{1/2} \rightarrow 1s$ and $2p_{3/2} \rightarrow 1s$ are dipole allowed. Their spectroscopic terms are $K_{\alpha 1}$ (17.48 keV) and $K_{\alpha 2}$ (17.37 keV), respectively. There are also transitions from the M-shell to the K-shell, labeled $K_{\beta 1}$ (19.61 keV) and $K_{\beta 3}$ ($\approx$ 19.5 keV). The energies in brackets refer to the Mo x-ray spectroscopic emission lines. The allowed x-ray transitions are graphically shown in Fig. 4.5. It is clear that the general transition process is the same for all atoms and the transition energies are specific for the atom in question.

Furthermore, as the transitions are within inner core levels, which do not take part in chemical binding, the x-ray energies are fingerprints for the atoms independent of the chemical environment in which they sit. This is at least true for heavy atoms. In light atoms, energy shifts due to the chemical bonding can be observed with high energy resolution x-ray spectroscopy. In any case, x-ray spectroscopy allows to identify atoms by their characteristic x-ray emission lines, which may be regarded as a "dry" chemical analysis (see below).

> **!** The x-ray characteristic spectrum consists of sharp spectral lines, corresponding to the energy difference between outer and inner electron energy states. The transmission occurs because a hole in an inner shell is filled by an electron from an outer shell.

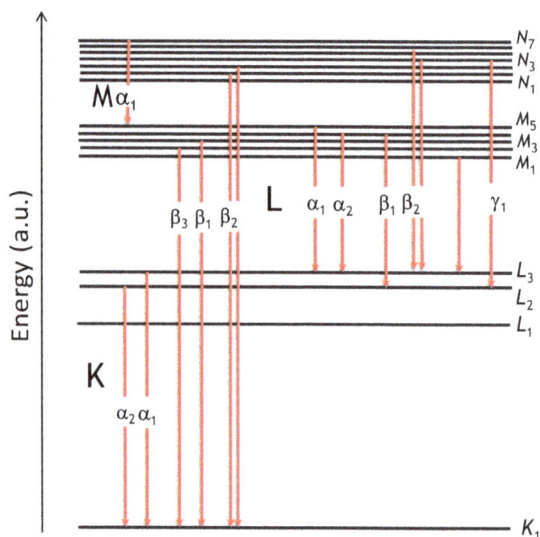

Fig. 4.5: Allowed dipole transitions for characteristic x-ray radiation. Adapted from *X-ray Data Booklet* (Center for X-ray Optics and Advanced Light Source Lawrence Berkeley National Laboratory).

Figure 4.6 shows the combined x-ray emission spectrum of a Mo target. It consists of the continuous bremsstrahlung and the characteristic spectrum. The bremsstrahl spectrum stretches from long wavelengths to the minimum wavelength (cutoff) at $\lambda_{min} = 0.35$ nm, corresponding to the maximum energy transfer at 35 kVp. Superimposed on the bremsstrahlung spectrum are sharp intensity lines due to characteristic $K_{\alpha 1}$ (17.48 keV), $K_{\alpha 2}$ (17.37 keV), and K_{β} (19.61 keV) x-ray transitions. The intensity of these lines is much higher than the bremsstrahlung over the same spectral range.

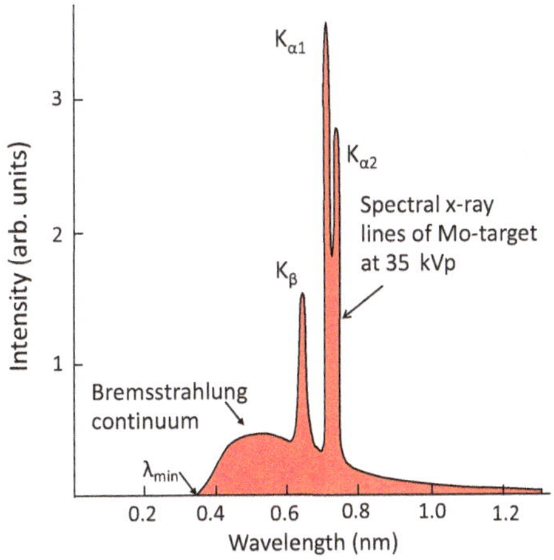

Fig. 4.6: Typical x-ray spectrum for a Mo target. The characteristic x-ray radiation is superposed on the bremsstrahlung spectrum.

Measured and calculated bremsstrahlungs spectra can be found in [2]. Some spectral emission and absorption energies are listed in Tab. 4.1.

Neglecting the spectroscopic splittings of the K, L, M, etc. lines due to finestructure effects, the x-ray photon energy can be estimated in analogy to the Bohr[7] model for the energy levels in hydrogen atoms. However, one has to take into account that after creating a hole in the K-shell by electron impact, the total charge of the remaining electrons is $Z - S$, where Z is the atomic number and S is a screening factor. With this modification, the empirical *Moseley's law*[8] for the photon energy from the K-shell reads:

$$E_{i \to f} = hf_{i \to f} = R_H (Z - S)^2 \left(\frac{1}{n_f^2} - \frac{1}{n_i^2} \right) \tag{4.12}$$

where n_i, n_f are the main quantum numbers of the atomic shells for the initial and final state of the electron transition, and $R_H (= 13.606$ eV) is the Rydberg[9] constant of the hydrogen atom. For the K_α transition, the screening factor can be approximated by $S = 1$ and the eq. (4.12) becomes:

$$E_{K_\alpha} = R_H (Z - 1)^2 \left(\frac{1}{1^2} - \frac{1}{2^2} \right) = R_H (Z - 1)^2 \frac{3}{4} \tag{4.13}$$

For Mo with $Z = 42$, we estimate a K_α-radiation energy of 17.146 keV, which is close to the observed energy of 17.480 keV.

---

7 Niels Bohr (1885–1962), Danish physicist and Nobel prize winner in phyics 1922.
8 Henry Gwyn Jeffreys Moseley (1887–1915), British physicist.
9 Johannes Robert Rydberg (1854–1919), Swedish physicist.

Tab. 4.1: X-ray absorption edges and characteristic emission lines for some selected elements.

Element	K_α absorption edge (keV)	Emission energies (keV)		
		$K_{\alpha 1}$	$K_{\alpha 2}$	K_β
Ni	8.333	7.478	7.460	8.264
Cu	8.979	8.047	8.027	8.905
Mo	20.0	17.479	17.374	19.608
Ag	25.514	22.162	21.990	24.942
W	69.52	59.318	57.981	67.244

Data from http://skuld.bmsc.washington.edu/scatter/AS_periodic.html and from the x-ray data booklet posted at http://xdb.lbl.gov/.

4.4.2 Energy dispersive x-ray chemical analysis

The element specificity of x-ray emission lines is a very useful property that can be utilized for chemical analysis. Let us assume that the electrons do not hit a target of well-known chemical composition but an unknown specimen. All atoms will be excited to emit characteristic x-ray radiation, supposing the incident electron energy is high enough to excite the inner K-shell electrons. We only need a detector to analyze the energy of the emitted characteristic radiation, and then we can identify different elements in the specimen. Furthermore, the intensities of the characteristic radiation lines are proportional to the concentration of the elements in the target. After proper calibration with standards, the chemical composition of specimens can be derived. Figure 4.7 shows an example where we recognize the K_α lines of Fe, Al, and to a lesser extent also of Si. In this example, a thin film of Fe and Al was deposited on a silicon substrate. The instrument, which provides such information, is called a *scanning electron microscope* (SEM). In contrast to x-ray tubes, an SEM works with a fine focus, high-energy electron beam that is scanned over the surface of a specimen. Secondary electrons emanating from the sample serve for imaging the sample surface. Simultaneously, the electrons excite atoms in the sample to emit characteristic x-ray radiation. These, in turn, can be used for a spatially resolved chemical analysis. SEMs belong to any modern analytic laboratory and they are often used in forensic medicine to identify inorganic chemicals such as those from gun powder or explosives [5].

4.4.3 Target material

So far, we have recognized that x-ray intensities can be produced by bombarding targets with electrons, or more general, with charged particles. The x-ray emission spectrum is a superposition of a continuous bremsstrahl spectrum and a characteristic

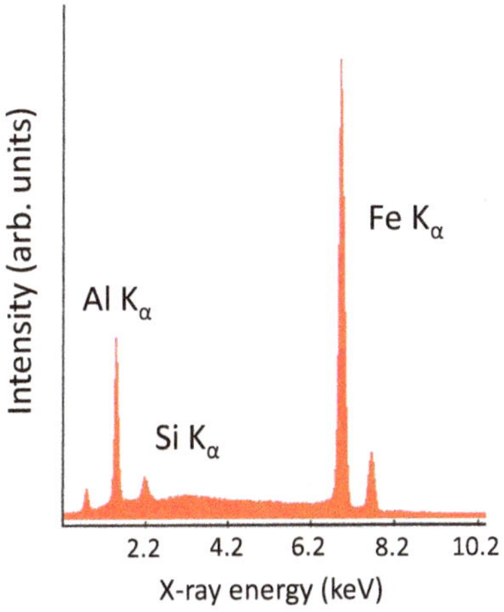

Fig. 4.7: X-ray spectrum recorded with a scanning electron microscope is used for chemical analysis of specimens.

spectrum featuring high-intensity sharp x-ray lines. For radiography and radiotherapy, only the bremsstrahlung spectrum is useful, while the characteristic spectrum may be regarded as a disturbance that cannot be avoided completely. However, to minimize its effect on radiographs, it is beneficial to shift the characteristic lines to the high-energy end of the bremsstrahlung spectrum. This can be achieved by coating the Cu-target with a tungsten film. Tungsten is a high atomic number ($Z = 74$) metal that features characteristic x-ray lines at about 60 keV. When using accelerating voltages of 60–80 kVp, the characteristic tungsten K_α and K_β x-ray emission lines are hardly excited and do not dominate the bremsstrahlung spectrum. One exemption is *mammography*, i.e., the radiography of the female breast. In mammography, lower energy x-rays are preferred because of their higher absorption in soft tissue. This yields better contrast for slight changes in electron densities due to cancerous tissue. In mammography, usually, molybdenum targets are employed. The characteristic Mo-K_α radiation at a wavelength of 0.071 nm and x-ray energy of 17.44 keV is used for imaging; the respective Mo x-ray spectrum is shown in Fig. 4.6.

In x-ray tubes, the x-ray intensity is composed of bremsstrahlung and characteristic radiation. Bremsstrahlung is due to the de-acceleration of electrons in the target, characteristic radiation is the result of core electron excitation. X-ray radiography utilizes mainly bremsstrahlung. Only for mammographic screening, often Mo K_α radiation is employed.

4.5 X-ray generators

4.5.1 X-ray tubes for radiography

In medical applications, the exposure time for taking a single x-ray image is rather short, on the order of a few milliseconds. Furthermore, there is a sufficient time gap between one exposure and the next. Therefore, heating or even melting of the anode is less likely than x-ray tubes in continuous operation mode. For this reason, x-ray tubes for medical applications have a different design than standard laboratory x-ray tubes, a typical cross section is shown in Fig. 4.8. The cathode filament has an off-center position and the electron beam hits the rim of a rotating wedge-shaped anode attached to a rotor of an electrical motor. The rotor is driven in a magnetic field of a stator outside the vacuum-sealed tube. Thus the "hot" spot on the anode is spread out over the target rim. The target wedge has an inclination angle of 12°, permitting an x-ray radiation cone with an opening angle of about 24°. This wide opening angle is useful for large area exposures applied for thorax radiographs. Otherwise, the beam can be shaped to the desired size with the help of diaphragms. Note that in medical x-ray tubes, the cathode is grounded, and the anode is at high potential, contrary to a commercial laboratory or industrial x-ray tubes. The physics of bremsstrahlung generation remains, though, the same.

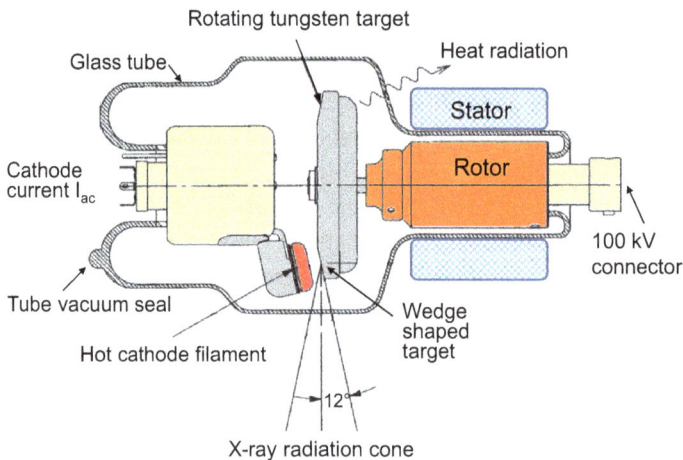

Fig. 4.8: Cross section of a typical x-ray tube for radiography. The glass tube is evacuated. An external stator drives the rotor with the target attached. The hot target emits radiation to the environment but is not actively cooled.

Figure 4.9 shows a real tube together with a housing and radiation shield. Such tubes can be operated up to about 150 kVp and are used for short exposure times. For

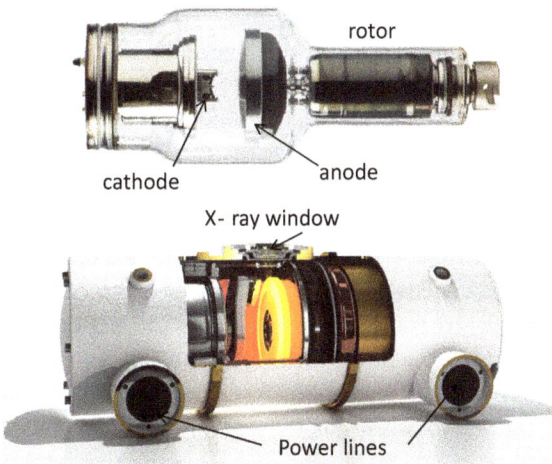

Fig. 4.9: Rotating anode x-ray tube for radiography. Top panel: The glass housing of cathode and rotating anode is powered by a stator and rotor. Bottom panel: metal case for protection, with openings for power lines and x-ray transparent windows. A cut-away opening shows an artist's view of the glowing target at about 2500 °C during operation. (Adapted from http://www.oem-prod ucts.siemens.com/x-ray-tube).

longer exposure times, the air cooling power would not be sufficient. From an accelerator point of view, these standard x-ray tubes are called *direct voltage accelerators*.

For taking computed tomography (CT) images (Chapter 8) or for x-ray videos (cine-radiography), often longer exposure times are needed. Considering the heating and cooling curves of medical x-ray tubes such as the one shown in Fig. 4.9, a maximum average power consumption can be specified for a finite exposure time. Let us assume that a power of 12 kW for a total exposure time of 10 s would be a safe operation. However, we need 40 kW during single exposures. This condition can be met by a pulsed time structure of x-ray exposure. If the pulse duration is taken as $\Delta t = 4$ ms and the repeat frequency or frame rate is 60 s^{-1}, then the duty time is 240 ms and the duty factor DF is 24% compared to continuous exposure. If the peak power is chosen as $P_p = 80$ kV × 500 mA = 40 kW, then the average power is $P_a = P_p \times \text{DF} = 9.6$kW, which is below the specified safety limit. Pulsing the x-ray power allows going to higher kVp, yielding sharper images in a shorter time. A frame rate of 60 s^{-1} is higher than required for a video but guarantees sharp pictures of moving organs like the heart. The estimate made here solely refers to the safe operation of the x-ray tube and does not take into account radiation safety aspects with respect to the patient. Radiation safety is an issue discussed in Chapter 7.

The manufacturer Siemens presented an innovative new concept of a medical rotating x-ray tube for single-shot x-ray radiographs and CT scans [6]. The so-called Straton rotary x-ray tube is an advance in the design of rotary x-ray tubes for medical applications and overcomes a number of limitations of the standard rotary x-ray tubes. The cross section of the Straton tube is shown in Fig. 4.10 (left picture), and a picture of the bare tube is shown in the right panel. The anode and cathode are fixed in an evacuated metal housing and rotate via an external motor. X-rays from the anode penetrate through a thin Al window, which also cuts off the soft end of the x-ray spectrum. The tube is immersed in an oil trough for cooling. The electron beam starts from the rotating cathode and goes to the rotating anode on a curved path. The electron beam is dynamically deflected by an external magnetic field, shaping the beam and controlling the focal spot size at the same time. Furthermore, with this construction, the beam can be flung back and forth between two focal points of different target materials at a high frequency. Due to the oil cooling, there is no limit to the exposure times. Although ingenious in construction, the Straton tube in practice faces serious problems regarding beam stability. This requires stabilization electronics that negate the advantages of a more compact x-ray tube. As a result, the Straton tube is not widely used in the clinical setting.

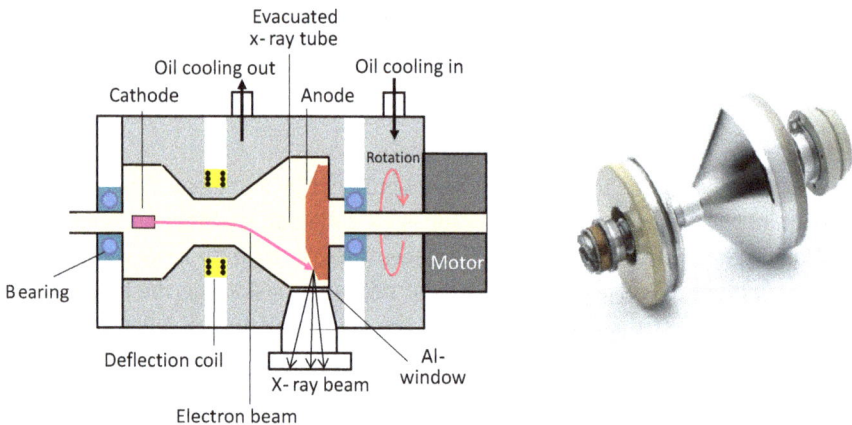

Fig. 4.10: Straton sealed x-ray tube immersed in an oil trough for cooling. Left panel: schematics of the working principle. The entire tube is rotated including the target. The cathode is located in the center and the electron beam is deflected to the target by an external magnetic field; Right panel: photograph of the rotating and sealed x-ray tube. (Adapted from www.siemens.com/press).

4.5.2 Linear accelerators for radiotherapy

XRT requires different types of x-ray machines to produce much harder x-ray bremsstrahlung than used in radiography. In the past, high-voltage x-ray tubes for peak voltages of up to 250 kVp were used for XRT. Since the 1960s, the sealed-off tubes have been replaced by linear electron accelerators with peak energies up to 25 MeV. These high MeV x-ray energies produce a completely different type of tissue irradiation than the lower energy x-ray tubes did in the past, as we will discuss in

Chapter 2 of volume 3. The technology of linear accelerators for medical applications originates from high energy physics and has been steadily improved over the
past years to make them shorter and more lightweight by using steeper voltage gradients in the accelerating cavities.

Figure 4.11 shows a block diagram of the most important components of a *linear
accelerator* (linac), consisting of a klystron, pulse modulator, electron source, microwave cavities for electron acceleration, bending magnet, target, collimators, and treatment couch. The power supply and vacuum pumps are omitted in Fig. 4.11 for clarity.
Klystrons produce microwave power for accelerating pulsed electron bunches, which
are additionally modulated in the pulse modulator. The klystron can be replaced by a
magnetron for lower power applications between 1 and 6 MeV, but for higher energies,
klystrons are preferred. In both cases, the aim is to produce electron bunches from a
dc electron source. The electron source is similar to x-ray tubes: electrons are emitted
into the vacuum by thermionic extraction from a hot filament.

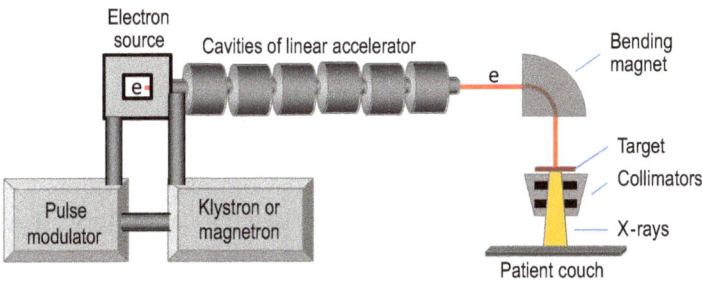

Fig. 4.11: Block diagram of a linear accelerator for radiation therapy.

The linac consists of a linear sequence of pill-box-like cavities (Fig. 4.12(c)) [7]. In each
cavity, several microwave standing modes can be excited. The one used most frequently is the so-called TM_{010} accelerating mode outlined schematically in Fig. 4.12.
Panel (a) gives a side view, and panel (b) is a view in the direction of the electron path.
The TM_{010} mode is characterized by a longitudinal electric field parallel to the cylinder
axis of the cavity, which accelerates electrons entering through a central hole. A curl-
ing magnetic field around this axis causes ohmic dissipation that needs to be cooled
off. The letters TM refer to the fact that the magnetic field is oriented perpendicular to
the accelerating current; the first index indicates the azimuthal dependence of the
magnetic field, the second index the radial dependence, and third index the longitudinal component. Thus in TM_{010} mode, the magnetic field has only a radial component.

The electron acceleration does not depend on the voltage difference as in a
standard x-ray tube. Instead, the acceleration depends on the phase of the voltage in
the cavity that the electron experiences when entering the cavity at a velocity v. Obviously, kicking the electron in a phase just beyond the maximum voltage amplitude is

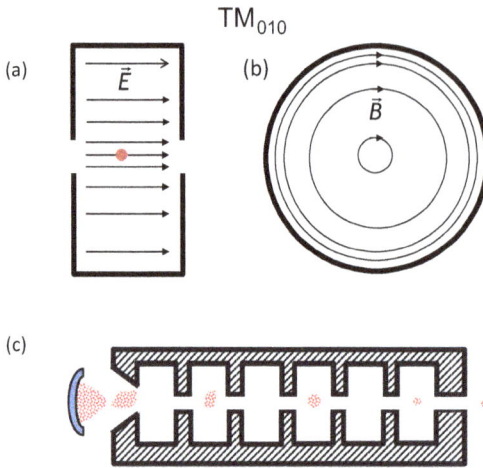

Fig. 4.12: Electrical and magnetic field components for the TM_{010} cavity mode: (a) side view of a cavity; (b) view along the electron path; (c) sequence of cavities in a linac with the electron source to the left.

most effective. This is equivalent to pushing a child on a swing just after the point of return from the maximum amplitude.

The microwave cavities convert the RF energy from the klystron/magnetron into high-energy electron bunches [8]. Once the high-energy electrons exit the linac, they may immediately hit a target for producing x-ray bremsstrahlung. However, In most cases, the horizontal electron beam is first turned by a bending magnet from horizontal to vertical orientation (see Fig. 4.11), and then the beam hits a tungsten target like in an x-ray tube. The bending of the electron beam is achieved by a sector magnet turning the electrons either by 90° or by 270°. Linacs are some 1–2 m long, depending on the maximum energy. For practical reasons, linacs are mounted horizontally inside a gantry with magnet sector and collimator head as seen in the artist's view in Fig. 4.13. The entire gantry can then be rotated about the patient couch at the center. In most advanced applications, the linac is short and comparatively lightweight producing 4–8 MeV electrons. The linac is integrated into a robotic arm that can be positioned anywhere around the patient. In this design, called cyberknife, the magnet sector for bending the electron beam is no longer required, making the linac even shorter and lighter. Cyberknife technology is described in more detail in Chapter 2 of Volume 3. By hitting a tungsten target, an intense high-energy photon beam is produced. The photon energy is so high that it can easily penetrate a thin tungsten target while the electrons are stopped. The x-ray beam is shaped by collimators to the specified physical target volume to be exposed to radiation.

Medical linac's high electron and photon energies are difficult to handle. For one, they require extensive radiation shielding. Moreover, the electron beam as well as the photon beam cause a number of side effects, such as neutron production.

Neutrons, in turn, cause further problems via activating materials when captured. In the end, when turning off a linac, there will be an "afterglow" of radiation from the material that was exposed to high energy radiation. This radiation consists of fast neutrons and gamma spectral lines. Therefore it is mandatory to take extra measures of radiation shielding to protect patients and clinical staff.

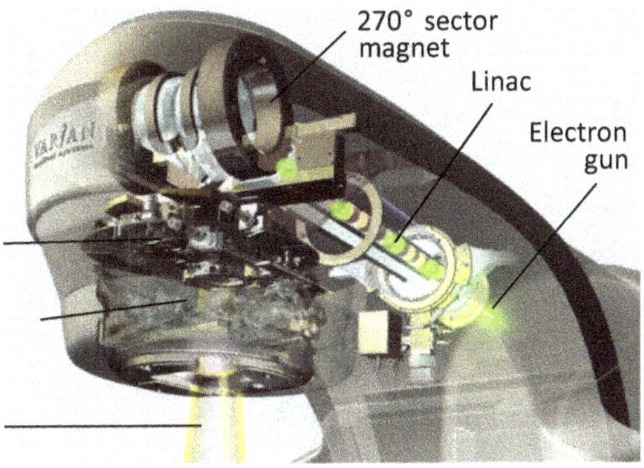

Fig. 4.13: Linac inside a gantry for x-ray production. (Courtesy Varian Medical Systems International AG, Switzerland; all rights reserved).

4.5.3 Synchrotron radiation

The third type of machine that produces x-rays is an electron *synchrotron* [9]. First electrons are brought up to speed with the help of a linac and a booster ring. Once the electron energy is ramped up, the electrons are injected as space-time separated bunches into a storage ring. A lattice of magnets keeps the electrons on track in a narrow circular tube, which is evacuated to pressures of less than 10^{-9} mbar. Ultra-high vacuum is required to avoid collisions of electrons with residual gas atoms that shortens the storage life time. A schematic outline of a synchrotron is shown in the inset of Fig. 4.15. While electrons in a circular orbit are constantly accelerated toward the center (radial acceleration), they emit EM radiation like in an antenna. At low nonrelativistic energies, the radiation profile has a typical dipole distribution shown in the left panel of Fig. 4.14. However, in storage rings, electrons have a speed close to the speed of light, i.e., they are highly relativistic. Then for an observer in the rest frame, the radiation profile becomes distorted, radiating into a narrow radiation cone with an opening angle:

$$\Delta\phi = \frac{1}{\gamma} \tag{4.14}$$

in the direction of the traveling electrons, as sketched in the right panel of Fig. 4.14.

Since

$$y = \frac{E_{kin}}{m_0 c^2},$$ (4.15)

where E_{kin} and $m_0 c^2$ are the kinetic energy and rest energy of the orbiting electron, respectively, and assuming for E_{kin} a typical energy of 6 GeV in a storage ring, we find for the opening angle:

$$y = \frac{E_{kin}}{m_0 c^2} = \frac{6\,\text{GeV}}{0.5\,\text{MeV}} = 12 \times 10^3; \qquad \Delta\phi = \frac{1}{y} = 5 \times 10^{-3°}$$

In this narrow radiation cone, the intensity is extremely high, orders of magnitude higher than in conventional x-ray tubes.

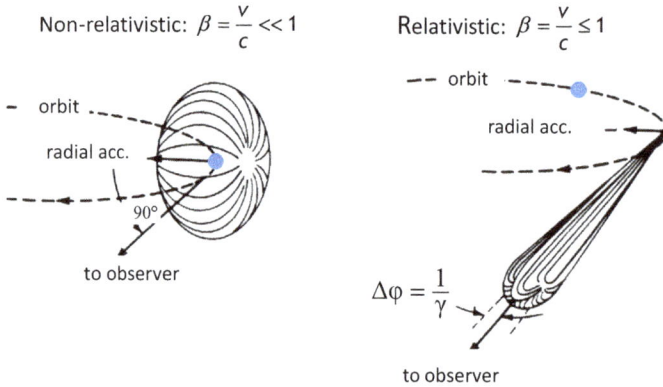

Non-relativistic: $\beta = \frac{v}{c} \ll 1$

Relativistic: $\beta = \frac{v}{c} \le 1$

$\Delta\phi = \frac{1}{y}$

Fig. 4.14: Radiation profile for a non-relativistic electron on a circular orbit (left panel) and for a relativistic electron in a synchrotron storage ring (right panel).

The energy spectrum of synchrotron radiation is continuous and very broad, similar to a bremsstrahlung spectrum with a cutoff that depends on E_{kin} and the radius R of the storage ring. The radius R of the orbit and magnetic field B of the bending magnets are related according to:

$$R[m] = 3.3 \frac{E_{kin}[GeV]}{B[T]}$$ (4.16)

In Fig. 4.15 a typical spectrum is shown for $E_{kin} = 1$ GeV and a magnetic field $B = 1.2$ T. The maximum photon intensity is reached at the critical photon energy E_c, indicated by a blue arrow. E_c is related to the kinetic electron energy and the magnetic field of the bending magnets via:

$$E_c[keV] = 0.665E_{kin}^2[GeV]^2 \times B[T] \tag{4.17}$$

Plugin in numbers from above, we obtain a critical photon energy $E_c = 0.8$ keV. The critical and cutoff energy increases with the square of E_{kin} and hard x-ray radiation ($E_{x-ray} > 10 keV$) of interest for medical applications can be gained at electron energies of $E_{kin} \approx 6$ GeV and higher.

The radiation spectrum of storage rings can be modified by using special magnetic insertion devices, which wiggle or undulate the electrons on their orbit to give them short periodic kicks for additional acceleration and increased intensity. However, a discussion of such devices is beyond the scope of this text. Synchrotron radiation is used sometimes for cancer irradiation, but more importantly, it has been used for x-ray radiography with phase contrast enhancement, presented in Section 8.8. A major use of synchrotron radiation is structural investigations of materials, including biomolecular crystallography. Further information on the basics, methods, and applications of synchrotron radiation can be found in [9].

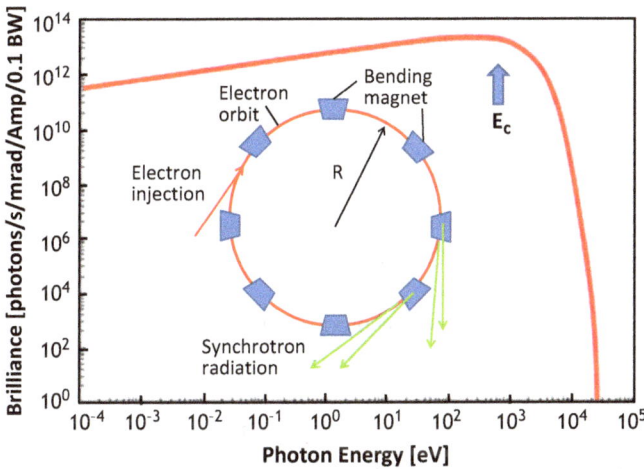

Fig. 4.15: Energy spectrum of synchrotron radiation for 1 GeV electrons and a 1.2 T field of the bending magnets. The inset shows an outline of an electron storage ring (synchrotron), where bending magnets keep the electrons on a circular orbit.

Continuous x-ray spectra are generated in x-ray tubes and in linear electron accelerators (linacs) when electrons hit a target or by electrons orbiting in a synchrotron storage ring.

4.6 Summary

S4.1 X-rays are produced by accelerating electrons in a static potential gradient between cathode and anode inside an evacuated tube, followed by a sudden de-acceleration in a target.

S4.2 X-rays produced by accelerating electrons and stopping in a target consist of a continuous bremsstrahlung spectrum and a characteristic spectrum of several discrete atomic emission lines.

S4.3 The continuous bremsstrahlung spectrum is cutoff on the high energy side when the kinetic energy of electrons is completely converted in EM radiation.

S4.4 The integrated intensity of the bremsstrahlung spectrum scales linearly with the cathode current and quadratically with the accelerating voltage.

S4.5 The characteristic spectrum is specific for the target material used. The emission lines correspond to differences in the binding energies of core atomic levels.

S4.6 The characteristic spectrum can be used for the chemical analysis of the target material.

S4.7 X-ray tubes for radiography contain rotating targets for distributing the heat load from the electron beam.

S4.8 X-ray tubes for radiography can only be used for short periods of time. When taking CT images or cinematographies, a pulsed time structure with low duty factor has to be applied.

S4.9 Medical x-ray tubes for radiography operate up to about 140 kV.

S4.10 For radiotherapy, very high-energy radiation is required in the range of several MeV.

S4.11 High energy electrons up to 25 MeV can be produced with a linear accelerator.

S4.12 Linear accelerators are mounted in gantries which can be rotated about a target at the isocenter.

S4.13 A continuous spectrum of x-ray photons with very high intensity can also be produced by electrons in a synchrotron storage ring.

S4.14 Synchrotron radiation is mainly used for x-ray scattering and x-ray spectroscopy of materials, including biomaterials.

S4.15 All beam shaping of x-ray beams is achieved by collimators.

? Questions

Q4.1 What is the relation between x-ray energies and x-ray wavelengths?
Q4.2 What types of x-ray radiation are usually distinguished?
Q4.3 What instrument can produce x-ray radiation?
Q4.4 How is bremsstrahlung generated?
Q4.5 How is characteristic x-ray radiation generated?
Q4.6 How can x-ray radiation be distinguished from γ-radiation?
Q4.7 What are typical x-ray energies for radiography and radiotherapy?

Q4.8   What are the essential parts of a medical x-ray tube for radiography?
Q4.9   The intensity of x-ray bremsstrahlung is a linear function of which energy difference?
Q4.10   What determines the cutoff energy E_{max}?
Q4.11   What is the purpose of the Al-filter used for radiography?
Q4.12   What is characteristic x-ray radiation useful for?
Q4.13   In what medical discipline is characteristic radiation used as an analytic tool?
Q4.14   Which transition corresponds to the $K_{\alpha 1}$ x-ray radiation?
Q4.15   The accelerating voltage of an x-ray tube may be 50 kV. With this voltage, is it possible to excite tungsten K_{α} radiation?
Q4.16   Which energy is higher, the excitation energy or the corresponding energy of the x-ray characteristic radiation? As an example, compare Mo-K_{α} excitation energy (absorption edge) with characteristic Mo-K_{α} radiation.
Q4.17   What are the differences between the Mosely law and Bohr's model for the hydrogen atom?
Q4.18   What are the characteristic design features of an x-ray tube for radiography?
Q4.19   Are x-ray tubes for radiography actively cooled?
Q4.20   What are the advantages of the Straton rotating x-ray tube compared to traditional ones?
Q4.21   What are the principle design features of linear accelerators?
Q4.22   Synchrotron radiation is radiation that is emitted from orbiting electrons in a storage ring. When these electrons radiate x-ray photons at high intensity, do they stop orbiting?
Q4.23   What is synchrotron radiation mainly used for?

Attained competence	+	0	−	
I know how x-rays can be produced				
I can name the three main sources for x-ray radiation				
I can distinguish between bremsstrahl spectrum and characteristic spectrum				
I know how to calculate the cutoff bremsstrahl energy of an x-ray tube				
I know the reasons why x-ray absorption spectra are different from x-ray emission spectra of a specific element				
I know that x-ray emission and absorption spectra can be used for chemical analysis of materials				
I know how very high x-ray energies are generated				
I know what synchrotron radiation is mainly used for				

Exercises

E4.1 **Threshold energy:** What is the minimum x-ray energy (threshold energy) required to excite the Mo-K$_\alpha$ characteristic x-ray radiation via x-ray absorption? How does this compare with the characteristic x-ray energy of the Mo-K$_\alpha$ radiation? Please explain the difference.

E4.2 **Energy resolution:** What energy resolution is required to distinguish between the Mo-K$_{\alpha1}$ and Mo-K$_{\alpha2}$ radiation? What is the wavelength resolution required for the same emission lines? Give the answer in absolute and relative terms.

E4.3 **Cutoff energy:** What is the cutoff Bremsstrahl wavelength if the accelerating anode voltage is 40 kV?

E4.4 **Absorption edge energy:** Mo-K$_{\alpha1}$ radiation has an energy of 17.48 keV. The Mo-K$_{\alpha1}$ absorption edge is at 20.00 keV. What is the absorption edge energy of the Mo-L$_{\alpha1}$ radiation?

References

[1] Eisberg R, Resnik R. Quantum physics of atoms, molecules, solids, nuclei, and particles. 2nd. New York, London, Sydney, Toronto: John Wiley and Sons; 1985.

[2] Birch R, Marshall M. Computation of bremsstrahlung X-ray spectra and comparison with spectra measured with a Ge(Li) detector. Phys Med Biol. 1979; 24: 505–517.

[3] Mendenhall MH, Henins A, Hudson LT, Szabo CI, Windover D, Cline JP. High-precision measurement of the x-ray Cu Kα spectrum. J Phys B At Mol Opt Phys. 2017; 50: 115004.

[4] X-ray Data Booklet (Center for X-ray Optics and Advanced Light Source Lawrence Berkeley National Laboratory). At: http://xdb.lbl.gov/

[5] Gentile G, Andreola S, Bailo P, Battistini A, Boracchi M, Tambuzzi S, Zoja R. A brief review of scanning electron microscopy with energy-dispersive X-ray use in forensic medicine. Am J Forensic Med Pathol. 2020; 41: 280–286.

[6] Schardt P, Deuringer J, Freudenberger J, Hell E, Knüpfer W, Mattern D, Schild M. New x-ray tube performance in computed tomography by introducing the rotating envelope tube technology. Med Phys. 2004; 31: 2699–2706.

[7] Wangler TP. RF linear accelerators. 2nd. New York, London, Sydney, Toronto: Wiley & Sons; 2008.

[8] Wiedemann H. Particle accelerator physics. 4th. Graduate Text in Physics. Berlin, Heidelberg, New York: Springer; 2015.

[9] Mobilio S, Boscherini F, Meneghini C editors, Synchrotron radiation: Basics, methods, and applications. Berlin, Heidelberg: Springer Verlag; 2015.

Further reading

Bushberg JT, Seibert JA, Leidholdt EM, Jr, Boone JM. The essential physics of medical imaging. 3rd.
    Philadelphia, New York, London: Lippincott Williams & Wilkins, Wolter Kluwer; 2012.
Dance DR, Christofides S, Maidment ADA, McLean ID, Ng KH editors, Diagnostic radiology physics –
    A handbook for teachers and students. IAEA; 2014.
Wille K. The physics of particle accelerators. Oxford, New York, Athens: Oxford University
    Press; 2000.

Useful website

http://hyperphysics.phy-astr.gsu.edu/hbase/hph.html

5 Nuclei and isotopes

Physical properties of nuclei

Atomic mass number	$A = Z + N$
Atomic number or nuclear charge number	Z
Number of neutrons in a nucleus	N
Atomic mass unit	u, $1\ u = 1.6605 \times 10^{-27}$ kg
Decay products	α, β^+, β^-, γ, v, $\bar{v}$
Neutron mean lifetime	882 s
Activity after n incremental decay times	$A_n = A_0 (A_1/A_0)^n$
Reaction rate for isotopes B from target nuclei A	$R_B = N_A \sigma_a J_a$
Isotope production	Neutron capture, proton bombardment

5.1 Introduction

Chapters 8–10 in this volume and 1 – 5 in volume 3 deal with ionizing radiation for imaging (x-ray radiography, scintigraphy, SPECT, and PET) and irradiation (x-ray radiation therapy, γ – knife, proton and neutron irradiation, and brachytherapy). Therefore, it is appropriate to first introduce some basic properties of nuclides and isotopes, especially of radioactive isotopes used in nuclear medicine. This chapter is not intended to replace textbooks on nuclear physics. But it provides enough background information to better understand the following chapters. Chapter 6 discusses the interaction of radiation with matter, knowledge which is a prerequisite for appropriate radiation protection measures, presented in Chapter 7.

Atoms have an electronic shell of orbiting electrons about the *nucleus* at the center. The term "nucleus" should not be confused with the nucleus of a biological cell. Although the names are identical, it should be clear from the context whether an atomic nucleus or a cell nucleus is meant. The atomic nucleus consists of Z *protons* and N *neutrons* bound together by a strong but short-ranged nuclear force. Another term for the nucleus is *nuclide*. Both terms are used in this chapter synonymously. The plural form of a nucleus is nuclei; the plural of a nuclide is nuclides. Individual protons and neutrons together are called nucleons. They are part of the hadron family, which are particles composed of *quarks*. More specifically, nucleons are *baryons*, as they always contain three quarks. Protons have two up quarks with charge $+2/3e$ and one down quark with charge $-1/3e$, where e is the elementary electrical charge. Therefore, protons have in total one positive elementary charge $+e$, which has the same magnitude but opposite sign as electrons in atomic shells. Neutrons have one up and two down quarks. Neutrons are neutral since their quark charge sums up to zero.

https://doi.org/10.1515/9783110757095-005

The sum of protons and neutrons in nuclides is the *mass number* $A = Z + N$. The number of protons Z in nuclides is also known as the *atomic number* or *nuclear charge number*. In charge-neutral atoms, the number of protons equals the number of electrons. Charged atoms are called *ions*. They are either *cations* if the number of protons is larger than the number of electrons and *anions* in the opposite case. Both protons and neutrons have angular momentum (spin) of $S = 1/2\hbar$ with an associated magnetic moment. The magnetic moment of protons is important for magnetic resonance imaging, as we have seen in Chapter 3. Neutrons also have a magnetic moment, which is utilized in research for probing magnetic materials. Protons and neutrons are by a factor of roughly 2000 heavier than electrons (exact: 1836 for protons and 1838 for neutrons). Their combined mass in nuclides determines the *atomic mass*.

5.2 Isotopes

In light nuclei, the number of neutrons N and the number of protons Z is roughly equal: $N \cong Z$. One exemption is hydrogen, which contains only one proton but no neutron. Nuclides with the same number of protons Z but a different number of neutrons N are called *isotopes*. The general nomenclature of isotopes is $^A_Z X$. Here X stands for the chemical symbol, A is the mass number and Z the atomic number. The chemical symbol X together with the atomic number Z is redundant and therefore the subscript Z is often omitted. The number of neutrons follows from $N = A - Z$. Isotopes have the same atomic number Z but different atomic mass numbers A.

As an example, we discuss the isotopes of hydrogen. Hydrogen $^1_1 H$ with $Z = 1$, $N = 0$ and deuterium (D) $^2_1 H$ with $Z = 1$, $N = 1$ are temporally stable isotopes. Hydrogen has another isotope: tritium (T) $^3_1 H$ with $Z=1$, $N = 2$. Tritium (or triton) is not stable but decays in time by converting one neutron into a proton, simultaneously emitting a negatively charged electron. The decay product is a stable $^3_2 He$ isotope with the same mass number as tritium but different Z. The decay is formally written similar to a chemical reaction:

$$^3_1 H \rightarrow {}^3_2 He + {}^{\ 0}_{-1} e^- + \bar{\nu} e.$$

Here $\bar{\nu}_e$ is an electron anti-neutrino, which we will not discuss any further. The half-life time of the decay is 12.3 years, i.e., after 12.3 years, one half of the tritium isotopes are converted into $^3_2 He$.

The isotopes of hydrogen are the only ones, which have their own chemical symbols: H for hydrogen, D for deuterium, and T for tritium. The bare nuclei of these three isotopes are called proton (p), deuteron (d), and triton (t). All other isotopes are only designated by their nuclear notation.

The example of the tritium decay teaches us many important facts about nuclei. First, isotopes become unstable if the number of neutrons is either too large or too small compared to protons. Unstable isotopes are called *radioisotopes*. Second, radioactive decay transforms a radioisotope into a temporally stable one, and the decay time is characteristic of the isotope. Third, radioactive decay, called *radioactivity*, occurs by the emission of charged particles that change the atomic number Z. As we will see later, radioactive decay may require a cascade of transitions before a stable isotope is finally reached. Often γ-radiation is emitted alongside, which brings isotopes from an excited nuclear state into the ground state.

Figure 5.1 shows the stable isotopes of light elements regarding the number of protons versus neutrons. The ratio N/Z is very close to 1. From this chart, we recognize that the elements Be, F, Na, and Al have only one stable isotope.

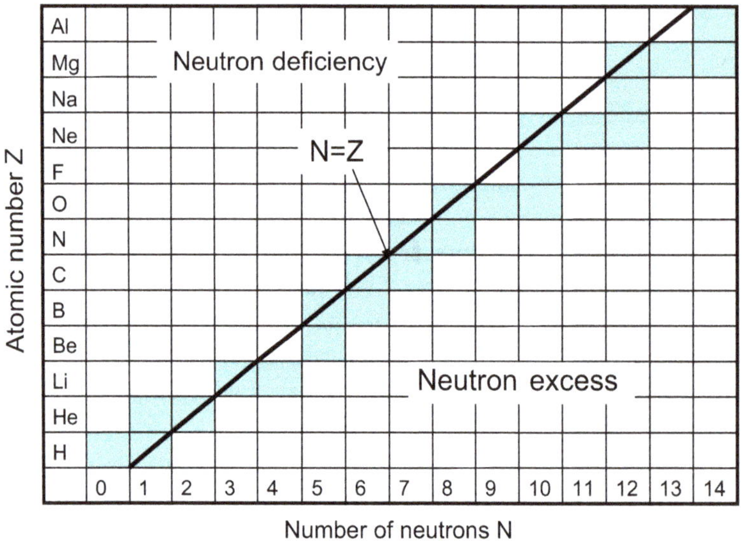

Fig. 5.1: Chart of stable isotopes for the light elements.

Figure 5.2 shows an extended nuclear chart of light and heavy isotopes, including stable and unstable isotopes, known as the Segrè[1] chart. We notice that the valley of stable isotopes, indicated by black dots, leans over to a neutron excess and ends at the element lead $^{208}_{82}$Pb. The ratio N/Z changes continuously from 1 for light isotopes to 1.5 for heavy isotopes. With an increasing atomic number, the repulsive Coulomb interaction between the protons weakens the nuclear binding energy, which is counterbalanced by a surplus of neutrons. All isotopes which are heavier than $^{208}_{82}$Pb are

---

1 Emilio Gino Segrè (1905–1989), Italian physicist, Nobel Prize in Physics 1959.

unstable and decay over time. The decay schemes are distinguished by blue for electron emission, red for positron emission, and yellow for α-particle emission.

There are two types of *radioactive isotopes:* those that occur naturally and those produced artificially. The naturally occurring radioisotopes can be further subdivided into those being around since the birth of the Earth, and those continuously reproduced by the environment. The former ones are known as *terrestrial radioisotopes*. They are long-lived with half lifetimes in the order of the age of the Earth. Typical examples are ^{40}K, ^{232}Th, ^{235}U, and ^{238}U. The shorter-lived *cosmic radioisotopes* are continuously generated by cosmic rays, such as ^{14}C, ^{3}H, and ^{7}Be. Radioisotopes can also be produced artificially by proton bombardment in an accelerator or by neutron capture (NC) in a fission reactor. These topics are discussed in more detail in Section 5.6.

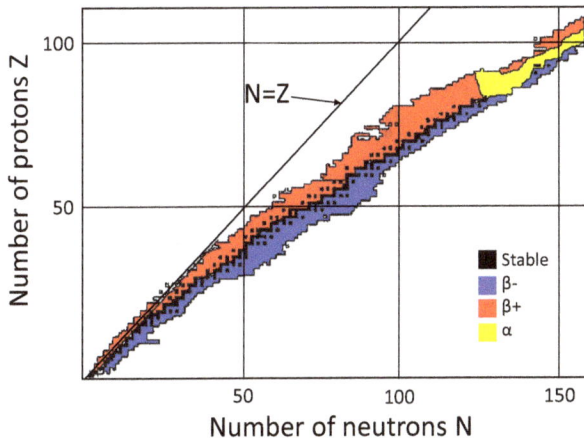

Fig. 5.2: Chart of stable isotopes (black dots) and radioactive isotopes colored in blue for electron emission, red for positron emission, and yellow for α particle-emission. Note that stable isotopes lie at or below the line $N = Z$, for heavier nuclei beyond $Z = 12$, $N > Z$.

> Nucleons with atomic mass A are composed of Z protons with charge $Z \times e^{+}$ and N charge neutral neutrons. For light stable nuclei, $N \cong Z$, for heavy nuclei, N>Z.

5.3 Atomic mass and atomic weight

A, N, and Z are integer numbers. When assigning a unit of mass for neutrons and protons, the total mass of a nucleus should also be an integer in this unit. However, this is not so because of the strong nuclear binding energy. The total mass of a nucleus is smaller than the sum of the individual neutron masses m_n and proton mass m_p:

$$m(A, Z) = Zm_p + Nm_n - B/c^2. \tag{5.1}$$

The *mass deficiency* is due to the binding energy B of nuclei. The binding energy varies with the mass number A. No distinction is made between protons and neutrons because the strong nuclear force is independent of charge. For light nuclei, the binding energy per nucleon B/A first increases and reaches a maximum of 8.6 MeV at the mass number $A \approx 58$ corresponding to Fe. For higher mass numbers, B/A drops off again. From the plot of B/A versus A in Fig. 5.3, we also recognize that in ${}^4_2$He the binding energy is particularly high compared to neighboring isotopes. This explains the formation of ${}^4_2$He embryos in heavy nuclei before α-decay takes place. Furthermore, this plot shows that energy can be gained by fusing light elements to heavier nucleons or splitting heavy elements into smaller fragments. The fusion process occurs in the Sun and all other stars; the fission process occurs in nuclear power plants and neutron research reactors.

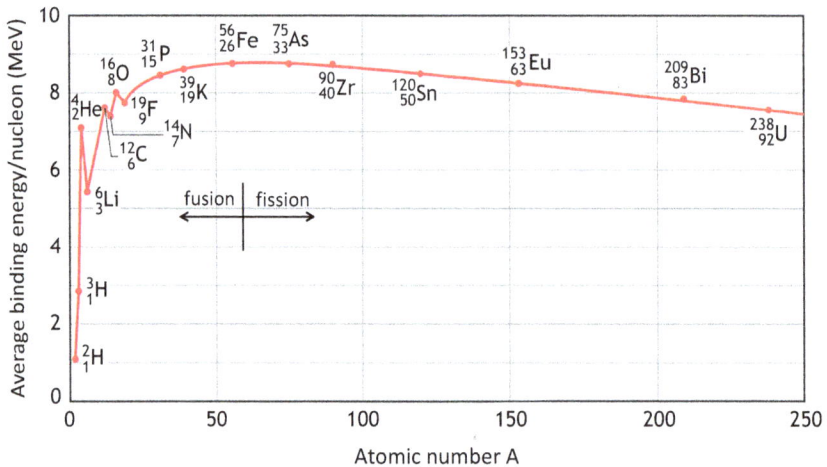

Fig. 5.3: Binding energy of nucleons as a function of atomic number A (adapted from physics. stackexchange.com and licensed under Creative Commons).

Because of the dependence of the nuclear mass on the mass number A, a unit for the atomic mass was defined independently of the energy variation. The *unified atomic mass unit* or short *atomic mass unit* with the symbol u is defined as the mass of the isotope ^{12}C divided by its mass number 12:

$$1u = 1m_u = m\left({}^{12}_6C\right)/12. \tag{5.2}$$

In SI units, u has a mass of 1.6605×10^{-27} kg and the energy equivalent of $m_u c^2 = 931.494$ MeV. In terms of atomic mass units, the mass of a proton is $m_p = 1.007276\ u$

and the mass of a neutron is $m_n = 1.008664\ u$. The free neutron is slightly heavier than the proton and therefore decays by $_{-1}^{0}e$-emission[2] with a mean lifetime of 882 ± 5 s into a proton, which is a stable particle.

By definition, $_{6}^{12}C$ has a unified atomic mass of 12 u. Taking this atomic mass in grams (here 12 g) provides a definition for the number of atoms per atomic mass, which is the Avogadro number N_{Av}. In words: 12 g of isotopically pure $_{6}^{12}C$ contains 1 mol of $_{6}^{12}C$ atoms or $N_{Av} = 6.022 \times 10^{23}$ $_{6}^{12}C$-particles. N_{Av} is known as the Avogadro[3] number. One mole of hydrogen or 1 g of protons contains the same number of particles: $N_{Av} = 1\,g/1\ u$. The mass of isotopically pure substances is always close to an integer multiple of u. For example, the atomic mass of ^{56}Fe is close to 56 u, etc.

The *atomic weight* is also quoted as the gram equivalent per mole of atoms. But in contrast to the atomic mass, the atomic weight takes an average over all naturally occurring and stable isotopes with their respective abundance. Typically, the atomic weight numbers are not close to multiples of u. For instance, the atomic weight of carbon is 12.011. From the 15 known isotopes of carbon, only 2 are stable: ^{12}C with an abundance of 98.9% and ^{13}C with an abundance of 1.1%. The weighted sum $12 \times 0.989 + 13 \times 0.011 = 12.011$ is the atomic weight of carbon. For an isotopic mixture, the atomic weight expressed in grams contains 1 mol of atoms. Atomic weights and isotopic compositions for all elements can be found in Ref. [1]. Tab. 5.1 gives an overview of symbols and terms used in nuclear physics.

Tab. 5.1: Overview of symbols for atomic and nuclear properties.

A	Atomic mass, atomic mass number
Z	Atomic number (number of protons)
N	Number of neutrons
u	Atomic mass unit
A_r	Atomic weight, relative atomic mass
N_{Av}	Avogadro number

5.4 Nuclear decay

Four types of radioactive decays can be distinguished that transform unstable radioisotopes into stable ones:

---

2 The $_{-1}^{0}e$-emission is also quoted as β^--emission in the literature. Here we use both expressions equivalently.

3 Lorenzo Romano Amedeo Carlo Avogadro (1776–1856), Italian physicist and chemist.

5.4.1 Electron emission (β^-)

Nuclei exhibiting a neutron excess are unstable. In the nucleus, one of the neutrons is converted into a proton by emitting a negatively charged electron, so-called
β^--particle, and an anti-neutrino. After emission, the positive charge of the nucleus
increases by 1 ($Z \rightarrow Z + 1$). Therefore, the daughter nucleus is a different chemical element placed on the right side of the parent nucleus in the chemical table with a different chemical symbol. The beta-minus decay has the general nuclear reaction form:

$$_Z^A\text{X} \rightarrow _{Z+1}^A\text{Y} + _{-1}^0\text{e}^- + \bar{\nu}_e + E_{e^-,\bar{\nu}_e}. \tag{5.3}$$

Decays by electron emission do not change the atomic mass number A. Those decays are called *isobaric transitions*. However, the ratio N/Z decreases, bringing the
daughter nucleus closer to the valley of stability. The anti-neutrino, together with
the electron, carries away the total binding energy $E_{e,\bar{\nu}_e}$. The partition of the total
energy between electron E_e and anti-neutrino $E_{\bar{\nu}_e}$ is not fixed; see Infobox on the β-
decay. Therefore, the energy spectrum of emitted electrons in β^--decay is very
broad, stretching from zero to a maximum energy $E_{e^-}^{\max}$. The interaction of neutrinos
with matter is essentially zero and the rest mass is much smaller than that of electrons. Neutrinos became noticeable mainly because of the broad electron energy
distribution, which required another particle due to energy and momentum
conservation.

Not only the energy is conserved during the decay but the electric charge, the
baryon number, and the lepton number are also conserved. Baryons are all heavy
particles composed of three quarks. Leptons are the light particles including electrons, positrons, and neutrinos. Leptons are associated with a lepton quantum
number $+1$ for particles and -1 for their antiparticles. In the β^--decay, the lepton
number is conserved. Before the decay, there are no leptons; therefore, the sum of
the lepton numbers is zero. After β^--decay, electron and anti-neutrino have opposite lepton numbers yielding again zero for their sum.

A prominent example for electron (β^-) emission is the decay of the terrestrial
radioisotope $_{19}^{40}\text{K}$:

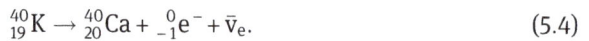

$$_{19}^{40}\text{K} \rightarrow _{20}^{40}\text{Ca} + _{-1}^0\text{e}^- + \bar{\nu}_e. \tag{5.4}$$

In the nuclear chart of Fig. 5.4, β^--decays are always directed from below upward
into the valley of stable isotopes. Because of the natural abundance of ^{40}K (0.012%),
its decay also occurs in our bodies, constituting a natural internal radioactive source.

5.4.2 Positron emission (β^+)

Nuclei exhibiting a neutron deficiency are unstable. In the nucleus, one of the protons
is converted into a neutron by emitting a positively charged electron and a neutrino.

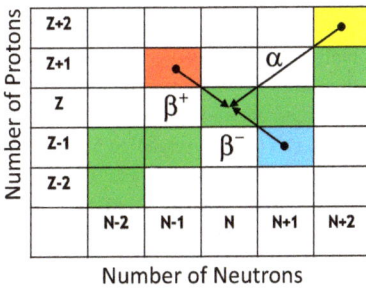

Fig. 5.4: Decay scheme for α- and β-radiation.

The positively charged electron is called a *positron*, denoted as β^+ or $_{+1}^{0}e^+$. After emission, the positive charge of the nucleus decreases by 1. The decay has the general nuclear reaction form:

$$_{Z}^{A}X \rightarrow {}_{Z-1}^{A}Y + \beta^+ + \nu_e + E_{e^+,\nu_e}. \tag{5.5}$$

Positron decay is also an isobaric transition, changing the chemical element but not the atomic number. The ratio N/Z increases, balancing the neutron deficiency. In the nuclear chart of Fig. 5.4, β^+-decay is always directed from above toward the valley of stable isotopes. Because of the participation of neutrinos in the decay, the energy distribution of positrons is very broad again, similar to the case of β^--decay (see Infobox I on β-decay). As expected, all conservation laws for energy, electric charge, baryon, and lepton number are fulfilled.

Positrons are antiparticles to electrons. They do not occur in nature naturally; they only occur as a decay product. Positrons can be stored for a long time in isolation (vacuum). As soon as they approach an electron, however, they combine and immediately annihilate each other by emitting two γ-photons in opposite directions, carrying away their rest mass energy. This correlated photon–photon emission is used in positron emission tomography (PET) to image tumor tissue in the body. The radioisotope ^{18}F is one of the preferred isotopes for PET, which is discussed in Chapter 10. The decay scheme of ^{18}F with a half-life of about 110 min is: $_{9}^{18}F \rightarrow {}_{8}^{18}O + \beta^+ + \nu_e$.

Infobox I: β-Decay

In all radioactive decays, energy and momentum as well as some other conservation laws regarding the symmetry of space and time must be preserved. During the detailed analysis of the β-decay it was noticed early on that the conservation of energy and momentum appears to be violated unless another charge-neutral particle is postulated.

Below is shown the energetically broad positron emission spectrum of ^{64}Cu in contrast to the sharp γ-emission lines of the same isotope. The comparison clearly shows that the decay via positrons and generally via β-particles fundamentally differs from the α- and γ-decays [4].

The mysteriously missing particle in the β-decay is the neutrino.[4] Although there is no doubt that this elementary particle exists, the interaction with matter is so weak that the rest mass could not yet be determined unambiguously. In addition, the parity is violated in the case of β-decay, so that a small difference in the decay rate can be discerned whether the β-emission occurs parallel or antiparallel to the nuclear spin. This violation can only be remedied by postulating a new force at work during the β-decay, which is called the *electroweak force*. This force must be of extremely short range, shorter than the range of the strong nuclear interaction that binds the nucleons in a nucleus. This short interaction controls the conversion of protons into neutrons and vice versa in nuclei as well as the β^--decay of free neutrons.

Thanks to the weak force, positrons exist and can be used for tomographic imaging.

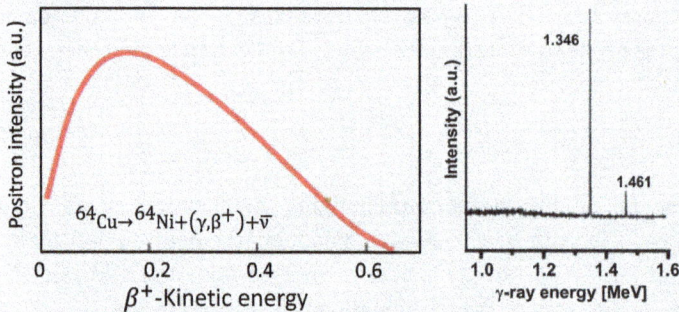

Left: Positron emission spectrum of ^{64}Cu. Right: γ-emission lines of ^{64}Cu.

5.4.3 Electron capture (EC)

In some cases, the conversion from a proton to a neutron is accomplished by capturing an electron from the K-shell of an atom. This creates a hole in the K-shell, similar to the excitation of characteristic X-rays by ejecting a K-shell electron. The hole in the K-shell is filled with an electron from the L-shell, and the energy difference is emitted as an X-ray photon. Therefore, isotopes that show electron capture (EC) decay are also known as x-ray emitters. Formally, the decay can be described by the reaction:

$$_Z^A X + e^- \rightarrow _{Z-1}^A Y + \nu_e + hf_{\text{x-ray}}. \tag{5.6}$$

An example is the already quoted decay of ^{40}K that may also proceed via EC:

$$_{19}^{40} K + e^- \rightarrow _{18}^{40} Ar + \nu_e.$$

---

4 The neutrino was postulated by Wolfgang Pauli ((1900–1958), Nobel Prize 1945); its existence was confirmed in 1956 by Frederick Reines and Clyde Cowan (Nobel Prize 1995).

EC is an alternative mode of decay for unstable isotopes if the energy is insufficient for the decay through positron emission. Positron emission requires at least 1.022 MeV binding energy difference between parent and daughter isotopes, corresponding to the rest mass of an electron-positron pair. When the energy difference is less, EC is an alternative pathway to the ground state. Therefore, the decay caused by EC is sometimes referred to as *inverse positron decay*.

5.4.4 α-Particle decay

Heavy nuclides often strongly bond subunits of two protons and two neutrons, forming a ^{4}He-nucleus. These subunits may escape from their parent nucleus by quantum tunneling through the Coulomb barrier of the nuclear potential. After successful tunneling, the α-particles $\left(\alpha = {}_2^4\text{He}\right)$ have a typical kinetic energy of about 5 MeV. α-particle decay changes the charge of the parent nucleus by $\Delta Z = -2$ and the mass number by $\Delta A = -4$. The decay has the general nuclear reaction form:

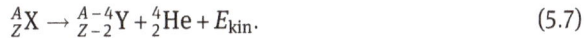

$$ {}_Z^A\text{X} \rightarrow {}_{Z-2}^{A-4}\text{Y} + {}_2^4\text{He} + E_{\text{kin}}. \tag{5.7} $$

In the nuclear chart of Fig. 5.4, α-decay always occurs along the diagonal or parallel to the valley of stable isotopes. An example is the decay of radium into radon:

$$ {}_{88}^{226}\text{Ra} \rightarrow {}_{86}^{222}\text{Rn} + {}_2^4\text{He} + E_{\text{kin}}. $$

The emission of α-particles is the most common radioactive decay mode for the heaviest nuclides. A whole series of decays is often observed, starting with uranium, thorium, or neptunium until a stable isotope is formed after a series of α-emissions. For example, ^{238}U needs eight α-emissions before it forms the stable isotope ^{206}Pb. These heavy nuclides were originally fused together in supernova explosions and can still be found in nature today due to their very long lifespan of about 5×10^9 a.

> There are stable and unstable isotopes. The unstable isotopes decay into stable isotopes with the ejection of α-, β^--, β^+- particles and emission of γ-radiation. **!**

5.4.5 Decay schemes

After ejection of α- or β-particles, the daughter nucleus often remains in an excited, high angular momentum state for a short time. The transition to the ground state occurs via the emission of electromagnetic radiation, termed *γ-radiation*, to distinguish it from characteristic x-rays. Physically, both radiations are identical. But the sources are different, atomic shells on the one hand and excited nuclei on the other.

The energy of γ-photons ranges from keV to MeV. A typical example is the decay of the isotope ^{137}Cs shown in Fig. 5.5. ^{137}Cs is a β⁻-emitter, i.e., after β⁻–emission, ^{137}Cs transforms into the isotope ^{137}Ba with the same mass number. There are two possibilities for the decay, called *branching*. Either the ^{137}Cs isotope emits a β⁻-electron with the maximum energy of 1.17 MeV, bringing ^{137}Ba immediately into the ground state; or ^{137}Cs emits a β⁻-electron with lower energy of 0.51 MeV, leaving ^{137}Ba* in an excited nuclear state that goes into a stable ground state after emission of a γ-photon of 0.66 MeV. The first transition has a probability of only 5%, whereas the majority of transitions (95%) occur via the intermediate excited ^{137}Ba* energy level into the ground state ^{137}Ba. The asterisk indicates a metastable state of the nucleus. The half-lifetime of ^{137}Cs is about 30 years, independent of the branching, whereas the γ-emission is relatively prompt with a $T_{1/2}$ of 2.5 min. The hard γ-radiation of 0.66 MeV is used among many other applications for radiation therapy.[5]

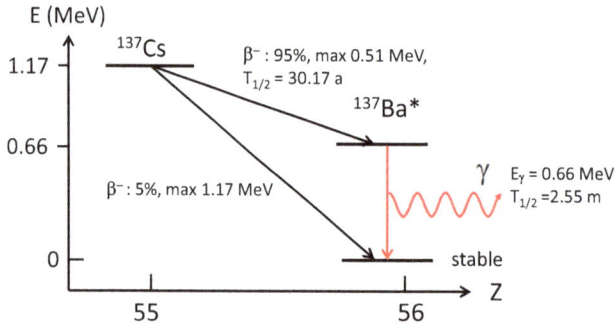

Fig. 5.5: Radioactive decay of ^{137}Cs and branching ratio.

Another interesting example for branching is the potassium isotopes $^{39-41}$K already introduced. ^{39}K and ^{41}K are stable isotopes; only ^{40}K is radioactive with a very long half-life of $T_{1/2} = 1.3 \times 10^9$ a. This isotope may either decay by EC with a probability of 11% or by electron emission with a probability of 89%, transforming the radioisotope into either stable Ar or Ca. The decay scheme is shown in Fig. 5.6.

The natural abundance of ^{40}K is only 0.012%. Nevertheless, we estimate that about 0.02 g of ^{40}K is in our body, decaying continuously with an activity of about 5000 events per second. As electrons do not have a long range, most of the electrons emitted remain in our body. But some escape and can be counted with full-body scintillation scanners.

---

5 ^{137}Cs is a fission product generated in nuclear reactors. ^{137}Cs was set free during the nuclear accident in Chernobyl 1986, contaminating large parts of Europe. As the lifetime of the isotope is about 30 years, more than half of the original amount has decayed meanwhile.

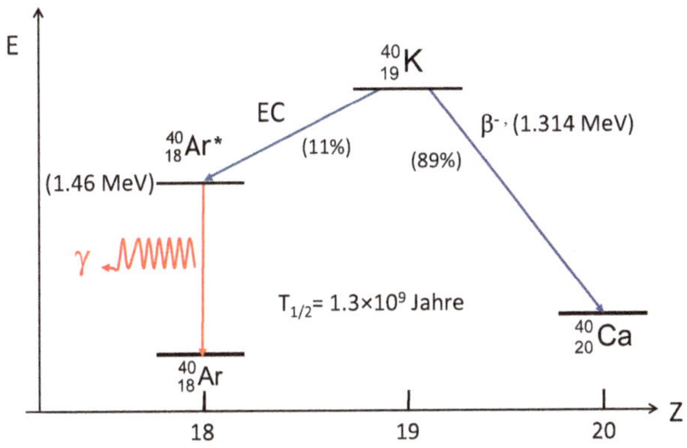

Fig. 5.6: Decay scheme of the isotope ^{40}K.

Another instructive example is technetium (Tc), the most commonly used isotope for scintigraphy. Since none of the Tc isotopes are long-lived, Tc does not occur in nature but has to be produced artificially. The isotope ^{99}Tc is particularly useful for scintigraphy because of its short lifetime and the emission of γ-radiation with a low energy of 140 keV, similar to the x-ray energies used for radiography. The starting isotope is ^{99}Mo, a by-product of ^{235}U nuclear fission in a nuclear research reactor. ^{99}Mo decays through β^--emission into an excited state of ^{99}Tc*; the metastable ^{99}Tc* state decays with a half-life of 6 h via γ-radiation into the ground state: $^{99}_{43}$Tc* → $^{99}_{43}$Tc + γ. The decay scheme is shown in Fig. 5.16. Finally, ^{99}Tc decays into the stable isotope ^{99}Ru through another β^--emission. Decay schemes and lifetimes of all radioisotopes can be found in [2]. A medical isotope browser is available in [3].

5.5 Radioactivity

5.5.1 Exponential decay law

Radioactive decay is a stochastic process of unstable, disintegrating isotopes. In the case of a large number of identical radioisotopes, we do not know when a particular nuclide decays. However, it is possible to state exactly how long it takes for half of the original radioisotopes to disintegrate, denoted as *half-life*. The decay law describes the radioactive decay over time. N_0 is the number of radioisotopes present at time $t = 0$ and $dN(t)/dt$ is the decay rate at time $t > 0$. Now we assume that the decay rate $dN(t)/dt$ is linearly proportional to the number of radioisotopes $N(t)$ still present at time t:

$$ -\frac{dN(t)}{dt} \sim N(t). $$

The negative sign is due to the fact that $dN(t)/dt$ decreases over time. Now multiplying by a proportionality constant λ, we have the equation:

$$\frac{dN(t)}{dt} = -\lambda N(t). \tag{5.8}$$

The proportionality constant λ is denoted as the *decay constant*. Separating the variables, we can integrate eq. (5.2) and obtain

$$N(t) = N_0 \exp(-\lambda t), \tag{5.9}$$

where we have set the integration constant $N(t=0) = N_0$.

The expression in eq. (5.3) is known as the *exponential decay law*, plotted in Fig. 5.7.[6] The decay constant λ is the probability that a particular nuclide will disintegrate in 1 s. The inverse $1/\lambda$ is the *average lifetime T* characteristic of the decay and the radioisotope in question. T is the average time a nucleus (nuclide) survives before it disintegrates by any of the particle emissions, including γ-radiation. The exponential decay law expresses the result of a statistical process. Therefore, fluctuations of T are expected. However, due to the large number of isotopes in a typical sample, fluctuations from the average λ or T are very small.

After one average lifetime T, the number of nuclei that still exist has decreased by the factor e: $N(T) = N_0/e$. This means that after time T, 63% of all nuclides in a sample have decayed, and 37% still remain. Often not the lifetime T but the half-life $T_{1/2}$ is quoted in the literature. According to its name, the half-life $T_{1/2}$ is the time after which half of the radioactive isotopes have survived, and the other half has decayed (Fig. 5.7). Half-life and average lifetime are related by

$$T_{1/2} = \frac{\ln 2}{\lambda} = T \ln 2 = 0.693 \times T. \tag{5.10}$$

Thus, the half-life is 30% shorter than the average lifetime.

> **!** Radioactive decay is a random or stochastic process. The disintegration of an individual nuclei cannot be predicted. Nevertheless, the decay constant of a large number of radioactive isotopes can be determined precisely.

Radioactivity in the body decreases over time via decay and excretion when radioactive substances were incorporated either by digestion, inspiration, or injection. Therefore, the effective half-life of incorporated isotopes is substantially shorter than the physical half-life. The biological half-life $T_{1/2}^{b}$ and the physical half-life $T_{1/2}^{p}$ add reciprocally:

---

6 Note that in this context, N refers to the number of radioisotopes in a sample; the letter N may also designate the number of neutrons in a nucleus. From the context, it should be clear whether the number or radioisotopes or the number of neutrons is meant.

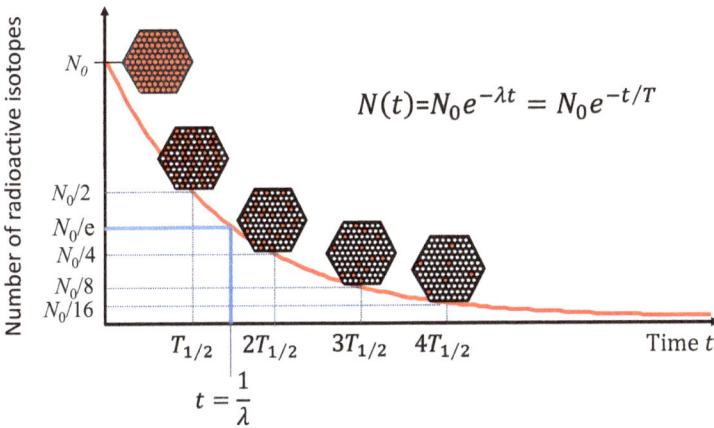

Fig. 5.7: Exponential decay law of radioactive isotopes. The red dots in the boxes symbolize the remaining radioisotopes at the respective times. The solid blue line indicates the number of isotopes after one lifetime; the number of isotopes after successive half-lives is indicated by dashed lines.

$$\frac{!}{T_{\text{eff}}} = \frac{1}{T^{\text{p}}_{1/2}} + \frac{\overset{\bullet}{1}}{T^{\text{b}}_{1/2}}, \tag{5.11}$$

and yield the effective half-life for incorporated radioisotopes:

$$T_{\text{eff}} = \frac{T^{\text{b}}_{1/2} \cdot T^{\text{p}}_{1/2}}{T^{\text{p}}_{1/2} + T^{\text{b}}_{1/2}}. \tag{5.12}$$

As an example, we discuss the radioisotope ^{131}I used for the scintigraphy of the thyroid gland. ^{131}I has a physical half-life of 8 days and a biological half-life of 80 days. The combined effective half-life is 7.27 days. The isotope ^{131}I is atypical as the biological half-life is usually much shorter than the physical half-life. In Tab. 5.2, half-life times of some relevant radioisotopes in clinical practice are listed. The isotopes ^{90}Sr, ^{210}Pb, ^{210}Po, and ^{233}U are not used for therapy or diagnostics, but they can get into the body by contamination through radioactive waste.

The radioisotope ^{14}C is constantly regenerated by cosmic radiation and naturally incorporated into the body via the food chain. Therefore, the density of the ^{14}C isotopes in our body and in all other living matter is constant. Only after death does the exchange of ^{12}C with ^{14}C stop, and the stored ^{14}C isotopes decay according to the physical half-life. Hence, the ^{14}C decay can be viewed as a biological clock used to date archaeological artifacts.

Tab. 5.2: Selected radioactive isotopes and their physical and biological half-life times.

Nuclide	Physical $T_{1/2}$	Biological $T_{1/2}$	Critical organs
^{90}Sr	28,1 a	11 a	Bones
^{210}Pb	22 a	730 days	Bones
^{210}Po	138 days	40 days	Spleen
^{233}U	$1,63 \times 10^5$ a	300 days	Bones
^{131}I	8 days	80 days	Thyroid
^{14}C	5570 a	35 a	Fatty tissue
^{99m}Tc	6 h	24 h	Bones
^{201}Tl	3.04 days	9.8 days	Heart, brain

5.5.2 Nuclear activity

The exponential decay law is not practical to handle in the laboratory since the number of radioisotopes in a particular sample is barely known. On the other hand, radioactivity is recognized by the emission of α-, β-, or γ-radiation and can be recorded with specialized detectors. The decay rate, i.e., the number of emissions per time $dN(t)/dt$, is also termed *activity A* and equals the count rate of a detector:

$$A(t) = -\frac{dN(t)}{dt} = \lambda N(t). \tag{5.13}$$

Hence according to eq. (5.9), we have

$$A(t) = \lambda N_0 e^{-\lambda t} = A_0 e^{-\lambda t}.$$

Therefore, it follows for the activity:

$$A(t) = A_0 e^{-\lambda t}. \tag{5.14}$$

According to eq. (5.14), the activity $A(t)$ is proportional to the decay constant λ. The larger λ is, the larger is the activity and the count rate, the shorter is the lifetime $T = 1/\lambda$. The activity has the same mathematical form as the exponential decay law but with the advantage of expressing a measurable quantity: the emissions per time interval. The SI unit of the activity is: $[A]$ = events/s = Becquerel[7] (Bq).

---

7 Antoine Henri Becquerel (1852–1908), French physicist and Nobel laureate in physics 1903.

1 Bq is one event per second, or one count per second if the detector is 100% efficient. An older unit is curie[8,9] (C), which corresponds to the activity of 1 g of radium. One curie (1 C) is equivalent to 3.7×10^{10} Bq. The curie unit is still sometimes used in clinics. While 1 C is a dangerously high activity, the typical activity levels of samples handled in clinics are closer to the μ C level.

As an example of the activity, we discuss again the isotope ^{40}K. A typical human body contains about 160 g of potassium, of which 0.012% or 0.02 g are radioactive. This corresponds to a number of $N_0 = (0.02/40) \times 1$ mol $= 3 \times 10^{20}$ radioactive ^{40}K isotopes in the body at any time. These isotopes decay via β^--emission with a half-life of $T_{1/2} = 1.3 \times 10^9$ a $= 4 \times 10^{16}$ s. Then we determine the activity as follows:

$$A = \frac{\ln 2}{T_{1/2}} \cdot N_0 = \frac{0.693}{4 \times 10^{16}} \cdot 3 \times 10^{20} \approx 5000 \, \text{s}^{-1} = 5000 \, \text{Bq.}$$

This is the decay rate quoted earlier. The specific activity a is the activity per mass unit. The SI unit is $[a] = $ Bq/kg. One kg of pure ^{40}K isotopes has an activity of 2.5×10^8 Bq/kg. More practical is the unit Bq/g. For instance, the specific activity of 1 g of radium is $a = 3.7 \times 10^{10}$ Bq/g. In brachytherapy discussed in Chapter 5 of Volume 3 it is desirable to have high specific activities, i.e., high activity in small amounts of material.

The exponential law allows simple relationships between activities after incremental decay times t, $2t$, $3t$, etc. The activity after time t is

$$A_1(t) = A_0 e^{-\lambda t} \tag{5.15}$$

or

$$\frac{A_1(t)}{A_0} = e^{-\lambda t}. \tag{5.16}$$

Then the activity after $2t$ is

$$A_2(2t) = A_0 e^{-\lambda 2t} = A_1(t) e^{-\lambda t} = A_1(t) \frac{A_1(t)}{A_0}. \tag{5.17}$$

For the activity after $n \times t$ we find the general expression:

$$A_n(n \cdot t) = A_{n-1} \frac{A_1(t)}{A_0} = A_1 \left(\frac{A_1(t)}{A_0} \right)^{n-1} \tag{5.18}$$

---

8 Pierre Curie (1859–1906), French physicist and Nobel laureate in physics 1903.
9 Marie Curie (1867–1934), Polish-French physicist and Nobel laureate in physics 1903 and in chemistry 1911.

or

$$A_n = A_0 \left(\frac{A_1}{A_0}\right)^n. \tag{5.19}$$

These expressions are useful in clinics when handling short-lived isotopes. For instance, when the present activity A_1 is known and the activity from an hour ago was A_0, the activity in a few hours can be predicted via the ratio A_1/A_0. Examples are given in Exercises E5.1–5.3.

5.5.3 Decay chains

After the decay of a nucleus A (parent nucleus) with the decay constant λ_A, often the daughter nucleus B is also radioactive with a decay constant λ_B. Then the daughter nucleus decays into C, which again disintegrates with decay constant λ_C into D, etc., until a stable isotope is formed. The successive decay of the parent nucleus into daughter nuclei B, C, etc. is known as a radioactive *decay chain*. Four of those chains occur naturally; for a discussion of those, we refer to textbooks on nuclear physics listed at the end of this chapter. Important for us is to recognize that the activities of parent and daughter nuclei are interdependent. The activity of the daughter nuclei depends on the production rate of the parent nuclei and the decay constant of the daughter nucleus. As long as A has not disintegrated, B is not born and cannot decay. The activity of the daughter nucleus A_B is, therefore, a function of the production rate of the daughter nucleus, which is equal to the decay rate of the parent nucleus $\lambda_A N_A$ minus the decay rate of the daughter nucleus $\lambda_B N_B$:

$$A_B = \lambda_A N_A - \lambda_B N_B. \tag{5.20}$$

Equation (5.14) can be integrated to yield the number of daughter nuclei $N_B(t)$ as a function of time [5]:

$$N_B(t) = \frac{\lambda_A}{\lambda_B - \lambda_A} N_A(0)\left(e^{-\lambda_A t} - e^{-\lambda_B t}\right) \tag{5.21}$$

Similar equations can be given for a sequence of decays from A to B to C, etc. The respective equations are called *Bateman equations*.[10]

An example is displayed in Fig. 5.8 for the case of $\lambda_A \cong \lambda_B$, i.e., the lifetimes of parent and daughter nuclei are about equal. Then the number of daughter nuclei has a maximum when most of the parent nuclei have decayed after one lifetime T,

---

10 Harry Bateman (1882–1946), British mathematician. The Bateman equations are also used in pharmacy for describing the invasion and elimination of drugs in the body.

and only at a later stage is the exponential decay of the daughter nuclei expressed. Finally, the decay from B to C forms a stable isotope.

In case that $\lambda_A \gg \lambda_B$, i.e., the activity of the parent nucleus is much higher than that of the daughter nucleus, the activity of the daughter isotope B remains low, even though the parent nuclei have already disintegrated. This situation is realized for the isotope ^{99}Tc, and the decay scheme is shown in Fig. 5.16. ^{99m}Tc decays fast to ^{99}Tc with a half-life of 6 h, whereas ^{99}Tc decays much slower to ^{99}Ru with a half-life of about 10^5 a. Thus, ^{99}Tc is quasi-"stable."

> In radioactivity, the number of decayable isotopes changes over time. But the probability of each individual isotope to decay stays constant. **!**

At the other extreme, when the parent isotope is long-lived but the daughter isotope decays much faster ($\lambda_A \ll \lambda_B$), then the number of daughter isotopes grows rapidly at the beginning and after some time, an equilibrium is reached where A and B decay at the same rate governed by λ_A. This condition is indeed realized for the decay of ^{99}Mo/^{99m}Tc and is shown in Fig. 9.11.

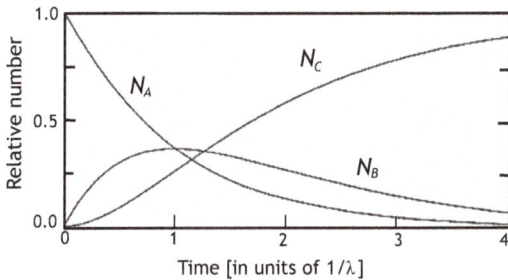

Fig. 5.8: Radioactive decay of successive parent (A) and daughter nuclides (B) with the same decay rates $\lambda_A = \lambda_B$. After the decay of isotope B into C, isotope C is assumed to be stable.

5.6 Radioisotope production

All isotopes used in nuclear medicine are produced artificially by nuclear reactions. There are three main methods for isotope production, which are explained later:
- charge particle irradiation;
- nuclear fission;
- neutron capture.

We start with a brief account of nuclear reactions and production rates in general terms that apply to all three methods and then discuss the three isotope production methods.

5.6.1 Nuclear reactions

Nuclear reactions require a target containing the starting nucleus A and a projectile a that transforms isotope A into another isotope B by emission of particle b. The standard terminology for the nuclear reaction is

$$a + A \rightarrow B + b$$

or equivalently:

$$A(a, b)B.$$

a and A are called the entrance channels, b and B are referred to as the exit channels. Nuclear reactions either require or set free some energy, which is expressed as the Q-value of the reaction:

$$Q = \Delta E_a + \Delta E_A - (\Delta E_b + \Delta E_B). \tag{5.22}$$

ΔE_x is the energy difference of particle x before and after reaction. If $Q > 0$, the reaction is exothermic, i.e., it releases energy; the released energy is transferred as kinetic energy to the final products.

If $Q < 0$, the reaction is endothermic and requires energy. The energy is converted from the kinetic energy of the entrance particle to the binding energy of the final particle B and the kinetic energy of b.

For both reactions, exothermic and endothermic, the complete nomenclature for nuclear reactions is, therefore:

$$A(a, b)B, Q \quad [\text{MeV}],$$

where Q is the mass difference between the parent and daughter nuclei.

5.6.2 Isotope production via irradiation

Now we want to estimate how long it takes to produce radioactive material by irradiation of a target with particles. The irradiation can be performed with protons, neutrons, or α-particles. The basic equations for the production rate of radioisotopes apply to all types of irradiation. We assume a constant incoming flux J_a of particles a, directed straight to the target, as indicated in Fig. 5.9. Then the flux (or intensity) is defined as the number of particles N_a in the beam per surface area S and time t:

$$J_a = \frac{N_a}{St} = \frac{N_a}{Sl}\frac{l}{t} = n_a v_a. \tag{5.23}$$

The incident flux equals the product of particle density n_a and their average velocity v_a in an imaginary tube of length l. The unit of J_a is $[J_a] = \text{m}^{-2}\,\text{s}^{-1}$.

target

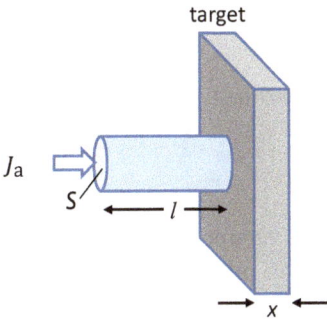

Fig. 5.9: Schematics for irradiation of a target.

Next, we consider the *cross section* σ_a for isotope production of particles a. In the simplest form, the cross section is defined as the ratio of successful events per unit time W divided by the incident particle flux J_a:

$$\sigma_a = \frac{W}{J_a}.$$ (5.24)

A successful event is, for instance, the capture of a neutron that converts a stable nucleus into a radioactive isotope.

The SI unit of the cross section $[\sigma]$ is m^2. It is usual to quote the cross section in terms of barns, where 1 barn $= 10^{-28}$ m$^2 = 10^{-24}$ cm^2. Small cross sections are also quoted in terms of mb or µb.

If N_A is the total number of isotopes A in the target hit by the particle beam flux J_a, and N_B are those isotopes, which have been transformed to B in time t, then W can be expressed as follows:

$$W = \frac{N_B}{N_A t} = \frac{R_B}{N_A}.$$ (5.25)

R_B is called the *reaction rate* for the production of isotope B. Reassembling these equations, we find the reaction rate:

$$R_B = N_A \sigma_a J_a$$ (5.26)

Thus, the reaction rate of isotope B is the product of the number of nuclei A in the target, the reaction cross section σ_a, and the incoming particle flux J_a. Cross sections for particular isotopes and reactions can be found in nuclear tables, such as in [6].

Now we are prepared to calculate the net *production rate* of isotope B. The production rate is the difference between the reaction rate R_B and the decay rate of isotope B:

$$\frac{dN_B}{dt} = R_B - \lambda_B N_B. \tag{5.27}$$

Integration over time yields the number of isotopes B after time t:

$$N_B(t) = \frac{R_B}{\lambda_B}\left(1 - e^{-\lambda_B t}\right). \tag{5.28}$$

$N_B(t)$ is plotted in Fig. 5.10. Initially, the increase of the isotope number B is controlled by the reaction rate R_B. But later, the increase levels off and saturates at the value $N(t \to \infty) = R_B/\lambda_B$. The activity of isotope B follows from the last equation:

$$A_B(t) = \lambda_B N_B(t) = R_B\left(1 - e^{-\lambda_B t}\right). \tag{5.29}$$

The activity of isotope B has the same time dependence as the number of isotopes $N_B(t)$. Certainly, after switching off the incident particle beam, the activity of the isotopes B will decay as usual.

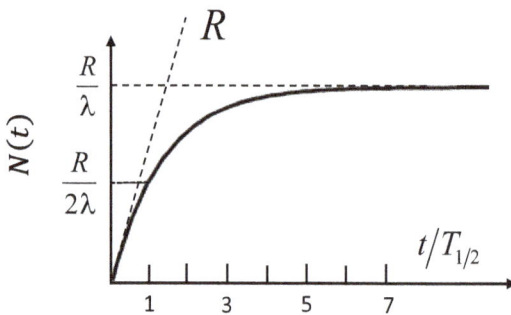

Fig. 5.10: Time dependence of the number of reactive isotopes produced. The timescale is in units of half-lives.

For isotope production, it is important to come close to the saturation value R_B/λ_B. After one lifetime $t = T_B = 1/\lambda_B$ already 63% of the saturation value has been reached, after two lifetimes 86%, etc. Knowing the experimentally available target parameters $R_B = N_A \sigma_a J_a$, the activity of the target after an irradiation time t by a particle flux J_a can easily be estimated. This is demonstrated in Exercises E5.7–E5.9.

> **!** Isotopes for nuclear medicine are produced artificially. The production rate depends on the reaction rate minus the decay rate.

5.6.3 Charge particle activation

The production of radioisotopes through bombardment with charged particles is known as *charge particle activation*. The charged particle is either a proton, deuteron, α-particle, or a carbon nucleus. In the simplest case, we talk about an $A(p, n)B$ reaction, where a proton is exchanged against a neutron, generating a radioisotope B.

A nuclear reaction between positively charged particles and nuclei is not the primary process. First, we expect that charged particles scatter elastically at the Coulomb potential of the nucleus, known as Rutherford[11] scattering. Rutherford scattering takes place as long as the center-of-mass energy is less than the Coulomb barrier. However, a nuclear reaction can take place if the charged particle has an energy greater than

$$E_p > E_{Coulomb} = \frac{Zze^2}{r},$$ (5.30)

where Z and z are the atomic numbers of the proton and the target nucleus, e is the elementary charge, and r is the distance between the particles. The Coulomb barrier can become very large as r approaches the nuclear dimension, such that nuclear reactions would be unlikely to occur. On the other hand, nuclear reactions do take place. Therefore, there must be a mechanism to overcome the Coulomb barrier. This mechanism is quantum tunneling, sketched in Fig. 5.11. The proton can penetrate the Coulomb wall by quantum tunneling forming a compound nucleus that lives for a very short time (~10^{-15} s) before decaying into the reaction products. Experiments show that the Coulomb barrier for protons on heavier target nuclei is on the order of 10 MeV. With a proton cyclotron, at least this proton energy should be reached. In the case of α-particles, the barrier height is about 25 MeV.

The reaction cross section for proton tunneling and nuclear reaction is expressed by:

$$\sigma_{r,i} = \sigma_{comp} \frac{P_i}{\sum_i P_i}.$$ (5.31)

Here σ_{comp} is the cross section for the compound formation, P_i is the probability for one particular reaction channel, and $\sum_i P_i$ is the sum over the probabilities of all reaction channels. The isotope production rate R then becomes with the use of eq. (5.23):

$$R = N_A \, J_a \left(1 - e^{-\lambda t}\right) \int_{E_s}^{E_0} \frac{\sigma(E)}{dE/dx} dE.$$ (5.32)

Here J_a is the incident proton flux, N_A is the number of nuclei exposed to the proton beam, t is the irradiation time, σ is the reaction cross section, and the integral is taken over the energy change along the distance traveled.

---

11 Ernest Rutherford (1871–1937), New-Zealand–British physicist and Nobel laureate 1908 in chemistry.

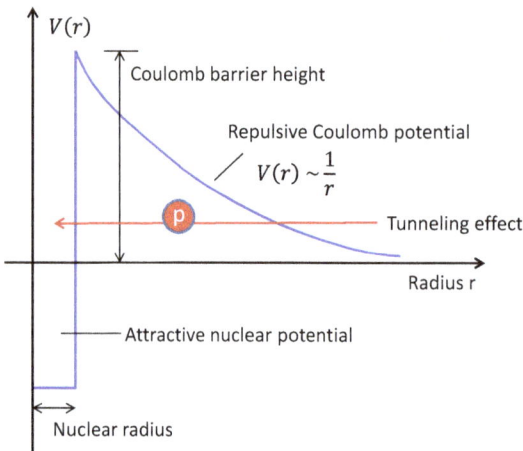

$V(r)$

Coulomb barrier height

Repulsive Coulomb potential

$V(r) \sim \dfrac{1}{r}$

p

Tunneling effect

Radius r

Attractive nuclear potential

Nuclear radius

Fig. 5.11: Sketch of the potential encountered by a proton approaching a heavy target nuclei. The proton will first experience the Coulomb potential and elastically scatter. In addition, there is a finite probability of the proton being trapped by the nuclear potential after penetrating the Coulomb barrier. This usually generates excited and unstable isotopes, which decay by neutron, gamma, or positron emission.

5.6.4 Cyclotron isotope production

Radioisotopes used in clinics sometimes have very short lifetimes, too short to be delivered from an external laboratory by a delivery service. Therefore, specialized radiation clinics have their own cyclotron for the production of short-lived isotopes. A cyclotron is an accelerator for light charged nuclei such as protons, deuterium, and α-particles.

The physical principle of cyclotrons is based on the Lorentz[12] force $\vec{F}_L$ of particles with charge q and velocity $\vec{v}$ in a magnetic induction field $\vec{B}$:

$$\vec{F}_L = q\left(\vec{v} \times \vec{B}\right). \tag{5.33}$$

We consider in Fig. 5.12 a charged particle beam with the velocity $\vec{v}$ along the x-axis. A homogeneous magnetic field $\vec{B}$ is oriented perpendicular to the xy plsnr along the z-axis. When the particle beam enters the magnetic field area, it is deflected due to the action of the Lorentz force. Since $\vec{v}$ and $\vec{B}$ are oriented perpendicular to each other, we can omit the vector product and simply write the Lorentz force in scalar form: $F_L = qvB$.

---

12 Hendrikus Albertus Lorentz (1853–1928), Dutch physicist and Nobel laureate in physics 1902.

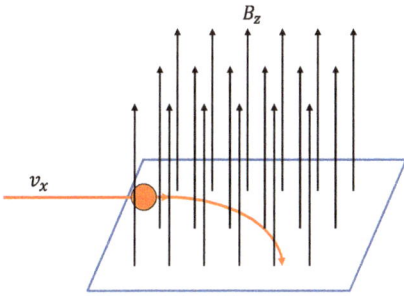

Fig. 5.12: The charged particle is deflected in a homogenous magnetic field due to the Lorentz force acting on the particle.

The deflection of the charged particle beam describes a circle because the Lorentz force pulls the particles inward and is balanced by the outward-oriented centrifugal force:

$$F_{centr} = mv^2/r. \tag{5.34}$$

In equilibrium, both forces balance:

$$qvB = \frac{mv^2}{r}, \tag{5.35}$$

and we have a stable orbit with the orbital radius:

$$r = \frac{mv}{qB}. \tag{5.36}$$

This effect is used in cyclotrons, invented by Lawrence.[13] The schematics of a cyclotron in the top and side view is shown in Fig. 5.13. It consists of two D-shaped metal cans called dees with a gap in between and placed in a vacuum chamber. The proton source is positioned at the center of the cyclotron between the dees. The vacuum chamber is inserted between the pole shoes of a large electromagnet forming a double C-shaped yoke. The coils are connected to a dc-current source. Whenever a charged particle has completed a half circle in one of the dees and crosses the gap, it gets a kick by an rf-generator connected with the dees, increasing the orbital velocity each half turn. Since the magnetic flux density B remains constant, the radius of the orbit increases continuously, forming an outward-going spiral. While the circulation time T_{cyc} is independent of the radius and remains constant:

Short lived radioisotopes are generated by charged particle bombardment of stable isotopes in cyclotrons.

[13] Ernest Lawrence (1901–1958), American physicist and Nobel Prize laureate 1938.

$$T_{\text{cyc}} = 2\pi \frac{m}{qB},$$

(5.37)

the kinetic energy increases and reaches a maximum when the radius of the particle orbit r comes close to the radius R of the dees. Then the maximum kinetic energy is

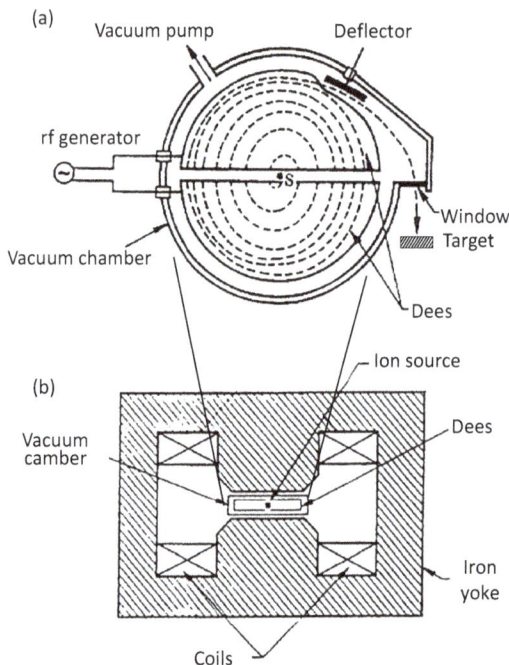

Fig. 5.13: Schematics of a cyclotron used for isotope production. (a) Top view of dees with ion source and target; (b) side view with magnet and coils. S is the source for protons being accelerated. The protons hit a target after penetrating a thin vacuum window.

$$E_{\text{kin, max}} = \frac{1}{2} \frac{(qBR)^2}{m}.$$

(5.38)

At the maximum energy, the accelerated particles are deflected by an electric field in a condenser, passing a vacuum window before hitting a target for isotope production via nuclear reaction.

Alternatively and often practiced is the use of negatively charged hydrogen molecules H_2^-. When the maximum energy $E_{\text{kin, max}}$ is reached, the molecules pass a graphite window which strips off the electrons. As the stripping changes the charge from minus to plus, the protons are automatically deflected without application of an electric field.

This classical treatment of the cyclotron kinematics should be corrected for relativistic mass increase when the particle velocity reaches more than 10% of the speed of light. Then in all equations, the mass m has to be replaced by $\gamma_0 m$, where $\gamma_0 = 1/\sqrt{1-(v/c)^2}$ is the Lorentz[14] correction factor of relativistic kinematics. With rf-frequencies between 10 and 30 MHz, kinetic energies of about 25 MeV are reached for protons and deuterons and about 50 MeV for α-particles.

With cyclotrons, mainly γ-emitters for scintigraphy (Chapter 9) and short-lived β^+-emitters for PET (Chapter 10) are produced. An example is the production of ^{18}F by proton bombardment according to the following nuclear reaction:

$$p + {}^{18}_{8}O \rightarrow {}^{18}_{9}F + n$$

Protons hit an ^{18}O-enriched target and convert the isotope ^{18}O by (p,n) exchange into ^{18}F. The cross section for this reaction has a maximum of 500 mb at about 5 MeV incident proton energy [7]. The natural abundance of the stable isotope ^{18}O is only 0.2%. Therefore, ^{18}O targets must be isotopically enriched to be effective. In most cases, ^{18}O-enriched water targets are utilized. After irradiation, fluorine is chemically extracted from the target and used for synthesizing fluorodeoxyglucose (^{18}F-FDG) required for PET scans. Using water targets has the additional benefit of thermalizing fast neutrons generated during the nuclear reaction. The thermalized neutrons are subsequently captured by boron-enriched protection shielding. The thermal neutron absorption cross section of boron is further discussed in Chapter 4 of Volume 3 because it is also important for the boron neutron capture therapy (BNCT). $^{18}_{9}$F is unstable as fluor has only one stable isotope containing 10 neutrons (see Fig. 5.1). Therefore, $^{18}_{9}$F decays by β^+ emission back to $^{18}_{8}$O: $^{18}_{9}F \rightarrow {}^{18}_{8}O + \beta^+ + \nu$. The subsequent $\beta^+ - \beta^-$ annihilation with the concomitant emission of two γ-photons flying in opposite directions is used for PET imaging, discussed in Chap. 10.

Table. 5.3 lists some typical γ-emitting radioisotopes commonly used for scintigraphy in clinics. Table 5.4 lists cyclotron produced short-lived positron-emitting isotopes used for PET and clinical research. It has been suggested that ^{99m}Tc may also be produced by proton bombardment of an Mo target via the reaction ^{100}Mo (p,2n)-^{99m}Tc as an alternative to the ^{235}U fission in a nuclear reactor [8]. However, the ^{99m}Tc yield by cyclotron production is much smaller than by the fission process such that the main source of ^{99m}Tc remains reactor-based for the foreseeable future.

Nowadays, compact cyclotrons are commercially available for radioisotope production in clinics using proton irradiation. The Eclipse cyclotron shown in Fig. 5.14 from the company Siemens is a 11-MeV, negative-ion, single-particle accelerator. It is specifically designed for clinical and commercial production of ^{18}F radiotracers and other positron-emitting radioisotopes such as ^{11}C, ^{13}N, ^{15}O, and ^{64}Cu.

---

14 See footnote no. 11.

Tab. 5.3: List of most common γ-emitting isotopes produced by proton reaction and used for scintigraphy.

Isotope	Reaction	Energy of γ-rad. (keV)	$T_{1/2}$ (h)
^{67}Ga	^{68}Zn(p, 2n)^{67}Ga	94, 184, 296	79
^{111}In	^{111}Cd(p,n)^{111}In	173, 247	67
^{123}I	^{124}Xe(p,2pn)^{123}I	159	13.1
^{201}Tl	^{203}Tl(p,3n)^{201}Pb → ^{201}Tl	x-Rays after EC 69–80	73

Tab. 5.4: Cyclotron produced β^+-emitters for PET and clinical research.

Isotope	Reaction	Emitter	$T_{1/2}$ (min)
^{11}C	^{14}N(p,α)^{11}C	β^+	20.4
^{13}N	^{16}O(p,α)^{13}N	β^+	10
^{15}O	^{14}N(d,n)^{15}O ^{15}N(p,n)^{15}O ^{16}O(p,np)^{15}O	β^+	2
^{18}F	^{18}O(p,n)^{18}F	β^+	110

All isotopes listed can be generated with proton energies up to 11 MeV.

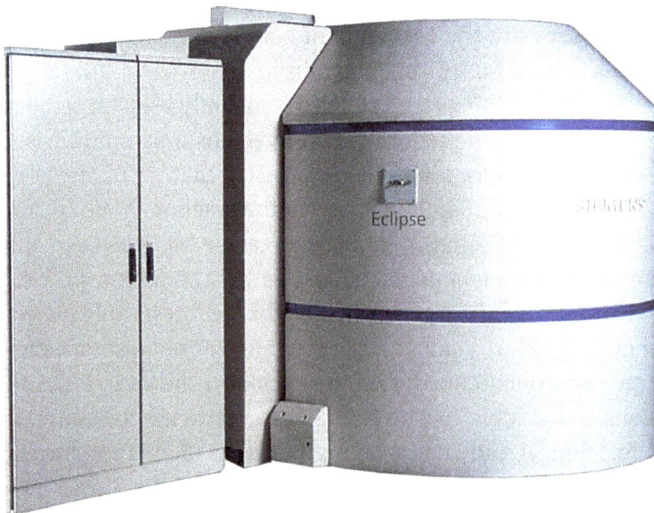

Fig. 5.14: Compact cyclotron "Eclipse" for production of short-lived radioisotopes in radiation clinics (reproduced with permission from www.siemens.com/press).

5.6.5 Radioisotope production by fission

Nuclear reactors specializing in research and isotope production "burn" isotope-enriched ^{235}U. By trapping thermal neutrons, the ^{235}U isotope is fragmented. In contrast to protons, neutrons do not have to penetrate a Coulomb barrier around the nucleon. Therefore, low-energy neutrons are sufficient for neutron capture (NC) and the subsequent fission reaction. The slower the neutrons, the longer the reaction time, and the more likely ^{235}U nuclei are excited to a degree of instability. The nuclear fission process splits the original nucleus into two unequal parts with the mass numbers centered at 95 and 135. The fragments of the original uranium nucleus also release further neutrons. Some of these neutrons are used for sustained fission reactions, and others can be used for neutron scattering experiments, irradiation, and isotope production.

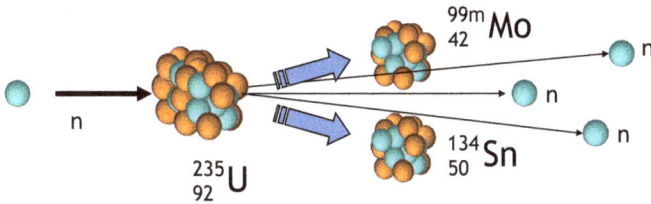

Fig. 5.15: Schematic illustration of the ^{235}U fission process.

There are several possibilities to produce ^{99}Mo/^{99m}Tc isotopes that are frequently used for scintigraphy: NC, proton irradiation, and fission. The last one has by far the highest cross section and is the most efficient process [9]. During the fission of ^{235}U, the isotope ^{99}Mo is released as one of the fission fragments, sketched in Fig. 5.15. About 6.1% of the fission products are ^{99}Mo. It has a half-life of 66 h and decays by β^--emission into the intermediate state ^{99m}Tc, which in turn has a half-life of 6 h and decays into ^{99}Tc by emission of 140 keV γ-radiation. This γ-radiation is utilized for scintigraphic imaging of the myocardium, and many other organs discussed in more detail in Chapter 9. ^{99}Tc is also a β^- – emitter, but with a very long half-life of 2.1×10^5 a, which can be regarded as stable on the timescale of a human life span. The biological lifetime is much shorter, on the order of 1 day via renal clearance. The decay scheme of ^{99}Mo/^{99m}Tc into ^{99}Tc/^{99}Ru is shown in Fig. 5.16.

Instead of extracting ^{99}Mo from nuclear fuel assemblies, it is more practical to irradiate a uranium metal containing about 20% enriched ^{235}U. The uranium metal sheet is placed in the reactor moderator to be exposed to thermal neutrons. After the ^{235}U fission process is saturated, typically after 6 days, the plate is removed and processed radiochemically in specialized isotope laboratories. The extracted ^{99}Mo/^{99m}Tc

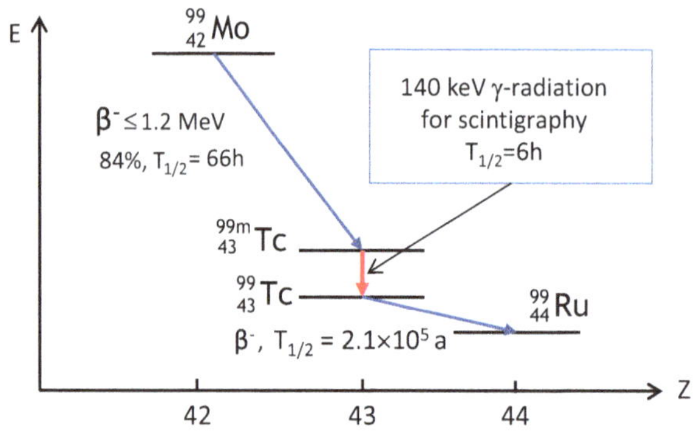

Fig. 5.16: Decay scheme of ^{99}Mo/^{99m}Tc into ^{99}Tc/^{99}Ru.

radioisotopes are divided into smaller activity quantities, safely deposited in generators, and finally sent to clinics for further elution of ^{99m}Tc (see Chapter 9 for details). Few nuclear reactors specialize in ^{235}U irradiation and few laboratories have facilities for chemical extraction of ^{99}Mo/^{99m}Tc. This is in stark contrast to the heavy use of ^{99m}Tc for scintigraphy, which is estimated at around 30 million scintigrams annually worldwide. The strong and increasing demand for this imaging modality would justify the establishment of additional centers for the production of ^{99}Mo/^{99m}Tc generators. One facility will soon be opened at the FRM II research reactor in Munich, Germany. On the other hand, it can also be argued that much of the imaging performed in the past by SPECT and PET has been taken over by specialized MRI techniques, such that the demand for ^{99}Mo/^{99m}Tc generators may decrease in the future.

5.6.6 Neutron activation

Stable isotopes can be converted into radioisotopes by NC. This process is called *neutron activation* and is characterized by an A(n, x)B reaction, where x stands for α-, β-, or γ- radiation. Neutron activation is used in many fields of science and technology, such as environmental science, geology, archeology, and materials sciences, to detect trace elements. Similar to the x-ray fluorescence analysis mentioned in Section 4.4.2, nuclear activation analysis enables a dry chemical analysis of isotopes and their abundances. This is because the emission spectrum, their energies, and half-lives are characteristic of the isotopes in the target [10]. Figure 5.17 shows a γ−emission spectrum as a function of photon energy and for two decay times after neutron exposure.

The cross section for NC is proportional to the reciprocal neutron velocity [12]: $1/v_n$. Thermal neutrons with energies of around 25 meV are predestined for such

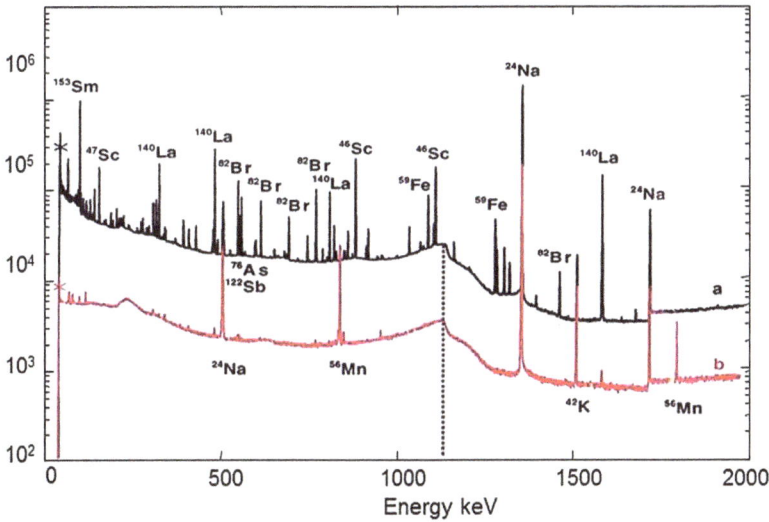

Fig. 5.17: γ-Emission spectra of a sample activated by thermal neutrons for 1 h. The spectra are recorded for isotopes with a decay time of 1 h (red) and a decay time of 1 day (black) (spectrum reproduced from MLZ at TUM, Munich, https://mlz-garching.de/naa/de).

reactions. Neutron activation is relatively simple and efficient. It only requires a "rabbit" system in a nuclear reactor. The samples are enclosed in carriers that are inserted into a tube leading close to a reactor vessel. After a standard activation time or activity level has been reached, the carrier is removed from the tube and further processed or analyzed. The relative ease of producing radioisotopes by NC also has a downside. Neutron activation is not particularly specific. Most of the heavier isotopes can be activated. For clinical applications, this poses a serious problem because of potential toxic contamination. Therefore, extreme care is required to activate only isotopically pure samples. A list of neutron capture cross sections is provided in [12].

Concerning clinical applications of neutron activation, we distinguish two variants: *in vivo activation* and *in vitro activation*. For in vivo activation, nonradioactive isotopes are first administered to the patient. The isotopes are then exposed to fast neutrons, which become moderated in the watery environment of the body and trigger (n, x) reactions locally in the tissue. The best known and successful application of this method is the BNCT via n, α-reaction presented in Chapter 4 of Volume 3.

In vitro thermal neutron activation is used for the production of a variety of nuclear decays [11], exemplified by the following reactions:

1. (n, γ) reaction; example: ${}^{98}_{42}\text{Mo} + \text{n} \rightarrow {}^{99}_{42}\text{Mo} + \gamma$; ($\sigma = 110$ mbarn)
2. $(n, \gamma) + \beta^-$-reaction; example: ${}^{130}_{52}\text{Te} + \text{n} \rightarrow {}^{131}_{52}\text{Te}^* + \gamma$; and ${}^{131}_{52}\text{Te}^* \rightarrow \beta^- + {}^{131}_{53}\text{I}$; ($\sigma = 67$ barn)
3. (n, α) reaction; example: ${}^{10}_{5}\text{B} + \text{n} \rightarrow {}^{7}_{3}\text{Li} + \alpha$ ($\sigma = 767$ barn)

The first example refers to the already presented ^{99}Mo/^{99m}Tc production. The cross section for this reaction is, though, rather small and therefore considered inefficient. Production via fission is preferred over thermal NC.

The second example is a sequential transition via a short-lived Te* radioisotope that decays further by β^--emission to the radioisotope ^{131}I. This isotope is also a beta-emitter with a half-life of 8 days, which is sometimes used for radiotherapy of the thyroid and less so for scintigraphy.

The third example shows the reaction already quoted for in-vivo BNCT in conjunction with thermal neutrons. This reaction is also used for detecting thermal neutrons by BF$_3$ gas-filled proportional counters or Geiger-Müller counters. The design of these counters is presented in Section 7.4.

Of clinical interest are some further isotopes produced by fission or NC. They are listed in Tab. 5.5 together with their main field of application. Here we distinguish between isotopes used for therapy and those for imaging. Radioisotopes of therapeutic interest should exhibit a high linear energy transfer (LET), high activity, and emit radiation of short range to affect mainly the targeted volume. In contrast, radioisotopes for imaging should be low LET radiation with a large range to be detected outside the body and to cause as little damage as possible in the body. All aspects concerning LET will be discussed extensively in later chapters.

Tab. 5.5: Radioisotopes generated either by nuclear fission or by neutron capture used in brachytherapy or scintigraphy.

Radioisotope	Reaction	Production	Application
^{137}Cs	^{235}U(n,γ)^{137}Cs→^{137}Ba	Fission	Brachytherapy
^{90}Sr	^{235}U(n,γ)^{90}Sr→^{90}Y	Fission	Brachytherapy
^{99m}Tc	^{235}U(n,γ)^{99}Mo→^{99m}Tc	Fission	Scintigraphy
^{133}Xe	^{235}U(n,γ)^{133}Xe	Fission	Scintigraphy
^{60}Co	^{59}Co(n,γ)^{60}Co→^{60}Ni	Neutron capture	Brachytherapy
^{192}Ir	^{191}Ir(n,γ)^{192}Ir→^{192}Pt	Neutron capture	Brachytherapy
^{125}I	^{124}Xe(n,e)^{125}I	Neutron capture	Brachytherapy
^{103}Pd	^{102}Pd(n,γ)^{103}Pd	Neutron capture	Brachytherapy
^{103}Pd	^{103}Ag(n,γ)^{103}Pd	Neutron capture	Brachytherapy

5.7 Summary

S5.1 Nucleons are composed of protons and neutrons.

S5.2 Atomic number A is the sum of protons Z and neutrons N; the nomenclature is ^{A}X, X stands for a chemical symbol.

S5.3 For light nucleons, the number of protons and neutrons is about equal. With increasing atomic number, nucleons tend to have more neutrons than protons.

S5.4   Isotopes are atoms with the same number of protons but different number of neutrons.

S5.5   Isotopes that have either too many or too few neutrons are unstable.

S5.6   All isotopes with atomic mass beyond lead (^{208}Pb) are unstable.

S5.7   Radioisotopes are unstable isotopes. They decay by emission of either α-, β^+, β^-, γ-radiation, or by EC from the K-shell, converting radioisotopes into stable isotopes.

S5.8   All radioisotopes decay according to the exponential decay law.

S5.9   Long-lived isotopes with half-lives in the range of the age of the Earth can still be found in nature.

S5.10   In decay chains, the production of daughter nuclei depends on the decay rate of mother nuclei.

S5.11   Radioisotopes can artificially be produced by proton irradiation in accelerators, by fission of heavy nucleons, and by thermal NC.

S5.12   The isotope ^{99m}Tc used for scintigraphy is extracted from fission products.

S5.13   Short-lived positron emitters for PET application are produced by proton bombardment in a cyclotron.

Questions   ?

Q5.1   Nuclei are composed of . . .?

Q5.2   How many quarks are in a nucleon?

Q5.3   How is the atomic number A defined?

Q5.4   What are isotopes?

Q5.5   When are isotopes stable, and when are they unstable?

Q5.6   What is the proper ratio of protons to neutrons for stable isotopes?

Q5.7   Unstable isotopes are radioactive isotopes. Name the two types of radioactive isotopes according to their production.

Q5.8   What is the definition of atomic mass unit?

Q5.9   Why is the atomic weight of chemical elements not an integer number of the atomic mass unit?

Q5.10   Which particles are emitted during radioactive decay of unstable isotopes?

Q5.11   Which of the decay products do not change the atomic number of the radioisotope?

Q5.12   What is the difference between lifetime of a radioisotope and half-life?

Q5.13   How is the activity defined?

Q5.14   What is the unit of the activity?

Q5.15   In the case of decay chains, how are the activities of parent and daughter radioisotopes related?

Q5.16   How can radioisotopes be produced artificially? Name at least three methods.

Q5.17   What method is used to produce the radioisotope ^{99m}Tc?

Q5.18   How do you explain the broad energy distribution for β-radiation?

⚡ Attained competence checker   +   0   –

I know how atomic nuclei are composed.

I appreciate the conditions that make nucleons unstable.

I am familiar with the three types of decay of radionuclides.

I know what the units are for nuclear activity.

I am aware of three methods of artificially producing radioactive isotopes.

I can distinguish between physical and biological lifetimes of radioisotopes.

I know how the half-life of radioisotopes can be calculated.

With a cascade of decays, it is clear to me what the activity of the daughter nuclei depends on.

ℹ Exercises

E5.1   **Activity:** The activity of a substance was 1000 Bq an hour ago. It is currently 900 Bq. How big will the activity be in an hour?

E5.2   **Biological half-life:** In a nuclear medicine examination, a radiopharmaceutical is used, which has an effective biological whole-body half-life of 6 h. The concerned patient wants to know how long it takes for the activity of the radioactive substance in their body to drop below 1%.

E5.3   **Radioactive oxygen isotope:** The radioactive oxygen isotope ^{15}O is often used in nuclear medicine. Its activity decreases to about 0.1% in 20 min. How long is its half-life?

E5.4   **Remaining isotopes:** A radioactive sample contains 10^{11} decayable isotopes. After their decay, stable isotopes are formed. The sample has an activity of 2000 Bq. About how many decayable isotopes are there after 5 h?

E5.5   **Reactor accident:** In the event of an accident in a nuclear power plant, small amounts of radioactive ^{134}Cs can be released, among other things. If this isotope gets into the human body, it is deposited there in the bones and can destroy the tissues through $\beta-$ and $\gamma-$ emission. The physical half-life of ^{134}Cs is 2 years. The biological half-life is 140 days.

     How long does it take for the body's radiation exposure to ^{134}Cs to drop to 12.5% or 1/8 of the initial activity?

E5.6   **Radioactive decay:** In nuclear reactors, radioactive nuclides N are generated by NC at a constant rate g (generated radioactive nuclei per second). The probability of decay per time unit (second) of the generated radioactive nuclide is λ.

     a. Derive an equation for the number of radioactive nuclei that are generated per unit of time.

     b. What is the maximum number of radioactive nuclei that can be generated under these conditions?

E5.7   **Activation of ^{60}Co:** A 1 g ^{60}Co sample with an activity of 100 MBq is requested for application in brachytherapy. Activation is performed in a nuclear reactor with a thermal neutron flux of 10^9 neutrons cm^{-2} s^{-1}. The target consists of isotopically clean ^{59}Co. The cross section for thermal NC in ^{59}Co is about 40 barn. For how long does the Co target need to be irradiated to reach the requested activity?

E5.8   **Cyclotron production rate:** A manufacturer of cyclotrons quotes that it can produce ^{18}F isotopes with an activity of 2000 mCi in 120 min at a proton beam current of 2 mA/cm^2.

a. Which nuclear reaction is used for the production of ^{18}F isotopes?
b. What is the most favorable projectile energy for this reaction? Consult the following graph.
c. What is the reaction rate used for this isotope production?
d. What is the reaction rate normalized to the proton beam intensity?

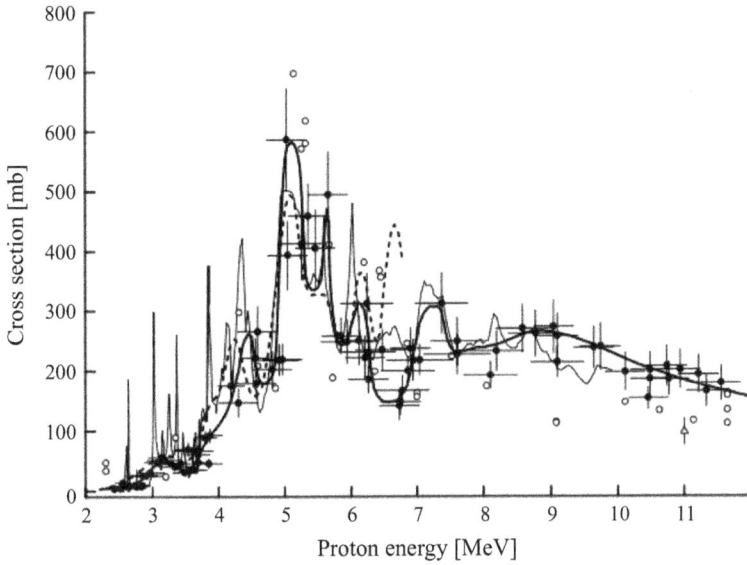

Cross section for the proton capture reaction ^{18}O(p, n)^{18}F (graph from [7]).

E5.9 **Isotope production in a neutron reactor:** A neutron facility is requested to produce a 100 μCi source of ^{36}Cl from a 1 g sample of natural nickel chloride. ^{35}NiCl$_2$ has a molecular weight of 129.6 and a natural abundance of 75.8%. The activation is performed by NC in a flux of 10^{14} cm^{-2} s^{-1}. The NC cross section for the reaction ^{35}Cl(n,γ)^{36}Cl is 43 barn and the half-life of ^{36}Cl is 3 × 10^5 years. How long will it take to produce such a radioactive source by NC?

E5.10 **Double decay:** In a decay chain, the nuclei A decays in the daughter nuclei B, which decays again in nuclei C. The decay probabilities of A and B are λ_A and λ_B respectively. Assume that $\lambda_B = 2\lambda_A$. Calculate the time when N_B has a maximum.

E5.11 **Prolonged time activity:** Show that when λ_B approaches λ_A, and after a prolonged time, the following relation holds:

$$\frac{N_B(t)}{N_A(t)} = \frac{\lambda_A}{\lambda_B - \lambda_A}$$

E5.12 **Multiple decays of radioactive substances:** A substance A disintegrates into another substance B, which is also radioactive and then decays into a stable substance C.
a. Set up the rate equation for the generation of substance B, where λ_A and λ_B are the decay constants of A and B, respectively; N_{A0} and N_{B0} are the number of nuclei at time $t = 0$.
b. Show that the rate equation leads to the inhomogeneous differential equation:

$$\frac{dN_B(t)}{dt} + \lambda_B N_B - \lambda_A N_{A0}\exp(-\lambda_A t) = 0$$

c. Show that the following equation is a solution of the inhomogeneous differential equation:

$$N_B(t) = \frac{\lambda_A}{\lambda_B - \lambda_A} N_{A0}(\exp(-\lambda_A t) - \exp(-\lambda_B t)) + N_{B0}\exp(-\lambda_B t)$$

d. Assume that the initial amount of substance B is zero. Sketch qualitatively the number of nuclei present in substances A, B, and C as a function of time for $\lambda_A = (1\ \text{h})^{-1}$ and $\lambda_B = (5\ \text{h})^{-1}$.

References

[1] Atomic weights and isotopic compositions for all elements: https://physics.nist.gov/cgi-bin/Compositions/stand_alone.pl
[2] https://www-nds.iaea.org/relnsd/vcharthtml/VChartHTML.html
[3] https://www-nds.iaea.org/relnsd/isotopia/isotopia.html
[4] Segre E. Nuclei and particles. An introduction to nuclear and subnuclear physics. New York, Amsterdam: W.A. Benjamin, Inc; 1965.
[5] Bateman H. Solution of a system of differential equations occurring in the theory of radioactive transformations. Proc Cambridge Phil Soc. 1910; 15: 423.
[6] https://physics.nist.gov/PhysRefData/Xcom/html/xcom1.html
[7] Hess E, Takacs S, Scholten B, Tarkanyi F, Coenen HH, Qaim SM. Excitation function of the $^{18}O(p,n)^{18}F$ nuclear reaction from threshold up to 30 MeV. Radiochim Acta. 2001; 89: 357–362.
[8] Schaffer P, Bénard F, Bernstein A, Buckley K, Celler A, Cockburn N, Corsaut J, et al. Direct production of ^{99m}Tc via $^{100}Mo(p,2n)$ on small medical cyclotrons. Phys Procedia. 2015; 66: 383–395.
[9] Organisation for Economic Co-operation and Development. The supply of medical radioisotopes. The path to reliability. Nuclear development. Paris: OECD Publishing; 2011.
[10] Ali MA. A brief overview of neutron activation analyses methodology and applications. 2nd conference on Nuclear and particle physics, Cairo; 1999. https://inis.iaea.org/collection/NCLCollectionStore/_Public/37/118/37118483.pdf
[11] Kim J, Narayan RJ, Lu X, Jay M. Neutron-activatable needles for radionuclide therapy of solid tumors. J Biomed Mater Res A. 2017; 105: 3273–3280.
[12] Kopecky J. Atlas of neutron capture cross sections. IAEA Nuclear Data Section; 1997.

Further reading

Segre E. Nuclei and particles. An introduction to nuclear and subnuclear physics. New York, Amsterdam: W.A. Benjamin, Inc; 1965.
Krane KS. Introductory nuclear physics. New York, London, Sydney, Toronto: John Wiley & Sons; 1988.
Lilley J. Nuclear physics, principles and applications. New York, London, Sydney, Toronto: Wiley & sons; 2013.

Cherry SR, Sorenson JA, Phelps ME. Nuclear medicine. 4th edition. Philadelphia, Pennsylvania: Elsevier; 2012.

Dance DR, Christofides S, Maidment ADA, McLean ID, Ng KH, eds. Diagnostic radiology physics – A handbook for teachers and students. Vienna: IAEA; 2014.

International Atomic Energy Agency (IAEA). Cyclotron produced radionuclides: Principles and practice. Technical Reports Series No. 465. Vienna: International Atomic Energy Agency (IAEA); 2008.

6 Interaction of radiation with matter

Physical properties of relevance for radiation - matter interaction	
Main interactions of photons	Photoelectric effect, Compton scattering, pair production
Range of photon beam	Not defined, exponential decay
Main interaction of protons with matter	Ionization, bremsstrahlung
Range of 5 MeV α-particles in air	5 cm
Interaction of electrons with matter	Ionization, Chernenkov radiation
Interaction of neutrons with matter	Collisions, kinetic energy exchange
Range of 2 MeV neutrons in water	4 cm
Compton wavelength $\lambda_C = h/m_e c$	2.43 pm
Treshold photon energy for pair production	1.022 MeV
Classical electron radius r_0	2.817 fm

6.1 Attenuation: Lambert-Beer law

In this chapter we consider the interaction of radiation with matter. This is a topic of fundamental physical interest and immense practical implications. Firstly, the analysis of the interaction of radiation with matter shows us many unusual physical processes such as photoelectric (PE) absorption, Compton scattering, and pair production. On the other hand, the analysis gives us information on how we can protect ourselves from dangerous radiation exposure, develop diagnostic procedures, and treat diseases, especially tumors. Here we mainly focus on the physical aspects of radiation interacting with matter. The clinical use of radiation is presented in the following chapters. Radiation protection is the subject of Chapter 7 on dosimetry. We start here with some basic definitions, which concern all types of radiation.

If a beam of α, β, γ-particles, neutrons, or electromagnetic (EM) radiation (photons) hits a target, three processes always take place regardless of the radiation type, sketched in Fig. 6.1(a):

1. Transmission;
2. Absorption;
3. Scattering.

The incident beam intensity I_0, also called flux, is defined as the number of particles N that impinge with perpendicular incidence on a slab of target material per cross-sectional area S and per time t:

https://doi.org/10.1515/9783110757095-006

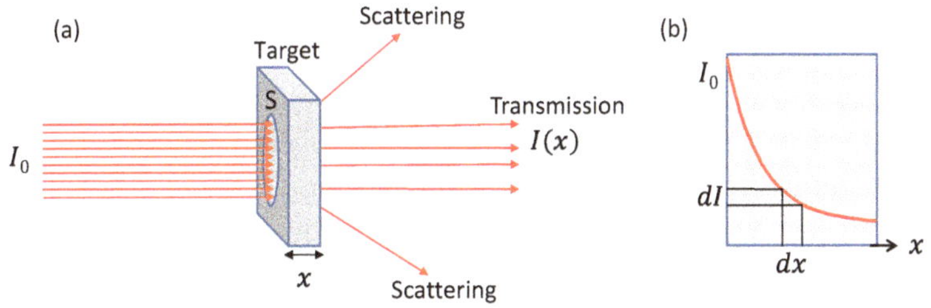

Fig. 6.1: (a) Incident particle beam becomes attenuated in a target by absorption and scattering; (b) exponential decay of the particle beam intensity by attenuation.

$$I_0 = \frac{N}{St} = \frac{\Phi}{t} . \tag{6.1}$$

The ratio

$$\Phi = \frac{N}{S} \tag{6.2}$$

is called particle *fluence*. The intensity (or flux) is therefore defined as fluence rate:

$$I = \frac{d\Phi}{dt} = \dot{\Phi} \tag{6.3}$$

$\dot{\Phi}$ is the first derivative of the fluence with respect to time. The SI unit of the intensity is $[I]$ = particle number $\times$ cm^{-2} $\times$ s^{-1}.

The definition of fluence and intensity applies to all types of radiation. The transmitted intensity $I(x)$ after passing a target of thickness x is, in general, attenuated. The attenuation is due to scattering or absorption. Scattering means that the particle trajectory has changed due to the interaction with the target. Absorption implies that the particle and its energy is converted to another particle and different energy (the total energy remaining conserved).

The incremental attenuation of the transmitted beam dI is proportional to the thickness of the target dx and the intensity $I(x)$ at that thickness (Fig. 6.1(b)):

$$dI = -\mu I(x)dx, \tag{6.4}$$

where the proportionality constant μ is called the linear attenuation coefficient (SI unit $[\mu]$ = m^{-1}). The negative sign expresses the incremental reduction of the intensity. Integration of (6.4) yields the *Lambert[1]–Beer[2] law* for the beam attenuation by a target of thickness x:

---

1 Johann Heinrich Lambert (1728–1777), Swiss-French mathematician and physicist.
2 August Beer (1825–1863), German physicist, chemist, and mathematician.

$$I(x) = I_0 \exp(-\mu x) \tag{6.5}$$

So far, so good. Any further analysis of the beam attenuation requires some knowledge about the nature of the particles impinging on the target and the target material. A distinction is made primarily between charged particles (electrons, positrons, protons, α-particles) and EM radiation (x-rays and γ-radiation). The interaction of neutrons with matter is again different and requires special consideration. In the following, we first consider the interaction of EM radiation with matter, followed by similar considerations for charged particles and neutrons.

> **!** attenuation = absorption + scattering

6.2 Interaction of EM radiation with matter

6.2.1 Attenuation coefficient of photons

We consider an EM plane wave with an electric field amplitude E_0 propagating in vacuum in the x-direction, described by:

$$E_{vac}(x) = E_0 \exp(ikx - \omega t) \tag{6.6}$$

$k = 2\pi/\lambda$ is the wavenumber. The time dependence of the oscillating field is not of interest and can be omitted in the following discussion. As the EM enters some material (gas, liquid, solid), the wavenumber changes due to refraction, such that in matter

$$E'(x) = E_0 \exp(ik'x - \omega t) \tag{6.7}$$

with $k' = kn$, and $n = k'/k$ are the refractive indexes. The refractive index of x-rays can be expressed by:

$$n = 1 - \delta + i\beta \tag{6.8}$$

The real part $1 - \delta < 1$ is smaller than 1 by only about one part in 10^{-6}. Therefore, lens optics with x-rays in the usual sense is not possible. However, the imaginary part that describes x-ray absorption can become rather large, in particular close to absorption edges. Now, we obtain for the EM–wave:

$$E'(x) = E_0 \exp(ik'x) = E_0 \exp(ik(1 - \delta)x) \exp(-k\beta x) \tag{6.9}$$

The intensity is the squared amplitude:

$$I'(x) = |E_0 \exp(ik(1-\delta)x)|^2 \exp(-2k\beta x) = I_0 \exp(-2k\beta x) \tag{6.10}$$

From the last equation, we find the attenuation coefficient:

$$\mu = 2k\beta = \frac{4\pi}{\lambda}\beta \tag{6.11}$$

β shows resonant behavior close to atomic absorption edges, as detailed further below in Section 6.2.3. Now we can derive again the attenuation coefficient:

$$-\frac{1}{x}\ln\frac{I}{I_0} = \mu, \tag{6.12}$$

which is the same expression as already stated in eq. (6.5).

For photons, the physics of absorption is contained in the linear attenuation coefficient μ. As the total attenuation depends on the density of the target material, μ is customarily normalized by the mass density for tabulation: μ/ρ_m (SI unit $[\mu/\rho]$ = m^2/kg), known as the *mass attenuation coefficient*. With this definition, the attenuation coefficient in a material with density ρ_x becomes:

$$\mu = \left(\frac{\mu}{\rho_m}\right)\rho_m \tag{6.13}$$

The relation between mass density $\rho_m = m/V$ and number density $\rho_N = N/V$ is:

$$\rho_N = \rho_m \frac{N_A}{A}, \tag{6.14}$$

where N_A is the Avogadro[3] number and A is the atomic mass. Now we may express the linear attenuation coefficient μ in terms of atomic units, i.e., as the product of the atomic number density ρ_N and the *atomic cross section* σ_a (SI units $[\sigma_a] = $ m^2):

$$\mu = \rho_N \sigma_a = \left(\rho_m \frac{N_A}{A}\right)\sigma_a. \tag{6.15}$$

and

$$\frac{\mu}{\rho_m} = \left(\frac{\rho_N}{\rho_m}\right)\sigma_a = \left(\frac{N_A}{A}\right)\sigma_a. \tag{6.16}$$

The atomic cross section σ_a is a measure of the range of interaction. As the name suggests, the cross section is the effective interaction size of an atom as seen by the

---

3 Lorenzo Romano Amedeo Carlo Avogadro (1776–1856), Italian physicist and chemist.

incoming particle. In the case of the strong nuclear force, the interaction range is very short, just limited to the size of a nucleus. However, in the case of Coulomb interaction, the interaction size is much larger and decays with the inverse distance.

In the following, we will discuss in more detail the mass absorption coefficients and the atomic cross sections for photons interacting with matter.

6.2.2 Mass attenuation coefficient of photons

When considering photon attenuation in matter, we distinguish four contributions:

$$\frac{\mu_{Ph}}{\rho_m} = \frac{1}{\rho_m}\left(\mu_{PE} + \mu_C + \mu_s + \mu_\pi\right) = \left(\frac{N_A}{A}\right)(\sigma_{PE} + \sigma_C + \sigma_\pi + \sigma_S) \tag{6.17}$$

The four different mass attenuation coefficients refer to the following effects:
- μ_{PE}/ρ_m stands for the PE effect, i.e., the complete absorption of x-ray photons and conversion into potential and kinetic energy of electrons.
- μ_C/ρ_m is the attenuation coefficient for Compton scattering; the Compton effect refers to inelastic scattering of x-ray photons at free electrons, where momentum and energy of the photon are partially transferred to electrons. Compton scattering is also called incoherent scattering and has to be distinguished from coherent scattering (Rayleigh or Thomson scattering).
- μ_s/ρ_m is the coefficient for coherent x-ray scattering. Coherent scattering is an elastic x-ray scattering process that does not change the energy of photons, only their direction. Coherent x-ray scattering is used for structural analysis.
- μ_π/ρ_m is the coefficient for pair production; high-energy x-ray photons with an energy exceeding the combined rest mass of electrons and positrons (≥ 1.022 MeV) can be converted into electron–positron pairs.

The PE effect and the pair production completely remove X-ray photons from the primary beam and convert their energy into another particle. We call this conversion absorption. In the Compton effect, the energy and direction of the X-ray photons change due to inelastic scattering, but the photon is not removed from the beam. The Compton-scattered photon is no longer in the transmitted beam because it is scattered in a different direction. Likewise, the elastic scattering stirs the beam in another direction but does not change its energy. Thus, the transmitted beam is weakened by two absorption processes and two scattering processes.

The x-ray mass attenuation coefficients μ_{Ph}/ρ are element-specific and depend on the photon energy. For pure elements, the coefficient as function of energy is listed in the x-ray data booklet [1] and is available in the internet [2]. In the case of compounds, μ_{Ph}/ρ_m is the sum over all components multiplied with appropriate weighting factors:

$$\frac{\mu_{Ph}}{\rho_m} = \sum_i w_i \left(\frac{\mu_{Ph}}{\rho_m}\right)_i \tag{6.18}$$

In the following subsections, we discuss the four contributions to the mass attenuation coefficients μ_{Ph}/ρ_m separately.

> The attenuation of photon beams (x-rays and γ-rays) is due to two absorption processes (PE effect and pair production) and two scattering processes (coherent Thomson scattering and incoherent Compton scattering). !

6.2.3 Photoelectric effect

The PE effect is a two-step process illustrated schematically in Fig. 6.2 and explained first by Einstein[4] in his famous publication dated from 1905 [4]. Here, the explanation of the PE effect is adapted to the inner core electrons. First, an x-ray photon is absorbed by an atom whenever the photon energy E_1 is at least the binding energy E_b of the electron: $E_1 \geq E_b$. Then the total x-ray energy is transferred to a core electron. The core electron may subsequently go to an unoccupied energy state in an outer shell or may leave the atom with a kinetic energy of:

$$E_{kin,e} = E_1 - E_b, \tag{6.19}$$

where $E_b = E_K$ is the binding energy of the K-shell electron. In the second step, the electron-hole left in the inner shell is filled by an electron from the next higher shell, as we have already discussed for the production of characteristic x-ray radiation via electron impact in Chapter 4. Assuming that a K-shell electron was kicked out and an electron filled the hole from the L-shell, the x-ray fluorescence radiation has the energy:

$$E_2 = E_K - E_L \tag{6.20}$$

E_2 always has lower energy than the initial photon energy: $E_2 < E_1$.

The linear x-ray mass attenuation coefficient for the PE effect $\mu_{ph}(E,Z)/\rho_m$ depends strongly on the energy of the incident x-rays and on the atomic number Z. The mass attenuation coefficient of Fe is plotted in Fig. 6.3 for illustration. Two aspects can immediately be recognized: (1) with increasing energy, the mass absorption coefficient decreases; (2) when the x-ray energy equals the binding energy of electrons in

---

4 Albert Einstein (1871–1955), German-US American physicist and Nobel Laureate in physics 1921.

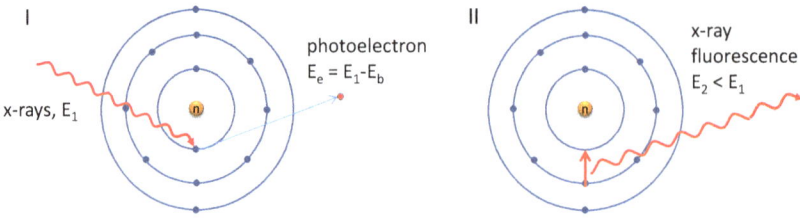

Fig. 6.2: The photoelectric effect takes place in two steps. First step: x-ray absorption and transfer of energy to an electron. Second step: filling of core-hole by an electron from the outer shell and emission of x-ray fluorescence radiation. n at the center stands for nucleus.

the K, L, or M–shell, the absorption coefficient shows a sharp, resonant-like increase. The mass attenuation coefficient of all chemical elements is tabulated in [1] and can also be found in the internet [2].

The absorption edges, as seen in Fig. 6.3, are important for contrast-enhanced x-ray radiography, which we will discuss in Chapter 8. In general, the dependence of $\mu_{\mathrm{PE}}(E, Z)/\rho_m$ on energy and atomic number Z of the target material is expressed in the following relation:

$$\frac{\mu_{\mathrm{PE}}}{\rho_m} \propto \lambda^3 Z^3. \tag{6.21}$$

Because of the strong Z dependence of the mass absorption coefficient, lead with a high Z ($Z = 82$) is used for protection against x-ray radiation, and barium ($Z = 56$) is used for contrast enhancement of x-ray radiographs of the gastrointestinal tract. In contrast, low Z beryllium ($Z = 4$) or alumininum ($Z = 13$) serve as x-ray "transparent" windows in vacuum chambers.

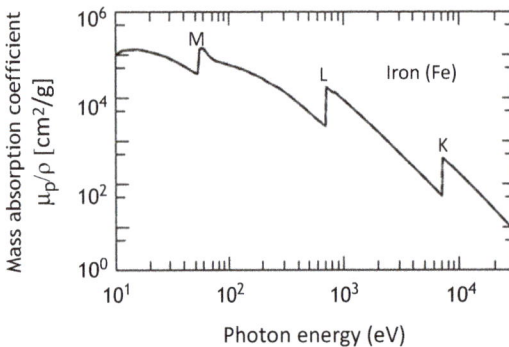

Fig. 6.3: Mass absorption coefficient for photons in Fe as a function of energy.

Next we want to discuss radiation damage caused by PE, in particular, in soft matter. To this end, we realize that the total energy of an x-ray photon is transferred to an electron in a single step. There is no partitioning of one photon energy to several electrons. Therefore, the kinetic energy of the photoelectron is determined by eq. (6.19): $E_{kin, e} = E_1 - E_b$.

In soft matter composed mainly of H, C, N, O, P, S, and a few light metals, the binding energies E_b of about 50–500 eV can be neglected compared to the high-energy x-ray photons used for radiography and radiotherapy. Therefore, when discussing radiation damage due to x-ray photons, we can set $E_{kin, e} \cong E_1$.

After PE energy conversion from photon to electron, the high-energy photoelectrons with some 100 keV and more kinetic energy will scatter strongly due to Coulomb interaction with ions in the material. Scattered photoelectrons will emit synchrotron radiation and excite other electrons in atoms from inner shells, generating new x-ray photons, etc. After a cascade of such events, including scattering, excitation, emission, and reabsorption, the photoelectrons will finally come to a stop. This is the same scenario that also β-particles experience in matter after radioactive decay, to be discussed further below. Because of the straggling process, high-energy photoelectrons in tissue have a range of a few tenths of millimeters. The biological damage done by x-ray absorption is entirely due to these high-energy photoelectrons and determines the quality factor Q of x-ray radiation in dosimetry, discussed in Chapter 7.

It should be noted that PE absorption in matter is a stochastic process that diminishes the x-ray intensity continuously and exponentially. However, it does not change the energy of those photons that have not been absorbed and are transmitted. This situation is illustrated in Fig. 6.4. Therefore, x-rays penetrating matter do not have a range in contrast to charged particles, but they have a half-thickness $x_{1/2}$ where the intensity has dropped to half of its original value:

$$x_{1/2} = \frac{\ln 2}{\left(\frac{\mu}{\rho}\right)\rho} \tag{6.22}$$

Attenuation of x-ray beams by the PE effect causes an exponentially decreasing intensity as a function of material thickness penetrated while the energy of the transmitted photons remains constant. The probability for PE absorption increases rapidly for x-ray energies close to absorption edges. !

6.2.4 Compton scattering

The second most important effect of x-rays interacting with matter is the *Compton effect or Compton scattering*. Compton scattering is an inelastic and incoherent scattering process of x-rays and γ-rays at free and resting electrons. Clearly, in matter there are no free or resting electrons. But viewed from the frame of a high-energy photon, a weakly bond electron is effectively free. With increasing energy between

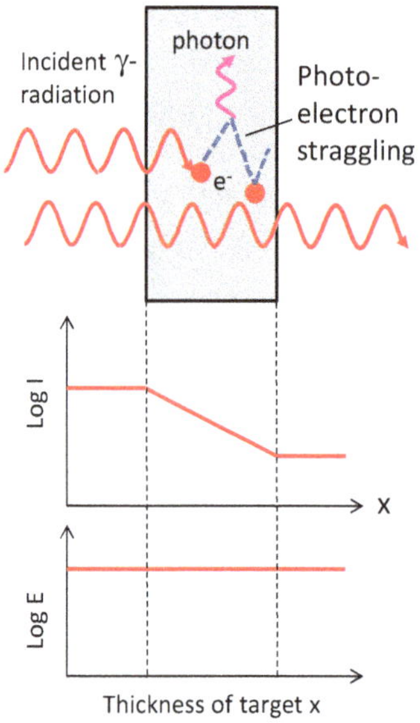

Fig. 6.4: Absorption of photons in matter via the photoelectric process. The photon intensity decreases exponentially with thickness x of the target, whereas the photon energy does not change. At the same time, high-energy photoelectrons are generated, which have a certain straggling distance in the material before coming to rest.

10^5 and 10^6 eV, Compton scattering becomes the dominating interaction process and is particularly important for scattering at light elements in soft matter.

In the classical experiment performed by Compton,[5] a monochromatic x-ray beam with a wavelength λ_i hits a target and the scattered beam is spectroscopically analyzed by a crystal analyzer for different scattering angles θ. At each scattering angle, two peaks are observed as function of wavelength: one that has the same energy and wavelength λ_i as the incoming beam and one that has a longer wavelength λ_f and a lower energy. The first peak is the elastically scattered beam, known as the Thomson[6] scattered beam, which is further discussed in the next section. The second peak is the inelastically scattered x-ray beam. During the scattering process, part of the photon energy has been transferred to the electron. The Compton scattering effect is clearly distinct from the PE effect. In Compton scattering, the photon is not absorbed as in the PE effect but rather scattered in a different direction. Nevertheless, it leads to an attenuation of the primary beam.

The conservation laws for momentum and energy describe the Compton effect completely. Conservation of momentum demands:

$$\vec{p}_{\lambda_i} = \vec{p}_{\lambda_f} + \vec{p}_e, \tag{6.23}$$

5 Sir Arthur Compton (1892–1962), US-American physicist and Nobel Laureate in Physics 1927.
6 Sir Joseph John Thomson (1856–1940), British physicist and Nobel prize recipient in Physics 1906).

(a)

(b)

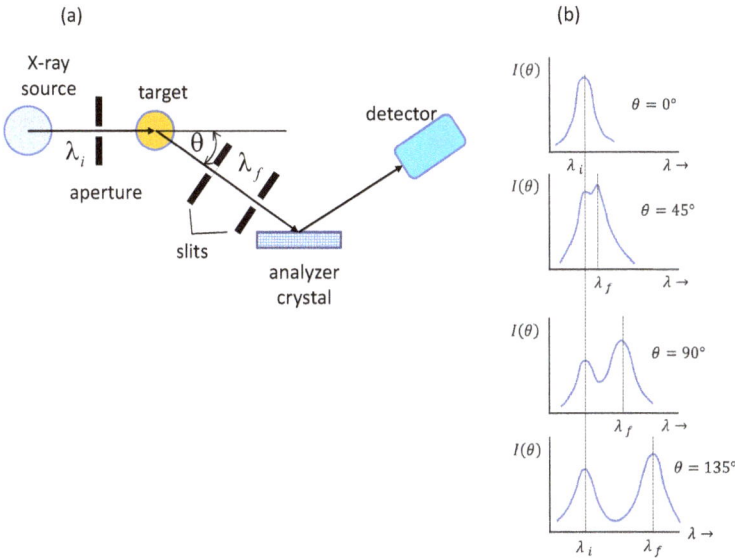

Fig. 6.5: (a) Experimental arrangement for the observation of the Compton effect; (b) x-ray spectra recorded at four scattering angles θ show two spectral lines: one at the unshifted wavelength of the incident beam λ_i and one at the longer wavelength λ_f. The separation between λ_i and λ_f increases with increasing scattering angle.

(a) $\alpha = 0.1$ (b) $\alpha = 5$

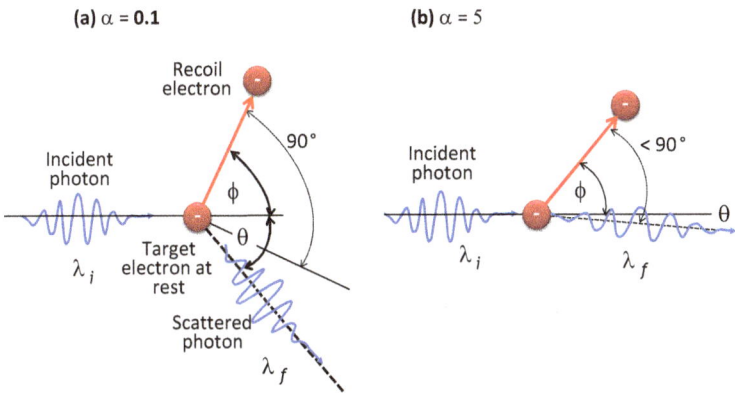

Fig. 6.6: Schematics of Compton scattering at low incident energy (panel (a)) and high energy (panel (b)). In panel (a) the scattering angle is arbitrarily set to $\theta = 50°$. The scattering angle ϕ of the recoil electron is determined by the conservation laws and is here 65°. Panel (b) illustrates the high-energy case where the scattering occurs mainly in the forward direction. Therefore, θ is set to 5°, and the recoil angle ϕ is 75°.

and conservation of energy requires:

$$E_i + E_{0,e} = E_f + E_e. \tag{6.24}$$

or

$$\frac{hc}{\lambda_i} + E_{0,e} = \frac{hc}{\lambda_f} + E_e. \tag{6.25}$$

Here $\lambda_{i,f}$ are the photon wavelengths before and after inelastic scattering, $E_{0,e} = m_e c^2$ is the rest energy of an electron, and E_e is the rest energy plus kinetic energy after scattering. Clearly, the electrons in atoms are never at rest. The condition is fulfilled, though, if the electron velocity is small compared to the photon speed of light c. Now working out the wavelength change $\Delta\lambda$ before and after scattering of the photons as a function of photon scattering angle θ yields:

$$\lambda_f - \lambda_i = \Delta\lambda = \frac{h}{m_e c}(1 - \cos\theta) = \lambda_C(1 - \cos\theta) \tag{6.26}$$

Here

$$\lambda_C = \frac{h}{m_e c} = 2.43 \text{pm}$$

is the *Compton wavelength*, which is a natural constant. The wavelength change implies an energy loss of the photon, and therefore the ratio of final to initial photon energy is:

$$\frac{E_f}{E_i} = \frac{\lambda_i}{\lambda_f} = \frac{1}{1 + \alpha(1 - \cos\theta)}. \tag{6.27}$$

The ratio:

$$\alpha = \frac{E_i}{m_e c^2} = \frac{\hbar\omega}{m_e c^2} \tag{6.28}$$

is also known as the *electron–photon coupling strength*.

The maximum energy loss of photons occurs at a scattering angle of $\theta = 180°$, causing a wavelength shift of $\Delta\lambda = 2\lambda_C$. At the same time, the energy transfer to the electron reaches a maximum. The electron energy follows from the difference $E_e = E_i - E_f$ and can be determined as:

$$\frac{E_e}{E_i} = \frac{\alpha(1 - \cos\theta)}{1 + \alpha(1 - \cos\theta)} \tag{6.29}$$

Both curves, eqs. (6.27) and (6.29), are plotted in Fig. 6.7 for $\alpha = 0.2$ and 10. The ratio $\alpha = 0.2$ corresponds to about 100 keVp x-rays used for radiography; $\alpha = 10$ corresponds to a photon energy of about 5 MeV, which is used for radiotherapy. We notice that for $\alpha = 10$ the photon energy is rapidly transferred to the electrons even at small scattering angles θ.

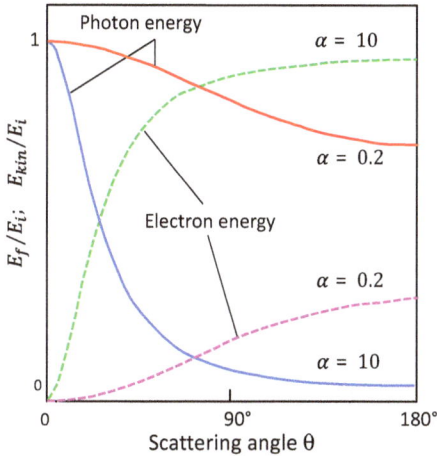

Fig. 6.7: Final photon energies (solid lines, red and blue) and electron energies (dashed lines, purple and green) plotted as function of the photon scattering angle θ and for two ratios $\alpha = E_i/m_e c^2 = 0.2$ and 10. $\alpha = 0.2$ is a typical value for radiography, and $\alpha = 10$ applies to x-ray radiotherapy.

For completeness, we also note the differential cross section for Compton scattering. This is the probability that a photon of energy $E_i = hc/\lambda_i$ is scattered into a solid angle $d\Omega = d\varphi d\theta$. The Compton scattering differential cross section, derived by Klein[7] and Nishina[8], is for an unpolarized incident x-ray beam:

$$\left(\frac{d\sigma}{d\Omega}\right)_{KN} = \frac{1}{2}Zr_0^2 \left(\frac{\lambda_f}{\lambda_i}\right)^2 \left(\frac{\lambda_f}{\lambda_i} + \frac{\lambda_i}{\lambda_f} - \sin^2\theta\right) \tag{6.30}$$

Here, $r_0 = e^2/m_e c^2$ is the classical electron radius and θ was defined before as the scattering angle of the photon in the scattering plane. The normalized differential Klein–Nishina cross section is plotted in Fig. 6.8 as a function of the scattering angle θ and for different coupling strengths α. The dip at 90° is a polarization effect. With increasing α, the angular distribution of the Compton-scattered photons is oriented in the forward direction.

The photon scattering angle θ and the electron scattering angle ϕ (Fig. 6.6) are related via:

---

7 Oskar Benjamin Klein (1894–1977) was a Swedish theoretical physicist.
8 Yoshio Nishina (1890–1951) was a Japanese theoretical physicist.

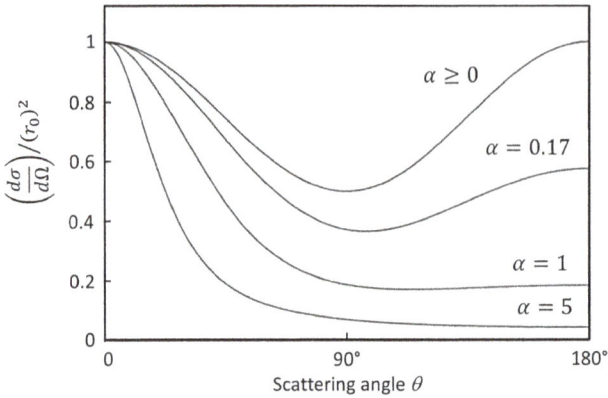

Fig. 6.8: Klein–Nishina differential cross section of Compton-scattered photons as a function of scattering angle θ and for different values of α. $\alpha = 0$ corresponds to elastic coherent Thomson scattering. $\alpha = 5$ refers to a photon energy of 2.5 MeV. Adapted from [9].

$$\tan(\phi) = \frac{\cos(\theta/2)}{1+\alpha} \tag{6.31}$$

For $\alpha \cong 0$ (Thomson scattering), these two scattering angles are simply related as:

$$\frac{\theta}{2} + \phi = 90° \tag{6.32}$$

However, with increasing α, the sum drops from 90° to lower angles.

Now we pull together all this information for a qualitative understanding of the Compton scattering process. We have seen above that Compton scattering produces "hot" electrons. According to eq. (6.30), high-energy photons scatter at small angles θ in a forward oriented cone with an opening angle of about 10°. Simultaneously, part of the photon energy is transferred to electrons (Fig. 6.7). The energy transfer produces "hot" electrons that scatter at angles of about 70° or less. These large angle recoil electrons are dose relevant, i.e., they cause biological damage by ionization processes in the tissue. The biological effects of radiation are discussed further in Chapter 7.

6.2.5 Coherent scattering of x-rays

We have seen in the previous section that the scattering of EM waves at free electrons has two components: inelastic or Compton scattering and elastic or Thomson scattering.[9] The scattering amplitude of Thomson scattering from a single electron is constant, isotropic and expressed by the *classical electron radius*:

$$r_0 = \frac{1}{4\pi\varepsilon_0}\frac{e_0^2}{m_e c^2} = 2.817 \times 10^{-15}\text{m}. \tag{6.33}$$

Here $\varepsilon_0\,(=8.854\ 10^{-12}\ \text{A}^2\text{s}^4/\text{kg m}^3)$ is the permittivity of free space, $e_0\,(=1.60210^{-19}\ \text{As})$ is the elementary charge, $m_e(=9.10910^{-31}\ \text{kg})$ is the electron mass, and c (=2.99 × 10^8 m/s) is the speed of EM waves in vacuum. The classical electron radius r_0 is not the true radius of an electron, which is unknown, but an effective radius that describes the interaction of EM radiation with electrons correctly.

Since x-ray wavelengths are in the order of atomic sizes, x-rays scatter from different parts of the electron distribution in an atom. The outgoing waves interfere, the more destructively the larger the scattering angle is (see Infobox on atomic form factor). The atomic form factor $f(\theta)$ takes into account the destructive interference with increasing θ, but for $\theta = 0$ all waves are in phase and therefore $f(0) = Z$, where Z is the atomic number. The total coherent scattering intensity into a cone of opening angle 2θ is:

$$I(\theta) = \left(\frac{r_0}{r}\right)^2 f^2(\theta)P(\phi). \tag{6.34}$$

Here r is the distance between target and detector, and $P(\phi) = (1 + \cos^2\phi)/2$ takes into account the polarization factor of the transverse EM wave.

Elastic coherent x-ray scattering is very important for the structural analysis of hard and soft materials and is the basis of crystallography. In the present context, coherent scattering is mainly manifested as small-angle x-ray scattering (SAXS). SAXS is an elastic and coherent scattering process of EM waves at density fluctuations in the target. The scattering occurs at particles with a different electron density ρ_p than the average electron density $\langle\rho\rangle$ of the surrounding medium in which these particles are embedded, illustrated in Fig. 6.9. For instance, if the target is homogenous water, no small-angle scattering occurs; but if water contains erythrocytes with a higher density than water, SAXS can be observed. The scattering intensity $I(\theta)$ as a function of scattering angle θ drops off with the fourth power of θ and scales with the number density of particles n_p in the matrix, the squared difference in electron density between particles (erythrocytes) and surrounding medium (water) $(\langle\rho\rangle - \rho_p)^2$, and the fourth power of the ratio: wavelength λ to gyration radius R_g of the particle $(\lambda/R_g)^4$. The SAXS intensity then approximates to [3]:

$$I(\theta) = \left(\frac{r_0 Z}{r}\right)^2 n_p \left(\langle\rho\rangle - \rho_p\right)^2 \frac{1}{\theta^4} \left(\frac{\lambda}{R_g}\right)^4 \tag{6.35}$$

---

9 This section is only of minor radiological interest and can be skipped. However, it is included for completeness and the interested reader.

In all cases relevant for x-ray radiography, the x-ray wavelengths range from 0.1 nm (10 keV) to 0.01 nm (100 keV), whereas the particle radius of cells is on the order of a micrometer. The ratio (λ/R_g) is therefore in the order of $10^{-4} - 10^{-5}$, indicating that SAXS drops off quickly with increasing x-ray energy (decreasing x-ray wavelength) and takes place at very small angles of about 0.05°. Nevertheless, SAXS can cause blurring of high-resolution x-ray images when all other geometric effects have been considered. SAXS leads to apparent attenuation of the primary incident beam intensity if the recording detector can discriminate between parallel beams and those which make a small angle θ with respect to the parallel orientation.

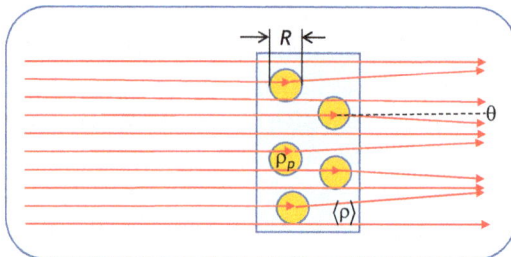

Fig. 6.9: Small angle x-ray scattering at particles of radius R that have a different electron density than the surrounding.

> [!] Coherent x-ray scattering changes the direction of the x-ray beam but not the energy of the x-ray photons. Small-angle coherent x-ray scattering may blur x-ray radiograhs.

> [i] **Infobox I: Atomic form factor**
>
> In general, the atomic form factor $f(Q)$ is expressed as an integral over the electron density distribution $\rho_{el}(R_e)$ in an atom, where $Q = (4\pi/\lambda)\sin\theta$ is the scattering vector. For small scattering angles $Q \approx (4\pi/\lambda)\theta$. The integral over the electron density is not a simple volume integral. A volume integral $\int \rho_{el}(R_e)dV$ would just recover the total number of electrons in an atom, i.e., the atomic number Z. Instead, the atomic form factor represents the Fourier transform of the electron density distribution within atoms or ions:
>
> $$f(Q) = \int \rho_{el}(R_e)e^{(iQ \cdot R_e)}dR_e.$$
>
> Here R_e is a radial vector of the electrons within an atom. The Fourier transform of the electron density yields the effective number of electrons that the x-rays are scattered from for increasing scattering vectors Q, starting from $Q = 0$ up to higher scattering vectors (or higher scattering angles θ). A schematic dependence of the scattered intensity $I(2\theta) = (f(Q))^2$ is shown below. Note that in the forward direction the intensity scales with Z^2, whereas in the backward direction $(2\theta = 180°)$ the intensity drops to zero.

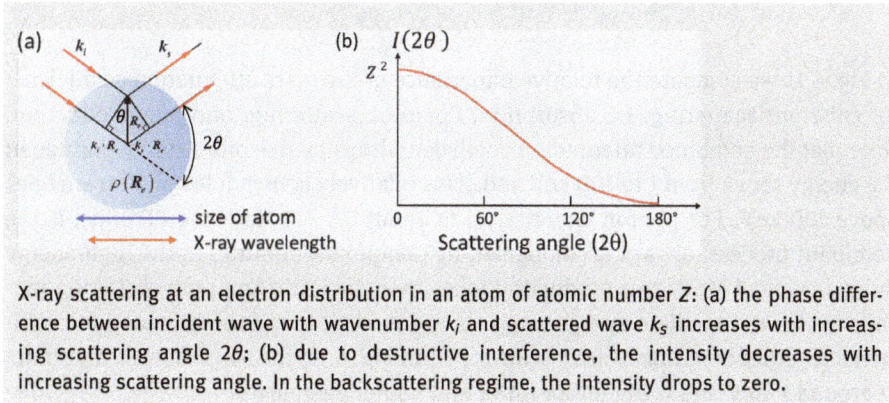

X-ray scattering at an electron distribution in an atom of atomic number Z: (a) the phase difference between incident wave with wavenumber k_i and scattered wave k_s increases with increasing scattering angle 2θ; (b) due to destructive interference, the intensity decreases with increasing scattering angle. In the backscattering regime, the intensity drops to zero.

6.2.6 Pair production

Pair production is a relativistic process. If the photon energy E_γ is larger than twice the rest mass of an electron, then in the Coulomb field of a nucleus the photon energy can be converted into a pair of particle and antiparticle, i.e., in an electron-positron pair. Electrons and positrons are equal in all physical aspects, but they have opposite charges and consequently opposite magnetic moments. The onset energy for pair production is

$$E_\gamma \geq 2m_ec^2 = 1.022\,\text{MeV}. \tag{6.36}$$

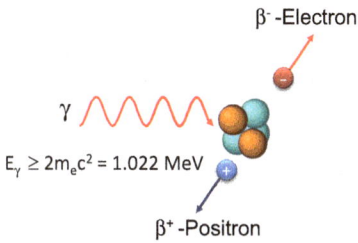

Fig. 6.10: In the field of a nucleus a high-energy photon may be converted into a pair of electron and positron. The threshold energy for pair production is 1.022 MeV.

Any photon energy in excess of the threshold energy is diverted in equal amounts to the electron–positron pair, which then fly apart at an angle of 180° for reasons of momentum conservation, illustrated in Fig. 6.10. The inverse process is the annihilation of positrons and electrons via converting their rest mass into two opposing γ – photons. Pair annihilation is utilized for imaging tumors and further discussed in Chapter 10 on positron emission tomography (PET).

6.2.7 Comparison of photon–electron interactions

In Fig. 6.11 we compare the relative importance of the mass attenuation coefficients for coherent scattering, PE absorption, Compton scattering, and pair production. Note that the combined attenuation coefficient drops by five orders of magnitude in the energy range from 1 to 100 keV and stays relatively constant for photon energies above 100 keV. For photon energies up to about 0.5 – 1 MeV, the PE effect is the dominant process, always accompanied by Compton scattering. In the high-energy range beyond 1 MeV, pair production takes over. Inelastic Compton scattering prevails in the intermediate energy range. If one puts all four mass attenuation coefficients together on a logarithmic scale for Fe in Fig. 6.11, the sum shows a minimum at around 1 MeV and is higher for lower and higher energies.

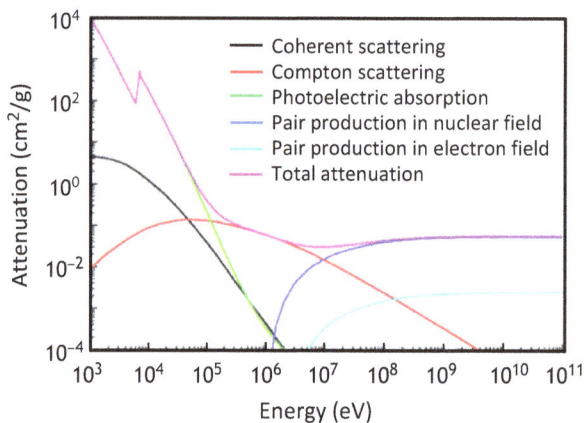

Fig. 6.11: Mass attenuation coefficients for photons plotted as function of energy. The sharp absorption edge at 7.1 keV refers to the K-edge of Fe. (Adapted from www.wikiwand.com, © creative commons).

The important point to realize is the fact that x-ray attenuation by any of the discussed means does not simply reduce the intensity. Attenuation implies that radiation is converted into other types of energy and often produces hot electrons that may have high biological effectiveness. For instance, when x-ray radiotherapy is performed with 20 MeV photons and higher, the photons are converted by equal amounts to hot Compton electrons and hot electron/positron pairs. The hot electrons are, subsequentially, of biological relevance.

In summary, the attenuation of x-rays in the range of 50 to 150 keV used for radiography is mainly determined by the PE effect and Compton scattering. For x-ray radiotherapy with energies beyond 4 MeV the PE effect plays no role. At these high photon energies, only Compton electrons and electron–positron pairs contribute to the therapeutic effect. The mass attenuation coefficients for the energy range from 1 keV to 10^2 MeV are tabulated and plotted in [8].

The attenuation of an x-ray beam passing through a slab of material is diminished by the PE !
effect at low photon energies (0–200 keV), by the Compton effect at intermediate energies
(100 keV–10 MeV), and by pair-production at high photon energies (1–20 MeV).

6.3 Interaction of charged particles with matter

6.3.1 Alpha particles

Alpha (α)-particles originate from the nuclear decay of heavy isotopes. After emission, these charged particles have a very high kinetic energy of around 5 MeV.
When penetrating matter, α-particles have, in contrast to photons, a well-defined
range. Figure 6.12 impressively shows the range of α-particles in air that are emitted
by a polonium source in the center. The picture was taken with a cloud chamber
that traces the straight paths of the α-particles until they suddenly came to a stand-
still at a very short distance. The total path length in the air is typically 5 cm and
the straggling at the end is no more than 5% of the total path length.

Fig. 6.12: Alpha–particle emission during decay of
^{212}Po to ^{208}Pb visualized with the help of a cloud
chamber. (Reproduced from W. Finkelnburg,
Introduction to Atomic Physics, Springer [5]).

First, we consider the energy loss of α-particles when penetrating matter. Because of
its charge, the primary interaction with matter is of the Coulomb type, which leads to
inelastic scattering with electrons. Since the mass ratio of α-particles to electrons is
very large (approx. 8000), each collision with an electron removes only little energy
from the α-particle, roughly 30 eV per collision. While electrons are scattered over a
wide angular range, the path of the α-particles remains more or less straight. This
situation is visible in Fig. 6.12 and sketched in Fig. 6.13.

In Fig. 6.14 the number of charged particles N in a particle beam is schematically plotted as function of distance x for a given material and the average energy
loss $\langle -dE/dx \rangle$ per distance x:

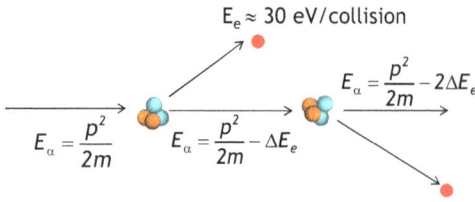

$E_e \approx 30$ eV/collision

$E_\alpha = \dfrac{p^2}{2m}$

$E_\alpha = \dfrac{p^2}{2m} - \Delta E_e$

$E_\alpha = \dfrac{p^2}{2m} - 2\Delta E_e$

Fig. 6.13: α-Particles collide with electrons and loose little by little energy until they are finally stopped after traversing a distance R.

$$\text{LET} = \left\langle -\frac{dE}{dx} \right\rangle \tag{6.37}$$

is called *linear energy transfer* (LET). LET is the rate at which energy is deposited by a particular type of radiation when passing through matter. The energy loss may take place by collision or radiation. The unit of LET is usually keV/μm. The minus sign is introduced so that LET is positive. LET is an essential property for radiotherapy, particularly for targeting tumors with photon or proton beams; more about that in Chapter 3 of Volume 3.

We notice that $N(x)$ in Fig. 6.14 stays relatively constant over a long distance and then suddenly drops to zero. The distance where the number of particles drops to half the original value is called the average *range* of charged particles $\langle R \rangle$. Over the same range, the LET $\langle -dE/dx \rangle$ is rather low and constant initially but then increases rapidly where the number of particles starts to decrease, followed by a sharp drop. The peak in $\langle -dE/dx \rangle$ close to the end of the particle track is called *Bragg peak*[10] and also defines the range $\langle R \rangle$ of charged particles. The full width of the peak at half maximum is the *straggling* distance σ over which the particles finally stop. A typical range for 5 MeV α-particles in air at 1 bar is 5 cm, and the straggling distance is about 2.5 mm.

The energy loss per distance traveled of monoenergetic charged particles by passing through matter with the atomic number Z is described approximately by the Bethe[11]-Bloch[12] equation. Assuming that energy loss takes place by ionization of the target material, this equation reads [6, 7]:

---

10 The Bragg peak here should not to be confused with the Bragg peak in coherent x-ray diffraction, but the discoverer is the same William Henry Bragg (1862–1942), who together with his son William Lawrence Bragg (1890–1971), received the Nobel prize in Physics for their x-ray diffraction work in 1915.

11 Hans Albrecht Bethe (1906–2005), German-US American physicist and Nobel laureat in physics 1967.

12 Felix Bloch (1905–1983 in), Swiss-Austrian-US American physicist and Nobel prize winner 1952.

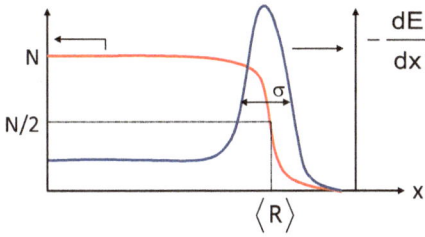

Fig. 6.14: Range of charged particles in matter. N is the number of particles in the beam and $-dE/dx$ is the energy loss per distance traveled, known as linear energy transfer (LET).

$$-\frac{dE_{ion}}{dx} = \frac{4\pi k_B Z_{eff}^2 e^4 n_e}{mc^2\beta^2} \ln\left(\frac{2mc^2\beta^2}{I(1-\beta^2)} - \beta^2\right).\tag{6.38}$$

k_B is the Boltzmann constant, $\beta = v/c$, Z_{eff}, m, v are the projectile effective charge, mass, and velocity, respectively. n_e is the electon density of the target material. I is the mean ionization energy of the light atoms in the target. The ionizing energy loss per distance traveled is also called *stopping power* S (units: MeV/cm):

$$S = -\frac{dE_{ion}}{dx}\tag{6.39}$$

Although the expressions look the same, there is a subtle difference between LET and S. LET includes all types of energy losses, while S excludes losses due to EM radiation. In table work, the *stopping power* of charged particles in matter is usually provided normalized by the mass density ρ_m of the material [8]:

$$\frac{S}{\rho_m} = -\frac{1}{\rho_m}\frac{dE_{ion}}{dx},\tag{6.40}$$

The SI units of S/ρ_m are [MeV cm^2/g]. The stopping power is a physical concept, assuming laboratory conditions. For the radiobiology, LET is more relevant than S as it is closer to biological conditions. On the other hand, integrating S yields the range of ionizing particles, whereas the range under LET conditions can only be determined empirically using phantoms or via Monte Carlo simulations. In the following, we will determine the average range $\langle R \rangle$ by integration of S.

The projectile continuously loses energy along its track by small amounts. When it has slowed down, the energy transfer to the target becomes most effective, and therefore, the energy loss per distance increases rapidly toward the end of the track. The total energy loss results from integrating over the projectile path from start to rest at $\langle R \rangle$:

$$E_{tot} = \int_0^{\langle R \rangle}\left(-\frac{dE}{dx}\right)dx.\tag{6.41}$$

The total energy is equivalent to the initial energy of the charged particle immediately before penetrating some slab of material with density ρ. Integration over the distance traveled before stopping yields the *range* $\langle R \rangle$. In general, the average range is obtained by integration of the stopping power from start to stop:

$$\langle R \rangle = - \int_{E_0}^{0} \left(\frac{dE}{dx} \right)^{-1} dE. \tag{6.42}$$

For the range of α-particles in the air at standard temperature, pressure, and humidity conditions, an empirical equation is quoted as follows [7, 9]:

$$\langle R \rangle_{\alpha, \text{air}} = 0.325 E^{3/2}. \tag{6.43}$$

Here the energy is in MeV and the range in centimeters. For scaling from air to another material, the following conversion is used, known as Bragg-Kleeman rule:

$$\langle R_{\alpha, 1} \rangle = \frac{\rho_{\text{air}}}{\rho_1} \frac{\sqrt{A_1}}{\sqrt{A_{\text{air}}}} \langle R_{\alpha, \text{air}} \rangle. \tag{6.44}$$

The symbols A_1 and ρ_1 are the respective average atomic numbers and mass densities for air and material 1. For air the effective numbers are: $A_{\text{air}} = 14.6$ and $\rho_{\text{air}} = 0.00127 \, \text{g/cm}^3$ (STP-conditions), and for tissue: $A_{\text{tissue}} = 9$ and $\rho_{\text{tissue}} = 1 \, \text{g/cm}^3$. The range equation can also be adapted to other charged particles, like protons or carbon ions. For conversion to protons we find [6.6, 6.7]:

$$\langle R_p \rangle = \frac{m_p}{m_\alpha} \left(\frac{Z_\alpha}{Z_p} \right)^2 \langle R_\alpha \rangle - k = 1.0072 \langle R_\alpha \rangle - k, \tag{6.45}$$

where k (~20 cm) is a constant accounting for differences in ionization energy of electrons via protons and α-particles. These empirical equations depend on the particle energy and have to be rescaled for specific energy ranges. In any case, the important point to recognize is the fact that heavy charged particles have a well-defined range $\langle R \rangle$ in materials of density ρ, which scales with the initial particle kinetic energy E:

$$\langle R \rangle \sim E^\delta, \tag{6.46}$$

where δ is an appropriate exponent ranging between 1.5 and 1.8. This relationship implies firstly that the range of charged particles in tissue can be tuned by the initial kinetic particle energy and, secondly, that most of the particle energy is deposited within this range. The ability to adjust the position of the Bragg peak in the target area using the incident particle energy is one of the great advantages of particle beam therapy over x-rays. This is discussed in more detail in Chapter 3 of Volume 3 on proton therapy.

Heavy charged particles (p, α, C) have a short range in matter and abruptly stop, when most of their energy is deposited within a narrow range known as Bragg peak. !

6.3.2 Beta-particles

Light β-particles are also charged. But their behavior is quite different from α-particles and protons due to their much smaller mass. They may collide with other electrons in the material elastically or inelastically, i.e., without or with energy transfer. In either case, scattering implies a large angular change of the trajectory. Deflection of the beam from the original path entails centripetal acceleration, which, in turn, generates synchrotron radiation. Fast β-particles lose their energy quickly through the emission of photons, either via Cherenkov[13] radiation or synchrotron radiation or by ionization of the target material. Cherenkov radiation occurs whenever the speed of charged particles v_p in a medium with refractive index n is higher than the phase velocity of light in the same medium: $v_p > c/n$. Cherenkov radiation is the light analog to supersonic sound propagation. Like sound waves, the Cherenkov radiation is emitted in a cone behind the traveling particle with an opening angle:

$$\cos\theta_c = \frac{c}{nv_p}.$$

(6.47)

The energy loss through emission of synchrotron radiation and Chernenkov radiation is proportional to:

$$-\frac{dE_\beta}{dx} \propto Z^2 E_\beta,$$

(6.48)

whereas energy loss through ionization has the form:

$$-\frac{dE_\beta}{dx} \propto Z \ln(E_\beta).$$

(6.49)

This shows that the energy loss through radiation is more dominant, particularly for high Z target materials.

Similar to α-radiation, β-particles have a maximum range $\langle R \rangle$. But in contrast to α-particles, the range is much longer and the straggling is broader. The range of β-particles depends on their primary energy and the mass density ρ of the

---

13 Pavel Alexesyevich Chernenkov (1904–1990), Russian physicist and Nobel laureate of physics in 1958.

target. An approximate formula for the *range of electrons* with energies between 3 and 20 MeV is [9, 10]:

$$\langle R_{50}[\text{cm}]\rangle = \frac{1}{\rho}(0.53\,E[\text{MeV}] - 0.106),\qquad(6.50)$$

where $\langle R_{50}\rangle$ is the distance at which the primary intensity has dropped to 50%. $\langle R_{50}\rangle$ is measured in centimeters and E in MeV. The range of 1 MeV electrons in air is about 3.5 m, whereas in tissues, the range is only 4 to 5 mm.

The greater straggling of electrons compared to the heavier charged particles (p, α) is partly due to the greater energy distribution of the β-particles since in β-decay the energy is shared with a neutrino. At the end of their range, β^+-particles annihilate with an electron from the target and emit their combined rest mass in the form of two high-energy photons traveling in opposite directions. PET uses the annihilation method for imaging tumors described in Chapter 10.

In summary, about the range of α-, β-, and γ-particles we can state the following: α-particles have a well-defined range; β-particles with the same energy have a much larger range and their straggling is more extensive; γ-particles and x-rays have no range, but a well-defined half-value thickness. Typical ranges are 5 cm in air for α-particles compared to 5 m for β-particles, but only a slight attenuation of γ-radiation in air. These different ranges are sketched in Fig. 6.15 and listed in Tab. 6.1. To determine the range of particles in tissue is more complicated. Usually, a water tank is used as a phantom target to simulate tissues. In the tank movable detectors are inserted to measure the range for different particles as function of incident energy. This is further discussed in Chapters 2 and 3 of Volume 3.

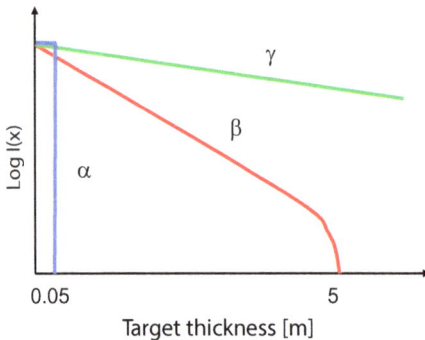

Fig. 6.15: Qualitative overview of the range of different particles with initial energy of about 1 MeV in air.

6.4 Interaction of neutrons with matter

Neutrons are charge neutral particles as the name indicates. Therefore neutrons lack Coulomb interaction and cannot lose energy by ionization of matter. From a radiation point of view, free neutrons have two possibilities to interact with matter: collision with other nuclei and absorption. Which interaction dominates depends on the neutron energy. Collisions prevail at high energies; resonant absorption becomes more likely in some nuclei at low energies. We distinguish between thermal neutrons, epithermal neutrons, and fast neutrons. Thermal neutrons have energy in the order of room temperature or about $1/40$ eV $= 25$ meV. Epithermal neutrons are all those with energy between 0.1 eV and 1 MeV, fast neutrons have energies beyond 1 MeV.

When fast neutrons with initial energy E_0 penetrate a target, they collide with other nuclei in the target and scatter. Assuming billiard-type collisions, the energy transfer solely depends on the scattering angle θ and mass number A of the target atom (see Fig. 6.16). In the forward direction ($\theta = 0$), no energy is transferred, i.e., the neutron energy does not change. For backscattering ($\theta = \pi$), the energy transfer is biggest and the neutron energy after collision is:

$$E_n^{\theta = \pi} = E_0 \left(\frac{A-1}{A+1}\right)^2, \tag{6.51}$$

Therefore the energy change per collision is:

$$\Delta E_n^{\theta = \pi} = E_0 \frac{4A}{(A+1)^2} \tag{6.52}$$

Since only part of the collisions lead to backscattering, the average energy loss per collision can be estimated as:

$$\langle \Delta E_n^{\theta} \rangle = E_0 \frac{2A}{(A+1)^2} \tag{6.53}$$

Obviously, the energy transfer to the target material is most effective for $A = 1$ (protons) and for atoms with small mass numbers A. After the first collision with the target atoms, the energy of the scattered neutrons is equally distributed between the maximal and the minimal value, i.e., no scattering angle is preferred. However, after further collisions, the neutron energy decreases step by step, reaching a value corresponding to the target atoms' thermal energy. In the end, the neutrons have a Maxwellian energy distribution centered at the thermal energy $3k_B T/2$. If the target consists of pure hydrogen, only six collisions are required before reaching thermal energy. Neutrons with energy of 2 MeV, typical for fission neutrons in nuclear reactors, need about 80 collisions in water and a range of about 40 mm before being thermalized. The process of slowing down fast neutrons by elastic collisions is called *moderation*. Obviously, hydrogen-rich materials such as water or paraffin are very effective neutron moderators. This is also the case when neutrons penetrate tissue. Radiation therapy with fast neutrons utilizes neutron moderation in tissue, a topic discussed further in Chapter 4 in Volume 3.

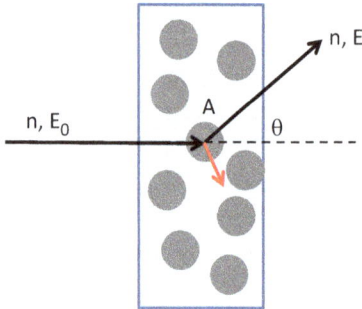

Fig. 6.16: Scattering geometry for neutrons hitting a target that consists of nuclei of mass number A. θ is the scattering angle.

As fast neutrons reach low energies by moderation, they are often captured by isotopes. The absorption cross section of neutrons as a function of neutron energy usually has three characteristic regions, illustrated schematically in Fig. 6.17 for a generic isotope: at $E < 1$ eV the absorption cross section decreases steadily with increasing neutron energy. In this region, the absorption cross section is inversely proportional to the neutron velocity v_n. This region is therefore referred to as the $1/v_n$ region. The epithermal energy range beyond 1 eV is called the *resonance region* [11]. In this region, the cross section raises sharply to high values in narrow energy bands. These sharp peaks in the absorption cross section are termed *resonance peaks*. They occur whenever the affinity of a nucleus for neutrons closely matches discrete quantum energy levels, i.e., when the combined binding energy plus kinetic energy of a neutron equals the energy difference between the ground state and an excited state. Absorption becomes negligible beyond the resonance regime, and elastic scattering of fast neutrons prevails, as discussed above. Neutron activation of isotopes by thermal neutrons is presented in Chapter 5 of this volume and plays an important role for brachytherapy, discussed in Chapter 5 of Volume 3.

Summarizing the various interactions of radiation with matter in an easy to remember comparison goes as follows: α-particles are stopped by a shirt, β-particles are slowed down within the thickness of the skin, fast neutrons are moderated to thermal energies within the top 4 cm of the body, and γ-radiation will penetrate the whole body, as illustrated in Fig. 6.18.

> **!** High-energy fast neutrons slow down by collisions with moderator atoms or molecules such as He or H_2O. Thermal neutrons are important for sustaining fission reactions in nuclear research reactors and for activating isotopes for materials analysis and medical treatment.

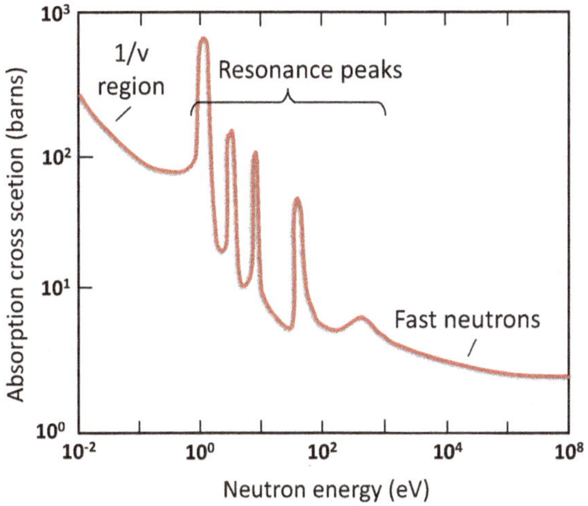

Fig. 6.17: Schematic absorption cross section of neutrons in a target material. Three regions can be distinguished: $1/v$ region, resonance region, and fast neutron region.

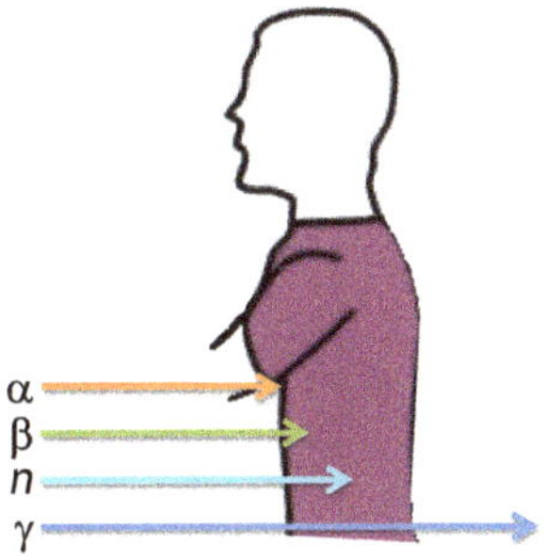

Fig. 6.18: Penetration of radiation in the body.

Tab. 6.1: Typical ranges for particles in water and lead. For γ-rays, the thicknesses quoted and indicated by an asterix are for the half-value thickness.

Particle	Energy	Range in water or tissue	Range in lead
α	4 MeV	5 μm	5 μm
β	1 MeV	5 mm	0.5 mm
γ	100 keV	5 cm*	0.1 mm*
γ	1 MeV	10 cm*	1 cm*
Neutrons	2 MeV	40 mm	~ m

6.5 Summary

S6.1   Particle beams that hit a target are characterized by an attenuation of the primary beam intensity, by transmission, and scattering.

S6.2   The attenuation of photon beams is governed by four effects of different importance in different energy regimes: coherent scattering, PE effect, Compton scattering, and pair production.

S6.3   SAXS is the scattering of EM waves at particles that have a different electron density than their surroundings.

S6.4   For x-rays in the energy region up to 0.5 MeV, the PE effect is most important.

S6.5   The PE effect consists of two steps: absorption of photons and emission of fluorescence radiation.

S6.6   The mass attenuation coefficient for the PE effect has pronounced edges at the binding energy of electrons in the K-, L-, or M-shells.

S6.7   The PE effect of hard x-rays generates high-energy photoelectrons.

S6.8   PE absorption decreases the intensity of the primary x-ray beam but does not change the energy of the x-ray photons.

S6.9   PE absorption causes an exponential decrease of intensity as a function of thickness penetrated.

S6.10 X-ray or γ-ray photons do not have a range but a half-thickness, where half of the intensity has diminished.

S6.11 X-rays and γ-rays produce hot photoelectrons whose range in tissue is similar to β-particles.

S6.12 The Compton effect involves the inelastic scattering of photons at electrons at rest.

S6.13 Compton scattering transfers energy from photons to electrons.

S6.14 Compton scattering changes the direction of the incident photon.

S6.15 Compton scattering becomes increasingly important with increasing photon energy.

S6.16 Pair production involves the materialization of photon energy into electron–positron pairs.

S6.17 Pair production has a threshold energy of 1.022 MeV photon energy.

S6.18 α-Particles have a defined range, which depends on the initial energy of the particle, the atomic number of the target material, and its density.

S6.19 The energy loss of α-particles is continuous. At the end of the range, the energy loss rate has a peak, called Bragg peak.

S6.20 The LET is defined as the energy loss per distance of charged particle traveled.

S6.21 LET for alpha particles is high at the end of their range.

S6.22 Electrons and positrons also have a range that is much longer than that of alpha particles.

S6.23 Positrons annihilate together with an electron at the end of their range.

S6.24 β-Particles lose energy mainly by Cherenkov and synchrotron radiation.

S6.25 The range of β-particles is greater than for α-particles. The straggling range is also more extended.

S6.26 Fast neutrons are slowed down by elastic scattering.

S6.27 Water is the most effective moderator for fast neutrons.

S6.28 The thermal neutron absorption cross section scales with $1/v_n$.

Questions

Q6.1 What are the two physical mechanisms that attenuate a particle beam interacting with a target?

Q6.2 What is the definition of fluence?

Q6.3 How is intensity related to fluence?

Q6.4 When photons interact with matter, what are the four different cross sections contributing to the attenuation of the intensity?

Q6.5 Coherent scattering has different names according to the size of the objects scattered from. Name a few of them.

Q6.6 What are the two steps that take place during photoelectric absorption process?

Q6.7 What is the difference between generating characteristic x-ray radiation using electrons and using photons?

Q6.8 What is the essential difference between the photoelectric effect and the Compton effect?

Q6.9 What is the difference between coherent x-ray scattering and Compton scattering?

Q6.10 Is pair production an important process?

Q6.11 Why do alpha particles suddenly stop?

Q6.12 What is the range of 5 MeV alpha particles in the normal atmosphere?

Q6.13 What is the Bragg peak for alpha particles due to?

Q6.14 Do photons have a range similar to alpha particles?

Q6.15 What is the range of β-particles?

Q6.16 What are the main interactions of fast and slow neutrons with matter?

Q6.17 What is the best moderator for fast neutrons?

Q6.18 In which energy range do neutrons experience resonance absorption?

Suggestion for home experiment

Take a white piece of paper and hold it in front of an LED light source. Describe the light absorption and scattering that you can observe.

Attained competence	+	0	–
I know what likely will happen when a particle beam hits a target.			
I can distinguish between fluence and flux.			
I know that the mass absorption coefficient for photons is the sum of four contributions.			
I am familiar with the basic physical concept of the photoelectric effect.			

I can distinguish between coherent elastic x-ray scattering and inelastic scattering.

I know that Compton scattering is an inelastic process.

I know that charged particles have a finite range in the matter.

I know that the LET depends on the distance traveled.

I know how to moderate neutrons from high energy after fission to thermal energies for further use.

I know what thermal neutrons are good for.

Exercises

E6.1 **Half-value thickness:** Compare photon energies of 100 keV typical for x-ray radiography and 500 keV of γ-radiation in PET. For both radiation sources, determine the half-value thickness $x_{1/2}$ in tissue of density 1 g/cm^3.

E6.2 **Compare the photoelectric effect and the Compton effect:** For both effects, state the conservation of momentum and energy.

E6.3 **Conservation of momentum and energy:** Show that for the Compton effect, energy and momentum are conserved simultaneously.

E6.4 **Violation of momentum and energy conservation:** Show that for the photoelectric effect at a free electron, momentum conservation and energy conservation are not possible simultaneously.

E6.5 **Comparison of PE and Compton effect:** Explain why in the photoelectric effect the electron cannot be free, whereas Compton effect is scattering of photons at free electrons.

E6.6 **Photon momentum:** What are the momentum, frequency, and wavelength of a photon that has an energy equal to the rest mass of an electron?

E6.7 **Compton effect:** Consider a γ-beam with the energy of the ^{137}Cs radioisotope. If the radiation is scattered from a free electron and the scattered photon beam is observed at right angle to the incident wave, determine:
   a. what is the energy and wavelength of the γ-beam?
   b. what is the wavelength shift?
   c. what is energy transfer to the electron, absolute, and relative?
   d. what is the energy loss of the photon?
   e. what is the scattering angle of the recoiling electron?
   f. what is the total angle $\theta+\phi$?

E6.8 **Range of α-particles:** What is the average range of 5 MeV α-particles in air and in tissue using. Use equations given in the text and international tables.

E6.9 **Confirm the eqs. (6.51) and (6.52)**

References

[1]   X-ray Data Booklet (Center for X-ray Optics and Advanced Light Source Lawrence Berkeley National Laboratory). At: http://xdb.lbl.gov/.

[2]   https://physics.nist.gov/PhysRefData/XrayMassCoef/tab3.html.

[3]   Hamley IW. Small-angle scattering: Theory, instrumentation, data, and applications. New York, London, Sydney, Toronto: Wiley&Sons; 2021.

[4]   Einstein A. Über einen die Erzeugung und Verwandlung des Lichtes betreffenden heuristischen Gesichtspunkt. Ann Phys. 1905; 322: 132–148.

[5]   Finkelnburg W. Introduction to atomic physics. Berlin, Heidelberg: Springer Verlag; 1958.

[6]   Livingston M, Bethe HA. Nuclear physics. Rev Mod Phys. 1937; 9: 245–390.

[7]   Leroy C, Giorgio-Rancoita P. Principles of radiation interaction in matter and detection. 2nd Edition. Singapore: World Scientific; 2009.

[8]   http://physics.nist.gov/PhysRefData/Star/Text/contents.html.

[9]   Segre E. Nuclei and particles. An introduction to nuclear and subnuclear physics. New York, Amsterdam: W.A. Benjamin, Inc; 1965.

[10]   Nordling C, Österman J. Physics handbook for science and egineering. Lund, Sweden: Studentlitteratur; 1996.

[11]   Krane KS. Introductory Nuclear Physics. New York, London, Sydney, Toronto: John Wiley&Sons; 1988.

Further reading

Khan FM, Gibbons JP. The physics of radiation therapy. 5th edition. Philadelphia, Baltimore, New York: Wolters Kluwer; 2014.

Lilley J. Nuclear physics, principles and applications. New York, London, Sydney, Toronto: John Wiley & Sons; 2013.

Krane KS. Introductory Nuclear Physics. New York, London, Sydney, Toronto: John Wiley & Sons; 1988.

Tavernier S. Experimental techniques in nuclear and particle physics. Berlin, Heidelberg: Springer-Verlag; 2010.

Useful website

Hyperphysics Webpage on Compton scattering provides a calculator for the incident and final photon and electron energy: http://hyperphysics.phy-astr.gsu.edu/hbase/hframe.html.

7 Dosimetry

Physical properties of radiation and radiation protection	
1 Gray (Gy)	= 1 J/1 kg
Annual average radiation exposure	3.5 mSv/a
Radiation sources	Terrestrial, cosmic, industrial, medical
Cumulative dose per transatlantic flight	30–40 µSv
ALARA principle	As low as reasonably achievable
Most important protections	Short time, long distance, no incorporation
Control area	>6 mSv
Lethal dose	5–8 Gy
Fractional dose during tumor irradiation	6–8 Gy

7.1 Introduction

Dosimetry combines precise radiation measurements, regulatory work, and surveillance. Dosimetry quantifies dose levels from any radiation source that produces ionizing radiation, such as x-ray generators, radioisotopes, and accelerators. Furthermore, dosimetry sets boundaries for radiation exposure to living organisms that are considered safe versus unsafe in the sense of causing radiation sickness. Finally, dosimetry determines adequate dose levels for radiography and radiotherapy. Specifically, dosimetry has to fulfill the following tasks in civilian areas and an additional one for radiology clinics:

1. Quantification of dose emitted by radiation sources
2. Designation of radiation control areas with increased dose above levels considered safe
3. Protection, safeguarding, and surveillance of people working in control areas
4. Guidance for the safe handling of radioisotopes and other radiation sources
5. Developing codes of practice for radiography and/or radiotherapy

Concerning these topics, international and legal binding conventions are established. The *International Atomic Energy Agency* (IAEA) with headquarters in Vienna, Austria, and the *International Commission for Radiological Protection* (ICRP) in Ottawa, Canada, frequently publish recommendations and guidelines on radiation protection and safeguarding [1].

All numbers provided in terms of dose have a statistical meaning based on a population rather than on individuals. For instance, if it is stated that an accumulated dose may lead to death within a certain number of years or may shorten the life span by so many years, this may or may not apply to an individual. Even for a large population, the numbers for low-dose exposure have a large variance. However, the expected

https://doi.org/10.1515/9783110757095-007

cumulative dose for serious radiation sickness and the lethal dose are rather well defined. Since radiation experiments cannot be performed on people, much of the experience with long-term radiation exposure comes from nuclear bombs on the cities Hiroshima and Nagasaki in 1945, nuclear tests, and major catastrophes that occured in Chernobyl 1986 and in Fukushima 2011.

7.2 Definitions of dose and dose rate

The radiation dose D_R is defined as the mean energy $\langle E \rangle$ of ionizing radiation imparted to a volume that contains a mass m of material, or, in short, the mean energy per mass:

$$D_R = \frac{\langle E \rangle}{m}. \tag{7.1}$$

This definition is independent of exposure time and the type of radiation, which can be from any radioactive source, nuclear reactor, or particle accelerator. The unit of dose is $[D_R] = J/kg$. $1\,J/1\,kg = 1$ Gray (Gy).[1] The definition of the dose is illustrated in Fig. 7.1; 1 J is the energy required to increase the temperature of 1 liter of water by 2.4×10^{-4} K. This does not sound much, but in terms of radiation, 1 Gy is a dose that already causes severe radiation sickness, and 5 Gy is lethal. This shows that the heating effect cannot be the cause for death, but the radiation damage of the cells and the genes inside, as we will see later in Chapter 1 of Volume 3.

To be more specific, the mean energy imparted to the volume with the mass m is the difference of radiant energy going into the volume, minus the radiant energy that the mass has not absorbed, plus the sum of all changes of rest energies $\sum_i E_{r,i}$ of nuclei and elementary particles i that may occur within the volume due to nuclear decay:

$$\langle E \rangle = \langle E \rangle_{in} - \langle E \rangle_{out} + \sum_i E_{r,i}. \tag{7.2}$$

The definition of the dose has some shortcomings. In particular, this definition does not account for the different ionization probabilities of photons and charged particles. Therefore, the dose by itself cannot express how dangerous particular radiation is for biological tissues. To correct for this missing information, a radiation weighting factor w_R was introduced, characterizing different types of radiation concerning their biolog-

---

1 Louis Harald Gray (1905–1965), British radiophysicist.

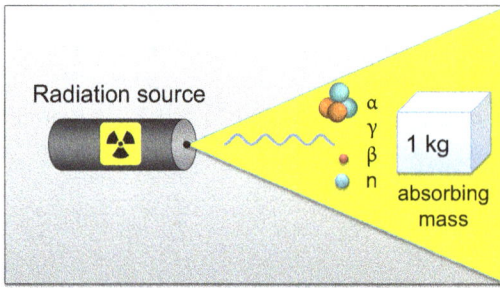

Fig. 7.1: The definition of a dose is independent of the type of radiation and relies solely on the energy absorbed by some mass.

ical effectiveness.[2] Table 7.1 lists the weighting factors w_R of different radiation types. They express how likely chemical bonds are broken by radiation, given the same energy of particles from a radioactive source. Obviously, α-particles are particularly dangerous as they are 20 times more efficient to break ionic bonds and form radicals on their track than γ-rays (through photoelectrons) or β-particles. Although the w_R factors listed in Tab. 7.1 appear to be constant, this is not the case. They depend on the *linear energy transfer* (LET = dE/dx) of particles introduced in Chapter 6.

Tab. 7.1: Radiation weighting factors for the dose of ionizing radiation [2].

Radiation source	Radiation weighting factor w_R
x-Rays, γ-rays, β-rays	1
Thermal neutrons	2.3
Fast neutron	10
Protons	2
α-Particles	20

Taking into account the radiation weighting factor w_R for certain types of radiation and the dose D_R of the same radiation, the *equivalent dose* H_T for specific tissue T is defined as follows:

$$H_T = w_R D_R \qquad (7.3)$$

---

2 There is currently some confusion regarding the use of the weighting factor w_R. In the past, this factor was referred to as quality factor Q or RBE. One might consider these factors to be equivalent. But that may not be correct. w_R is an approximate "administrative" factor used in recent ICRP reports to avoid short-term adjustments. The experimentally determined in vivo and in vitro values can be differently and more precisely defined for certain energy ranges.

The unit of the equivalent dose is $[H_T]$ = sievert,[3] $1\,\text{Sv} = 1\,\text{J}/\text{kg}$. Equivalent dose H_T is a measure of potential health effects on the human body due to low levels of ionizing radiation.

If the tissue is exposed to several different types of radiation, the sum has to be taken over all radiation sources i:

$$H_T = \sum_i w_{R,i} \times D_{R,i}. \tag{7.4}$$

Different organs in the body have different levels of sensitivity to radiation. To take these sensitivities into account, weighting factors $w_{T,j}$ have been introduced for different body parts when evaluating a full-body equivalent doses [3]. Table 7.2 gives a few examples. The tissue-weighted sum of equivalent doses in all tissues and organs of the body is then expressed by

$$H_{\text{eff}} = \sum_j w_{T,j} H_{T,j} = \sum_j w_{T,j} \sum_i w_{R,i} D_{R,i}. \tag{7.5}$$

The organ weighting factors have to fulfill the condition:

$$\sum_j w_{T,j} = 1. \tag{7.6}$$

The first sum goes over all tissues and organs exposed to radiation; the second sum goes over all types of radiation the body is exposed to. H_{eff} is called the *effective dose*.[4] According to the recommendations of the ICRP, weighting factors are highest

Tab. 7.2: Weighting factors w_T for organs and tissues [2].

Organs and tissues	Weighing factor $w_{T,j}$
Gonads	0.08
Red bone marrow	0.12
Colon	0.12
Lung	0.12
Stomach	0.12
Breast	0.12
Liver	0.04
Thyroid	0.04
Brain	0.01
Further organs and tissues	0.23
Total	1

---

3 Rolf Maximillian Sievert (1896–1966), Swedish radiophysicist.
4 In the literature, H_{eff} is denoted by the letter E. Here we use the notation H_{eff} to avoid any confusion with the symbol E reserved for energy in this chapter.

for breast, colon, lung, stomach, bone marrow (0.12) and lowest for bone surface, brain, salivary gland, and skin (0.01) [2].

Next, we define the *dose rate*, which is the dose per time of exposure:

$$\dot{D}_R = \frac{D_R}{\Delta t}, \qquad (7.7)$$

measured in Gy/a. Similarly, we define the equivalent dose rate:

$$\dot{H}_T = \frac{H_T}{\Delta t} \qquad (7.8)$$

measured in Sv/a. Here "a" stands for annum or year. Any appropriate time unit can be used in these definitions, but mSv/a is the most frequently used unit.

The dose rate makes it easy to calculate the dose. The dose rate can be expressed as follows:

$$\dot{D}_R = A\frac{E}{m}, \qquad (7.9)$$

where A is the activity of a radioactive source, i.e., the number of decays per second. If the activity is known, the integrated dose over time is determined via

$$D_R = \dot{D}_R \Delta t = A\frac{E}{m}\Delta t. \qquad (7.10)$$

The same arguments hold for the equivalent dose:

$$H_T = w_R \dot{D}_R \Delta t = w_R A\frac{E}{m}\Delta t. \qquad (7.11)$$

The units for the activity (becquerel, defined in Section 5.5.2), dose (Gy, Sv), and dose rate (Gy/a, Sv/h, etc.) are SI units. They are the only ones used in this text. The older unit rad (radiation absorbed dose) is replaced by gray, rem (roentgen equivalent man) is replaced by sievert, and curie is replaced by becquerel. The conversion factors are as follows:

$$1\,\text{Ci} = 3.7 \times 10^{10}\,\text{Bq},$$
$$1\,\text{rad} = 10^{-2}\,\text{Gy},$$
$$1\,\text{rem} = 10^{-2}\,\text{Sv}.$$

> ⚠ Dose is the absorbed radiation energy per mass; equivalent dose is the dose weighted by the biological effectiveness of radiation; equivalent effective dose is the equivalent dose weighted by the radiation sensitivity of organs exposed.

7.3 Kerma

Kerma is the dose that a beam of photons (x-rays or γ -rays) delivers when absorbed by *a* body. Before explaining kerma more precisely, we first need to introduce a few more definitions.

7.3.1 Flux and fluence

Fluence ϕ is defined as the number of particles N accumulated over time per unit area a:

$$\Phi = \frac{N}{a}. \tag{7.12}$$

The particle flux is defined as the number of particles N per unit time:

$$F = \frac{N}{t}. \tag{7.13}$$

Flux and fluence are connected via

$$\Phi = F\frac{t}{a}. \tag{7.14}$$

At constant flux F, the fluence increases linearly in time t.

7.3.2 Energy fluence

In the following, we consider only photons. The task is to determine the dose that a photon beam of fluence Φ_{ph} deposits when passing through matter with a mass attenuation coefficient μ/ρ_m. For estimating the radiation dose, it is essential to consider the photon beam energy. But the photon energy is not explicitly expressed in the definition of fluence in eq. (7.12). Therefore, we define another quantity, the *energy fluence* Ψ_{ph}. For a monoenergetic beam of photons, the energy fluence Ψ_{ph} is the product of the photon fluence Φ_{ph} and the photon energy $E_{ph} = \hbar w$:

$$\Psi_{ph} = \Phi_{ph}E_{ph}. \tag{7.15}$$

The SI unit of $[\Psi]$ is J/m². If a beam has a broad spectral distribution like the bremsstrahlung spectrum, either the integral over the energy distribution has to be taken or an effective average energy $\langle E_{ph} \rangle$ can be used in eq. (7.15):

7.3.3 Mass energy transfer coefficient

As an x-ray beam passes through matter, the beam becomes attenuated by four processes expressed in the mass attenuation coefficient for photons and discussed in detail in Section 6.2:

$$\frac{\mu_{Ph}}{\rho_m} = \frac{1}{\rho_m}\left(\mu_{PE} + \mu_C + \mu_s + \mu_\pi\right). \tag{7.16}$$

However, dose relevant are only those processes that transfer photon energy to charge particle kinetic energy. Photons deposit energy in a two-step process. First, photons transfer energy to charged particles either via the photoelectric process, Compton scattering, or pair production. Second, the energy-enhanced charged particles lose energy in matter by ionization and excitation of atoms and molecules over some straggling distance before coming to rest. The energy transferred from photons to charged particles is called *kerma* (*K*), which stands for *kinetic energy released in matter*. Kerma is determined by the product of energy fluence and the *mass energy transfer coefficient* μ_{tr}/ρ_m to be defined next.

The *mass energy transfer coefficient* μ_{tr}/ρ_m considers only those photon absorption processes, which result in kinetic energy of charged particles. Photoelectrons are the main contributors to photon-charge particle conversion. Compton scattering converts only part of the incident photon energy in kinetic energy of electrons, and coherent scattering does not convert any photon energy. In the case of pair production, only the excess energy beyond the threshold energy of 1.022 MeV is dose relevant, as only the excess energy can be converted into kinetic energy of electrons and positrons.

Therefore, not all four terms in eq. (7.16) contribute to the kinetic energy conversion to the same extent. The fraction f that contributes to the energy conversion is called *mass energy transfer coefficient*:

$$\frac{\mu_{tr}}{\rho_m} = \frac{1}{\rho_m}\left(f_{PE}\mu_{PE} + f_C\mu_C + f_S\mu_s + f_\pi\mu_\pi\right) = \frac{1}{\rho_m}\sum_i f_i\mu_i. \tag{7.17}$$

The fractions f_i are smaller than one or zero ($f_{PE}, f_C < 1, f_S = 0$), and they depend on the photon energy.

7.3.4 Mass energy absorption coefficient

After primary photons have transferred part of their energy to charged particles, the charged particles not only lose energy by collisions but also suffer radiative loss. A typical example is the bremsstrahlung, which is a radiation produced by de-acceleration of charged particles. The radiation losses due to these secondary processes are collected in the *mass energy absorption coefficient* μ_{abs}/ρ_m. The mass energy absorption coefficient is proportional to the mass energy transfer coefficient according to

$$\frac{\mu_{abs}}{\rho_m} = (1-g)\frac{\mu_{tr}}{\rho_m}, \qquad (7.18)$$

where g is the fraction of secondary charged particle energy lost in radiative processes. Clearly, these secondary radiative losses are not dose relevant. For photon energies used in x-ray diagnostics, g can be taken as zero.

7.3.5 Definition of kerma

Now we return to the definition of kerma. Kerma is defined as the photon energy fluence of an x- or γ-ray beam converted into kinetic energy of charged particles. Quantitatively kerma is given by

$$K = \Psi_{ph}\left(\frac{\mu_{tr}}{\varrho_m}\right). \qquad (7.19)$$

The SI unit of kerma is $[K] = J/kg$. The ratio μ_{tr}/ρ depends on the primary energy and ranges from 0.7 m²/kg for 20 keV x-rays to 0.2 m²/kg for 60 keV x-rays. This is because at higher energies, Compton scattering becomes more relevant in comparison to the photoelectric effect (see Fig. 6.11), contributing less to hot electrons. But this picture changes again at very high photon energies of up to 25 MeV used for tumor radiation treatment, when Compton scattering and pair production dominate. The values for μ_{tr}/ρ are plotted and tabulated in Ref. [3] and reproduced in Tab. 7.3 and Fig. 7.2 for water and dry air.

The definition of kerma can be rephrased as follows: kerma is the sum of all initial charge particle energies $dE = \sum_i dE_i$ that have been produced by photons (more general: by uncharged particles) per unit of mass dm:

$$K = \frac{dE}{dm}. \qquad (7.20)$$

As we have seen before in eq. (7.1), the *dose* (D_R) is defined as energy deposited by ionizing radiation per unit mass m of irradiated material:

$$D_R = \frac{E}{m} \qquad (7.21)$$

Dose has the same unit as kerma $[D] = J/kg$, where $1\ J/kg = 1$ Gy. But the definition of dose is broader. Dose considers all kinds of ionizing radiation of charged and uncharged particles: photons, protons, neutrons, kaons, etc. In contrast, kerma is limited to the fraction of charged particles produced by uncharged particles. Accordingly, the dose is calculated as follows:

$$D_R = \Psi \left(\frac{\mu}{\varrho_m} \right), \tag{7.22}$$

where Ψ is the energy fluence of any ionizing radiation and μ/ϱ_m is the corresponding mass absorption coefficient. In Tab. 7.3 and Fig. 7.2, both absorption coefficients are compared for water and dry air and for different energies. Further information on the photon mass attenuation coefficient can be found on the NIST webpage in Ref. [3].

Tab. 7.3: Comparison of mass energy absorption coefficient μ/ϱ_m and mass energy transfer coefficient μ_{tr}/ρ_m for water and dry air in units of cm^2/g.

E (MeV)	Water		Dry air	
	μ/ϱ_m	μ_{tr}/ρ_m	μ/ϱ_m	μ_{tr}/ρ_m
0.01	5.33	4.95	5.12	4.74
0.1	0.171	0.025	0.154	0.023
1.0	0.07	0.031	0.063	0.028
10	0.022	0.016	0.02	0.015
100	0.0173	0.016		

Note: Values are taken from Ref. [3, 4].

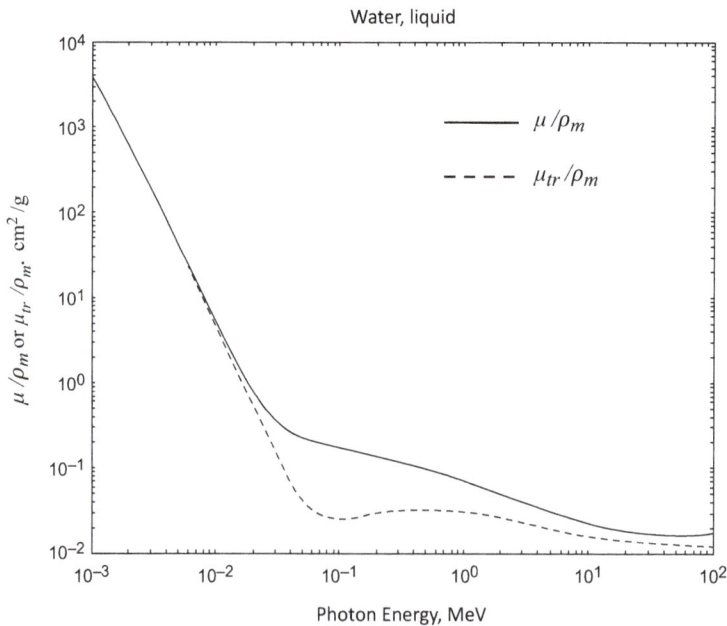

Fig. 7.2: Plot of the mass energy absorption coefficient μ/ϱ_m and mass energy transfer coefficient μ_{tr}/ρ_m for water (reproduced from Ref. [3]).

Kerma is defined as the energy fluence of an x-ray beam converted into kinetic energy of charged particles. Radiation dose is defined as the energy fluence of all charged and uncharged particles converted to kinetic energy of charged particles absorbed by the body.

7.3.6 Examples

In radiography, it is important to know the dose for a single image, a complete CT, a scintigraphic body scan, or the dose administered to a cancerous volume. We can calculate the dose for an x-ray radiograph relatively straightforwardly because all parameters are well defined and known: x-ray intensity determined by the power of the x-ray tube, exposure time, average mass absorption coefficient for tissues and bones, and quality factor for x-rays. In other cases, the dose calculation may be more difficult. We discuss four different cases:

1. **Flat panel x-ray radiography.** We assume the following values: x-ray intensity $I = 10^{10}$ photons/cm^2 s, mean x-ray photon energy of 60 keV for a bremsstrahlung spectrum with a kVp = 100 keV, exposure time $t = 50$ ms, average mass absorption coefficient for tissues and bones at 60 keV radiation is about 5 cm^2/g according to Fig. 6.11, w_R for x-rays is 1, which yields a dose according to

$$D_{\text{x-ray}} = I\Delta t \, E w_R \times \left(\frac{\mu}{\rho}\right) = \Phi \, E w_R \left(\frac{\mu}{\rho}\right) \qquad (7.23)$$

   of 24 mSv.

2. **Radioactive source.** Here we assume that the radioactive source of activity A is pointlike and that we determine the dose at a distance r on a spherical surface surrounding the source. If the radioactive source is a photon source (γ-rays), the dose at the radius r can be determined as follows:

$$D(E_{\text{ph}}, r) = \frac{A E_{\text{ph}} t}{4\pi r^2} \left(\frac{\mu_{\text{tr}}}{\rho_m}\right), \qquad (7.24)$$

   where μ_{tr}/ρ_m is the mass energy transfer coefficient. For the energy range between 0.02 and 2 MeV photon energy, the coefficient μ_{tr}/ρ_m is approximately constant and has the value 0.25 cm^2/g. For instance, a 0.5 MeV γ-source with an activity of 100 MBq produces in 1 h at a distance of 1 m a dose of 60 µGy.

3. **Scintigraphic imaging.** It is more difficult to estimate the dose for scintigraphic imaging because of many unknowns. The administered radioisotopes with a known activity spread out over the body and accumulate at hot spots such that the fluence is ill-defined. The time of exposure can only roughly be estimated by the effective lifetime of the radioisotopes used. Nevertheless, the following effective dose values have been estimated for scintigraphy [4]:

- 6 µSv for a 3 MBq ^{51}Cr measurement of glomerular filtration rate of the kidneys;
- 37 mSv for a 150 MBq ^{201}Tl cardiac stress test;
- 3.5 mSv for a common bone scan with 600 MBq of ^{99m}Tc.

In the last example, the activity of the source is much higher than in the other two cases. But the activity is spread over a much larger area, such that the dose is not exceedingly high. The quoted dose values for some medical procedures are listed in Tab. 7.4.

4. **Charge particle radiation.** For estimating radiation treatment with charged particles (cp), a slightly different equation is used [1, 5]:

$$D_{cp} = \Phi \cdot \text{LET} \cdot w_R \cdot \left(\frac{1}{\rho_m}\right) \cong \Phi\left(\frac{S}{\rho_m}\right). \tag{7.25}$$

Here LET is the energy loss dE of charged particles per distance dx traveled (dE/dx) within the Bragg peak and ρ_m is the mass density of the target material. The radiation weighing factor w_R is equivalent to the relative biological effectiveness (RBE), introduced in Chapter 1 of Volume 3. S is the stopping power (eq. (6.39)). The second relation in eq. (7.25) is identical if the dose due to photon radiation can be neglected. The dose rate is defined accordingly:

$$\dot{D}_{cp} = \dot{\Phi} \cdot \text{LET} \cdot w_R \cdot \left(\frac{1}{\rho_m}\right) = I \cdot \text{LET} \cdot w_R \cdot \left(\frac{1}{\rho_m}\right), \tag{7.26}$$

where I is the beam intensity. More on dose and LET will be discussed later in the chapters on radiography and radiotherapy using various radiation sources.

Finally, it should be reminded that all dose calculations are nothing but estimates. The actual dose to a patient cannot be determined and furthermore varies from person to person, depending on many individual factors such as bone density and body weight. However, to experimentally confirm that the calculated dose is reasonable, phantoms are used with similar density than tissue in combination with radiation detectors presented later. Furthermore, software packages are available to simulate the dose for different particles and energies to body organs [6, 7].

Tab. 7.4: Equivalent dose for different diagnostic medical radiation procedures with x-rays and y-rays.

Medical procedure	Dose in mSv
X-ray projection radiograph of the thorax	0.1–0.3
Full-body x-ray CT	20
Renal filtration with y-rays	0.006
PET/CT whole body protocol	23
SPECT – cardiac stress test with y-rays	37

7.4 Dosimeters and radiation monitors

For measuring the dose and dose rates of ionizing radiation, a number of detectors (dosimeters) are available. Dosimeters can be classified as

1. personal dosimeters carried by individuals while staying in a radiation control area; and
2. radiation monitors for monitoring radiation levels in control areas.

Personal dosimeters are important for estimating the radiation dose deposited over a period of time to a person wearing the dosimeter. In fact, wearing personal dosimeters is mandatory in all control areas with an average dose of more than 5 mSv/a. Figure 7.2 shows personal film badge dosimeters in panels (a) and (b) and a pocket dosimeter in panel (c). All of these dosimeters are used to record an accumulated dose over time. Film dosimeters are typically read once a month, while pocket dosimeters give the wearer an immediate read after any radiation exposure.

In film badge dosimeters, the dose is indicated by the blackening of an x-ray film that becomes visible after the film has been developed. The radiation type and its energy can be roughly estimated with various absorbers in front of the film. In panel (a) of Fig. 7.3, a film badge dosimeter is opened to show different compartments for monitoring different types and levels of radiation. Finger dosimeters (Fig. 7.3(b)) are used by personnel who have to work with their fingers in the vicinity of an open radiation source.

Direct reading pocket ionization dosimeters resemble in shape and size of fountain pens. The dosimeters contain a small ionization chamber filled with gas. In the

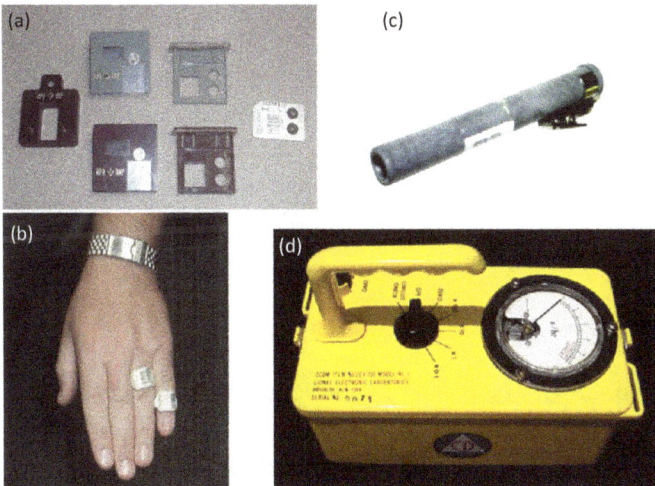

Fig. 7.3: (a) Film dosimeter; (b) finger dosimeter; (c) pocket ionization dosimeter; and (d) handheld radiation monitor.

center of the ionization chamber is a split anode wire connected at the base and the other end being free. When charged to a positive potential, electrostatic repulsion deflects these wires apart. Radiation causes ionization of the gas in the chamber; the electrons are attracted to the positive potential and at the same time, discharge the wire. This, in turn, reduces the deflection, which can be read out with the help of a magnifying lens. Digital electronic dosimeters more and more replace pocket dosimeters. The output of a radiation detector is collected and amplified. When a predetermined charge from radiation has been reached, the collected charge is discharged by simultaneously triggering an electronic counter. The counter then displays the accumulated exposure and dose rate in digital form.

Many types of handheld radiation monitors are available for monitoring the radiation level of an area. An example is displayed in Fig. 7.3(d). They are all based on discharge tubes with a central anode wire, schematically outlined in Fig. 7.4. Depending on the voltage applied, they can be categorized as ionization chambers, proportional counters, or Geiger-Müller detectors. When ionizing radiation enters through a thin window, the gas in the chamber becomes ionized and the free charges are separated in a voltage gradient: negative charges drift to the anode wire, and positive charges go to the grounded cathode chamber wall. The processes that occur after ion pairs have formed depend on the voltage applied. The charge versus voltage is plotted in Fig. 7.5.

7.4.1 Ionization chamber

At low voltage, all charges that have been produced by incoming ionizing particles are collected without amplification. The operation of the discharge tube is referred to as *ionization chamber* in this first plateau region (Fig. 7.5, IC area). No special requirement of the gas is specified and even air can be used, in which case no window is needed. Since the charge from a single particle is too small, ionization chambers are not useful as particle counters. Instead they are used in current mode, where the detected current is proportional to the count rate. In this mode, ionization chambers are robust and can be used to detect very high activities without dead-time problems.

7.4.2 Proportional counters

Proportional counters contain gases that are easy to ionize by charged particles. With the higher voltage gradient in proportional counters, the ion pairs become accelerated on the way to the electrodes, thereby strongly multiplying the charge collected. Therefore, the proportional counter can be used as a single event counter after electronic amplification of the charge signal. The total charge collected per particle is proportional to the applied voltage and the detected particle's energy.

Detector chamber

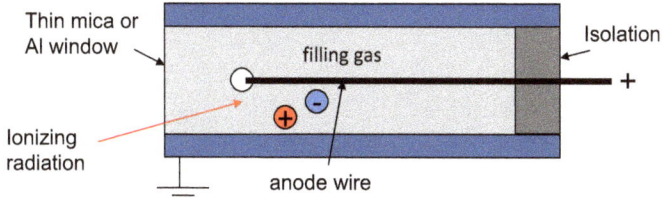

Counting electronics

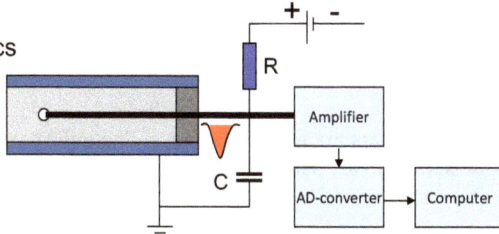

Fig. 7.4: The chamber for the ionization detector, for the proportional detector, and for the Geiger–Müller detector holds detecting gas, an anode wire at positive potential, and a thin window for entrance of the radiation. The counting electronics is shown in the bottom panel.

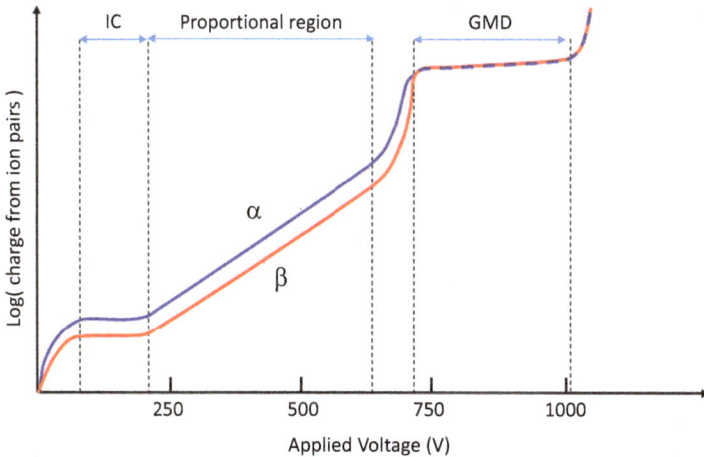

Fig. 7.5: Charge versus voltage characteristics for the different regions of operation. The ionization chamber (IC) operates at low voltage in the plateau region. The proportional counter operates in the region where the charge collected is proportional to the voltage applied. The Geiger-Müller detector (GMD) operates again in the plateau region, but at much higher voltage. The blue curve is the charge–voltage characteristics for α-particles and the red curve for β-particles.

This latter property makes the proportional counter interesting for spectroscopic investigations, as the activity and the particle's energy can be recorded.

7.4.3 Geiger-Müller detectors

Geiger[5]-Müller[6] detectors [8] work at an even higher voltage in the plateau area of the charge-voltage curve in Fig. 7.5. All other properties are the same as for the proportional counter. At this higher voltage, the charge amplification can be viewed as an avalanche effect independent of the exact voltage and independent of the particle energy. Geiger-Müller detectors are therefore clear event counters.

> **!** Dosimeters and radiation monitors are used for safeguarding individual radiation exposures and radiation levels in control areas. Badge dosimeters accumulate radiation over time, and counters measure the intensity.

7.4.4 Dead time

In both proportional counters and Geiger-Müller detectors, the discharge would continue if it was not quickly quenched by a special additive to the gas so that the chamber is ready to detect the next particle. Discharge and quenching require some time during which no other particle can be recorded. This time is called the dead time τ. The dead time is not important for low count rates, but dead time corrections must be applied for higher count rates. The dead time corrected count rate of recorded events N over a measuring time t is

$$\dot{N}_{\text{corr}} = \frac{N}{t\left(1 - N\frac{\tau}{t}\right)} = \frac{\dot{N}}{1 - \dot{N}\tau}, \tag{7.27}$$

where $\dot{N} = N/t$ is the uncorrected count rate. For instance, if the dead time τ is about 10 μs and the uncorrected count rate is 10^4 counts/s, the corrected count rate is 1.1×10^4 counts/s or 10% higher.

All three detectors are primarily charged particle detectors. They have difficulty detecting x-rays or γ-rays. This limitation can be partially overcome by adding noble gases with a high Z value, such as Ar or Xe, to the gas mixture of the detector. Anyway, scintillation counters are much more suitable for the detection of high-energy photons. They are used for scintigraphic imaging and described in Chapter 9.

---

5 Johannes Wilhelm Geiger (1882–1945), German physicist.
6 Walther Müller (1905–1979), German-American physicist.

7.5 Radiation exposure

People are exposed to many types of radiation, voluntarily or not. In general and as illustrated in Fig. 7.6, we distinguish between three different sources of radiation:

a. cosmic;
b. terrestrial; and
c. medical/industrial.

Isotopes produced by cosmic radiation are mainly due to the activity of our Sun or other solar systems. They eject mainly π-mesons, protons, and α-particles, which interact with the earth atmosphere and produce isotopes of carbon (^{14}C), tritium (^{3}H), and to a lesser extent ^{10}Be, ^{26}Al, and ^{36}Cl. ^{14}C finds its way into our body through the food chain. At sea level, the radiation received per person due to cosmic radiation is on average 0.3 mSv/a. This dose increases dramatically when going to higher altitudes. In normal flight altitude of 10 km, the dose rate is already 10 µSv/h and with one transatlantic flight, we accumulate about 0.1 mSv. Dose and dose rates are listed in Tab. 7.5.

Terrestrial radioactivity is due to long-lived isotopes, mainly from the radioisotopes ^{40}K, ^{232}Th, ^{235}U, ^{238}U, and their decay products. One important radioisotope is the noble gas radon (^{222}Rn), which decays with a half-life of 10.5 years into ^{218}Po. Since Rn as a noble gas atom is not bound in molecules, it emanates from uranium-rich rocks and accumulates in closed spaces, like in poorly vented basements. When inhaled, Rn reaches our lungs. As α-emitter, the biological effectiveness is particularly high. After decay, the daughter isotope ^{218}Po is not only radioactive but also toxic. The dose rate from terrestrial sources fluctuates strongly and depends on the geological conditions. Another vital contribution to terrestrial radiation is the decay of the natural isotope ^{40}K in our body, causing about 5000 Bq from 1.3 MeV β-radiation. This internal source of radiation accumulates to 0.3 mSv/a. Altogether, on average, one can assume that the dose rate from terrestrial sources is about 1 mSv/a.

Industrial activity and medical treatment are further sources of radiation exposure. Coal-burning power plants set free ^{14}C that would otherwise be bound. Nuclear power plants contribute a negligible amount to the environmental radioactivity unless there is a major accident, such as in Chernobyl 1986 and Fukushima 2011. Much of the radiation from the Chernobyl disaster came from ^{137}Cs fallout. This is a strong β^- and γ-emitter with a half-life of 30.2 years. Moreover, in addition to the high level of radioactivity, Cs$^+$ ions replace K$^+$ ions and disturb the cell function when incorporated with food. The radiation exposure from normal x-ray imaging without tumor treatment is estimated to be about 1 mSv annually. In total, the dose rate from industrial/medical applications per person is estimated to be about 1.2 mSv/a.

Summing up the three main sources of radioactivity (cosmic, terrestrial, and medical), we estimate a total dose rate of about 3.5 mSv/a any person receives on average. This is below the radiation control level starting at 5 mSv/a according to Tab. 7.5.

> ❗ The annual exposure of people to radiation from cosmic, terrestrial, or medical sources amounts on average to 3–3.5 mSv/a. This radiation level is considered safe.

Tab. 7.5: Typical dose per person and dose limitations in various countries. For dose limitations in specific countries or specific facilities, the local regulations should be consulted.

Dose and dose rate	Radiation source
0.01 mSv/a	Nuclear power plants
0.3 mSv/a	Cosmic sources
0.1 – 1 mSv/a	Cosmic ray exposure during flights
1 mSv/a	Terrestrial sources
1 mSv/a	Medical/industrial sources
3.5 mSv/a	Average radiation exposure
< 1 mSv/a or < 0.1 μ Sv/h	Maximum tolerable additional radiation exposure without further radiation control
1 – 6 mSv/a	Survey area
> 6 mSv/a	Control area
> 3 mSv/h	Off-limit area
20 mSv/a	Occupational dose limit
250 mSv	Maximum lifetime occupational dose
1000 mSv	Radiation sickness
3000 – 4000 mSv	50% death toll within 3–6 weeks
5000 – 8000 mSv	Lethal

7.6 Radiation protection

The basic guideline of radiation safety and radioprotection is expressed in the ALARA principle:

As Low As Reasonably Achievable.

The three most important measures for radiation protection are:
1. Reduce the exposure time
2. Get away from the radiation source
3. Choose the right and adequate shielding

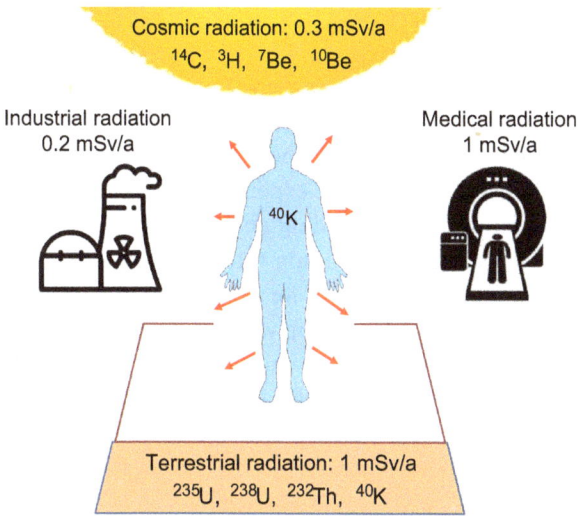

Fig. 7.6: Humans are exposed to cosmic, terrestrial, industrial, and medical radiation sources. A major source of radiation is the radioactive isotope ^{40}K, amounting to 0.3 mSv/a. The total dose: full body dose is on the average 3.5 mSv/a.

The first point is obvious and needs no further explanation.

The simplest and most effective measure against radiation is distance. For a radiation source that emits in all directions (4π), the inverse square law $1/r^2$ applies. If the distance to the source is doubled, the intensity drops by a factor of 4, because the exposed area has quadrupled. Figure 7.7 shows how this works. However, the inverse square law of distance does not apply to directed beams such as x-ray tubes and radiation from accelerators. Also radiation that enters the body through food or inhalation is beyond the inverse square law rule and should be avoided at all costs.

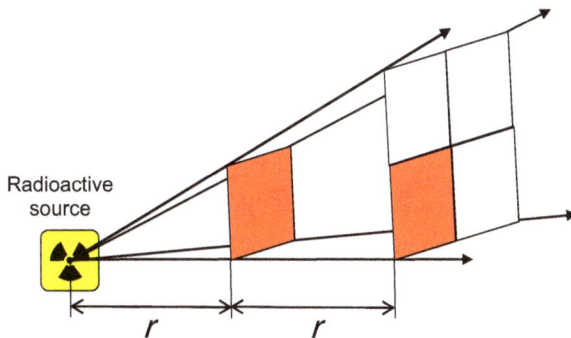

Fig. 7.7: Illustration of the quadratic distance rule. The intensity of a radioactive source drops by a factor of $(\frac{1}{2})^2 = 1/4$ at double distance $2\,r$ from the source.

Appropriate shielding against radiation can be an art. The range of particles in different environments gives a first hint to the best choice of shielding. Some of them are listed in Tab. 6.2 and many more can be found in the internet [5]. X-rays and γ-radiation can be shielded most effectively by high Z materials such as lead sheets or lead bricks. The required thickness can be determined as follows. First, the half-value layer (HVL) thickness follows from the exponential attenuation law according to $HVL = 0.693/\mu$ (see Section 6.2). Depending on the acceptable radiation level, the total thickness of the required shielding then is given by $(1/2)^n$, where n is the number of HVL to be used. For instance, for a reduction of the intensity to 1%, $n = 6$ is necessary. α-particles are easy to stop within 5 cm of air or by a sheet of paper. But any skin contact with an α-source should be strictly avoided. β-particles can be shielded with a thin aluminum metal sheet. It is considerably more difficult to shield fast neutrons. First, fast neutrons need to be slowed down to thermal energies by a sequence of collisions in water or paraffin. Then thermal neutrons can be captured by highly neutron-absorbing materials, such as boron, cadmium, or gadolinium. But thermal neutron capture does not come for free. For instance, the absorption reaction $_5^{10}B(n,\alpha)_3^7Li$ releases α-particles, and the other reactions release β- and γ-radiation. Therefore additional shielding is required for capturing secondary particles.

> **!** The most effective protection against hazardous radiation sources is shortening exposure time and enlarging distance from the source.

7.7 Summary

S7.1   Dose D is a measure of radiation absorbed by matter. The dose of 1 Gy equals the absorption of 1 J of radiant energy in 1 kg of matter.

S7.2   The unit sievert accounts for the relative biological effectiveness (RBE) of radiation.

S7.3   Equivalent dose is defined as the dose multiplied by a quality factor w_R specific for each type of radiation.

S7.4   At the same energy α-particles are 20 times more effective in destroying biological tissue than γ- or β-particles.

S7.5   Fluence is the number of particles N impinging on a defined surface area S.

S7.6   Energy fluence is the fluence of photons multipled with the energy per photon.

S7.7   The mass energy transfer coefficient considers only those absorption processes, which result in kinetic energy of ionizing charged particles.

S7.8   Kerma is defined as the energy fluence of an x-ray beam converted into kinetic energy of charge particles.

S7.9 The annual exposure of people to radiation from cosmic, terrestrial, or medical sources amounts on average to 3–3.5 mSv.

S7.10 For monitoring radiation in controlled areas, personal dosimeters and various detectors are available and must be carried.

S7.11 The three most important measures against radiation are time, distance, and shielding.

S7.12 Dosimetry quantifies dose levels and sets boundaries for radiation exposure considered safe versus hazardous to people.

Questions

Q7.1 What is dosimetry good for?

Q7.2 How is dose defined?

Q7.3 What factors characterize radiation weighting factor w_R and tissue weighting factor w_T?

Q7.4 What is the difference between dose and equivalent dose, and what are their units?

Q7.5 What is the difference between the weighting factor w_R and weighting factor w_T?

Q7.6 What is energy fluence in comparison to fluence?

Q7.7 What does the acronym kerma stand for, and how is kerma defined?

Q7.8 Does the definition of kerma hold for all types of radiation?

Q7.9 How can the mass energy transfer coefficient be distinguished from the mass attenuation coefficient?

Q7.10 What dosimeters are available for measuring the dose?

Q7.11 What is the difference between proportional counters and Geiger–Müller detectors?

Q7.12 What is meant by "dead time," and how does it affect the count rate?

Q7.13 What are the main sources of radiation to which people are exposed?

Q7.14 What is the tolerable dose without further control?

Q7.15 What is the ALARA principle?

Q7.16 What is the best protection measure against radiation?

Suggestions for home experiment

You may test the ALARA principle with a spherically shaped loudspeaker. Moving away from the source, the loudness should drop by a factor of 4 when doubling the distance.

Attained competence	+	0	–
I know what the units for the dose D are.			
I can distinguish between dose and equivalent dose.			
I know what kerma stands for.			
I know how to apply the weighting factors w_R and w_T.			
I know how to monitor radiation over longer periods of time.			
I know at which dose level personal dosimeters are required.			

I realize that counters have a dead time, and I know how to correct it.

The term ALARA is known to me and I follow the rule.

I can name the three main sources of radiation in daily life.

I know what the quadratic distance rule implies.

Exercises

E7.1   **Dose:** Confirm the dose of 24 mSv quoted below eq. (7.23).

E7.2   **Radiation exposure through ^{131}I therapy:** For a thyroid treatment, a patient is given an injection with an activity of 1.5×10^8 Bq of the isotope ^{131}I. The isotope decays to 89% with emission β_1 – particles with a maximum energy of 610 keV, and 11% with emission β_2 – particles with a maximum energy of 350 keV. The subsequent γ-decay is therapeutically not relevant. The radiation therapy development can be followed with the emitted γ-rays. The γ-radiation represents a burden for people in the vicinity (see E7.3). Assume that after a short time, this isotope completely accumulates in the thyroid gland with a mass of 20 g. The physical half-life of ^{131}I is 8 days, and the biological half-life is 80 days.

    a.   What is the dose rate at the beginning after the injection? Assume the "worst"-case scenario that all β-particles emit with their maximum energy and are completely absorbed in the thyroid gland.

    b.   Determine the effective half-life.

    c.   What is the dose absorbed by the body after one effective half-life?

E7.3   **Radiation exposure of bystanders:** Assume that a person weighing 70 kg is in the vicinity of the patient who was previously treated with ^{131}I at an activity of 1.2×10^{13} Bq. The patient was discharged from the hospital after an effective half-life (see Task E7.2). What is the dose that a person receives on the day of discharge at a distance of 1 m over a period of 1 h if the radiation exposure is caused exclusively by the γ-radiation of the patient with an energy of 580 keV γ – rays?

E7.4   **Count rate:**

    a.   An ionization chamber measures a source activity of 5×10^4 counts/s. The dead time of the detector is known to be 1 μs. What is the true source activity?

    b.   A very fast detector is known to measure the correct rate of 5×10^5 counts/s. An older and less expensive counter measures a source activity of 4×10^5 counts/s. What is the dead time of the older counter?

E7.5   **Dead time:** If $\dot{N}_t = 1000$ is the true count rate and $\dot{N}_m = 900$ is the measured count rate, what is the dead time τ of the detector?

E7.6   **Dip in the μ_{tr}/ρ curve:** In Fig. 7.3, a dip at 0.1 MeV can be seen in the μ_{tr}/ρ_m but not in the μ/ρ_m curve. Try to explain and rationalize the dip.

References

[1] International Commission on Radiological Protection. The 2007 recommendations of the international commission on radiological protection. ICRP Publication; vol. 203: ICRP 2007.
[2] https://www.icrp.org/.
[3] https://www.nist.gov/pml/x-ray-mass-attenuation-coefficients.
[4] Kato H. Photon mass energy transfer coefficients for elements z=1 to 92 and 48 additional substances of dosimetric interest. Nihon Hoshasen Gijutsu Gakkai Zasshi. 2014; 70: 684–691. (in Japanese language).
[3] Jacobi W. The concept of the effective dose a proposal for the combination of organ doses. Radiat Environ Biophys. 1975; 12: 101–109.
[4] Brindhaban A. Effective dose to patients from SPECT and CT during myocardial perfusion imaging. J Nucl Med Technol. 2020; 48: 143–147.
[5] Keller AM, Hahn K, Rossi HH. Intermediate dosimetric quantities. Radial Res. 1992; 130: 15–25.
[6] Ding A, Gao Y, Liu H, Caracappa PF, Long DJ, Bolch WE, Liu B, Xu XG. VirtualDose: A software for reporting organ doses from CT for adult and pediatric patients. Phys Med Biol. 2015; 60: 5601–5625.
[7] https://www.nist.gov/pml/stopping-power-range-tables-electrons-protons-and-helium-ions.
[8] Geiger H, Müller W. Elektronenzählrohr zur Messung schwächster Aktivitäten [Electron counting tube for measurement of weakest radioactivities]. Die Naturwissenschaften (In German). 1928; 16: 617–618.

Further reading

IAEA. Dosimetry in diagnostic radiology: An international code of practice. Technical Reports Series No. 457. IAEA; 2007.
Khan FM, Gibbons JP. The physics of radiation therapy. 5th edition. Philadelphia, Baltimore, New York: Wolters Kluwer; 2014.
Dance DR, Christofides S, Maidment ADA, McLean ID, Ng KH eds, Diagnostic radiology physics –A handbook for teachers and students. Vienna: IAEA; 2014.

Part C: **Radiography**

8 X-ray radiography

Physical parameters of radiography	
Contrast bone-tissue	+0.26
Contrast tissue-air	−0.2
Contrast muscle-blood	−0.003
HU for bones	+1000
HU for water	0
HU for air	−1000
Spatial resolution film	100 μm
Spatial resolution TFT	200 μm
Spatial resolution CMOS flat panel	100 μm
CT voxel size, typically	$0.5 \times 0.5 \times 0 \times 5$ mm^3
Dose: chest projection radiograph	0.1 mSv
Dose: mammography	3 mSv
Dose: CT chest	5–8 mSv
Dose: CT cardio-angiogram	7–13 mSv
Exposure time for a CT chest scan	60 s
X-ray contrast methods	Attenuation, contrast agents, dual energy, K-edge, phase contrast
Peak accelerating potential for radiography	80–140 kVp

8.1 Introduction

X-ray radiography is the most common imaging modality for the noninvasive representation of internal organs and foreign objects. It is also the oldest method, which
began in 1895 with the famous x-ray picture that Röntgen[1] took of his wife's hand.
In the early days, the radiation hazards associated with radiation exposure were
not recognized. Until the 1970s, it was common practice to check the fit of shoes
with an x-ray machine in the shoe store, called a shoe-fitting fluoroscope!

X-rays penetrate opaque materials because the energy (frequency) of x-ray photons is much higher than any energy bandgap in materials or any plasma frequency
of electrons in metals. On their way through gases, liquids, or solids, x-rays are attenuated but not refracted. Therefore, there are no x-ray lenses in the conventional
sense. Only the attenuation offers an imaging modality via shadow projection of
partially x-ray transparent and inhomogeneous media.

In Chapter 6, we learned about x-ray attenuation under the assumption of well-
defined conditions typical for physics laboratory experiments. The incident x-ray
beam was assumed to be monochromatic, the absorbing material had a well-defined

---

1 Wilhelm Conrad Röntgen(1845–1923), German Physicist, first Nobel Prize in Physics 1901.

https://doi.org/10.1515/9783110757095-008

thickness, and the density was taken to be constant. In medical practice, the situation is much more complex. Radiographs are recorded with a broadband x-ray source (modified bremsstrahlung spectrum), the absorbing tissue is highly inhomogeneous, and the thickness varies with the shape of organs. Nevertheless, with x-ray radiographs, we try to gain information about inner organs from transmitted x-ray beams that expose an appropriate x-ray detector. The magic word is contrast. A homogeneous sample would not offer any contrast. In this chapter, we learn about three possibilities of contrast formation: attenuation contrast, spectroscopic contrast, and phase contrast. We start our discussion with the attenuation contrast and discuss the more specialized contrast modalities in later sections.

Conventional x-ray radiography provides a two-dimensional (2D) projection image of the internal organs. While this is very useful for many clinical applications, it is also clear that three-dimensional (3D) x-ray imaging would be even more revealing. Indeed, 3D imaging, known as computed tomography (CT), can be performed. This imaging modality, introduced in the early 1970s, has undergone rapid development and today delivers detailed 3D images with a remarkable spatial resolution. CT is briefly discussed in the last part of this chapter. Basic information on x-ray generation and x-ray energy spectra relevant for radiography is presented in Chapter 4 and is recommended for review before continuing with this chapter.

X-ray radiography, including CT, is covered extensively in the literature on various levels of complexity, from introductory to professional. Here the text is adjusted to an introductory to intermediate level, explaining the most important physical aspects, but is not suitable for professional x-ray use. X-ray radiography has many different aspects concerning many fields of physics. It would be impossible to cover all these facets in this short chapter. The textbooks by Dance et al. [1] and Bushberg et al. [2] are recommended for comprehensive reviews on all physical aspects of medical imaging, including x-ray radiography. More specialized reviews are cited throughout this contribution. Recently, two excellent textbooks have appeared on biomedical imaging by Salditt et al. [3] and Maier et al. [4]. More recommendations can be found under "Further reading" at the end.

8.2 Standard x-ray radiography

8.2.1 Beam delivery and beam hardening

Standard radiography uses an x-ray generator like the one shown in Fig. 4.8. The x-rays are emitted cone-like in the forward direction emanating from a small focal spot on the anode. The x-ray beam is incoherent, covers a broad energy spectrum, and drops off in intensity with distance according to the $1/r^2$ law, where r is the distance to the focal spot on the anode. Collimators control the x-ray beam divergence. The divergence is wide for a large field of view (FOV) to image, for instance,

the entire upper chest, or narrow for imaging a specific organ. For the image formation, it is important that the focal spot is small. However, a small spot limits the
intensity. Therefore, a proper balance is required between intensity and image quality. This will be the subject of further discussions. A typical arrangement for taking
an x-ray image is depicted in Fig. 8.1: x-ray generator, filter, and collimator on one
side, radiation detection on the opposite side, and the patient in between. In contrast to nuclear imaging methods discussed in Chapters 9 and 10, the x-ray dose to
the patient occurs only during the short exposure time. There is no radiation before
the x-ray generator is turned on, nor after turning it off. The x-ray exposure does
not cause any afterglow in the body.

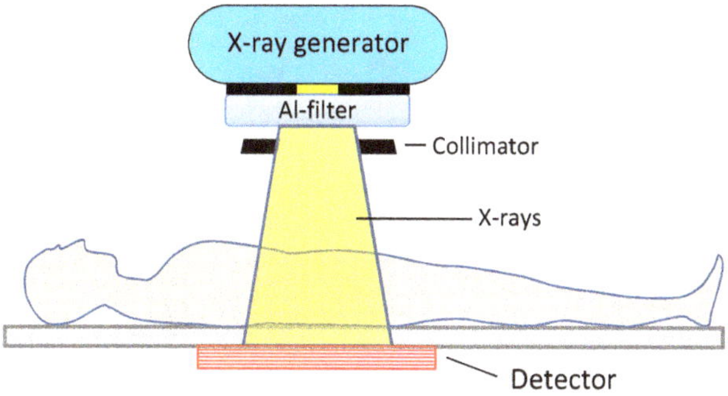

Fig. 8.1: Standard arrangement for taking x-ray radiographs. The x-ray source is outside the body.
A collimated beam penetrates the body and is attenuated by the body. The remaining intensity is
recorded by an appropriate detector.

Radiographs use the entire bremsstrahlung spectrum for imaging at operating voltages
between 80 and 140 kVp. However, the low-energy "soft" part of the x-ray spectrum is
highly absorbed in the skin, causing biological damage without contributing to the
image contrast. Therefore, for safety reasons, the spectrum is modified by passing the
x-rays first through an Al absorber plate (foil). This eliminates low-energy x-rays, while
the Al-foil hardly attenuates higher energy x-rays. A schematic x-ray spectrum before
and after filtering with an Al foil is reproduced in Figs. 4.3 and 8.2. Hence, radiographs
are usually taken with a filtered broad x-ray bremsstrahlung spectrum. In general, by
employing various types of filters (Al, Cu, etc.), the average bremsstrahlung spectrum is shifted to higher energies. This often serves for contrast variation or contrast
enhancement and is referred to as x-ray *beam hardening*. Therefore, filters have two
effects: they reduce the primary flux preferentially in the soft x-ray regime and shift
the average energy of the bremsstrahlung spectrum to higher energies. But not only
hardware filters contribute to beam hardening. Also the tissue itself causes beam
hardening simply by beam attenuation. As the beam propagates through the body,
lower energy photons are absorbed more readily than higher energy photons due to

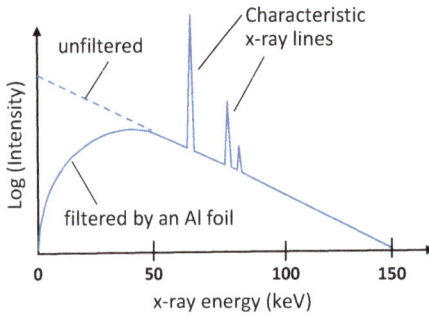

Fig. 8.2: Bremsstrahlung spectrum and characteristic spectrum of an x-ray tube before and after filtering the radiation with a thin Al foil.

the mass attenuation coefficient depending on the inverse third power of the photon energy $\mu/\rho \sim 1/E^3$ (see eq. (6.21)).

8.2.2 Magnification and penumbra

When exposed to x-rays for taking a radiograph, bones form an enlarged x-ray shadow, shown schematically in Fig. 8.3. This is due to two effects. First, x-rays are attenuated in bones more strongly than in the surrounding tissue. The stronger attenuation is because of bones' high calcium content (atomic number 20). Second, the shadow's magnification factor depends on the geometry, i.e., the distance between the focal spot and bone, and bone to the recording detector. Radiographs by x-ray shadow formation are called *projection radiographs* because they convert a 3D object into a 2D shadow image. On the recording x-ray film or flat panel detector, bones appear in light gray (low x-ray exposure of the film), while the surrounding tissue appears in dark gray (high x-ray exposure). When the radiograph is taken with a point-like x-ray source, boundaries between light and dark regions appear sharp (umbra), forming a high contrast image. If the x-ray source is extended, a halftone (*penumbra*) separates the bright and dark regions, causing a blurry appearance of the radiograph. There are two options to reduce but not remove the penumbra effect: shrink the source size and/or decrease the distance between film and body. The smaller the penumbra, the sharper is the contrast on the recording media.

8.2.3 Compton scattering and grids

Another potential source of fuzziness is inelastic Compton scattering (Fig. 6.6) and elastic small-angle x-ray scattering (Fig. 6.9). Small-angle scattering[2] occurs from any object with a density different than the surrounding matrix. Small-angle scattering is very useful for analyzing nano- to micro-size objects and it is extensively applied to investigate the atomic structure of soft matter. However, the relevant scattering angles are rather small when using hard x-rays (on the order of <0.05°). Therefore, small-angle scattering can safely be ignored as a source of fuzziness in x-ray projection radiographs.

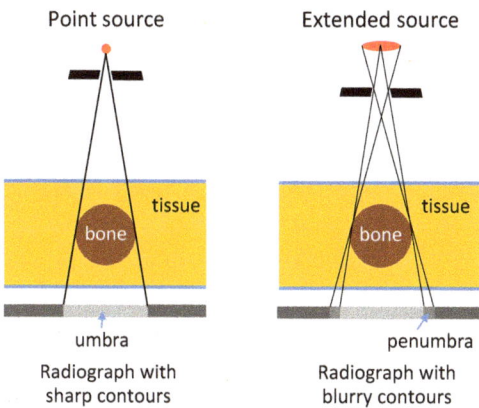

Fig. 8.3: Umbra and penumbra formed by absorption of x-rays in objects (bones) using a point source and an extended source.

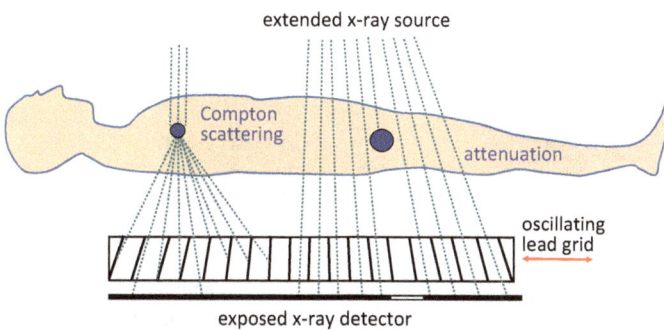

Fig. 8.4: A static or moving lead grid in front of the exposed x-ray detector suppresses stray x-rays and improves the contrast.

---

2 In medical literature, elastic x-ray scattering is often referred to as coherent scattering.

On the other hand, Compton scattering becomes increasingly important with higher x-ray energies (see diagram of the cross-sections versus the energy in Fig. 6.11) and contributes noticeably to the fuzziness. In order to solve this problem, a lead-containing grid with parallel lamellas is placed in front of the detector, allowing only straight x-rays to reach the recording detector, as shown schematically in Fig. 8.4. The grid also solves the penumbra problem when using x-ray tubes with an extended focal spot. But the grid creates another problem: it causes an indistinct shadow of itself on the x-ray image. In order to overcome this obstacle, the grid can be made to oscillate. The oscillating grid removes the shadow problem but leads to an overall reduced intensity. Grids are characterized by height, width, inclination angle, and septa thickness between the openings. If the septa walls are too thin, stray x-rays can penetrate, and their purpose is lost; thicker walls reduce the intensity. Hence, there is a trade-off between higher resolution and lower intensity. The ratio of x-ray exposure time with and without grid is termed the bucky factor F:

$$F = \frac{\text{exposure time with grid}}{\text{exposure time without grid}}.$$

In fact, grids improve contrast, increase resolution, and reduce background noise. However, they are not an essential part of x-ray radiography, in contrast to grids in SPE or SPECT, as we will see in Chapter 9.

Figure 8.5 compares the three main factors affecting the contrast of radiographs: (a) halftone from an extended source; (b) small-angle scattering and Compton scattering; (c) insufficient contrast due to similar attenuation coefficients of an object and surrounding matrix, such as muscles and fat.

Figure 8.6 shows radiographs from the pelvic space taken with and without an oscillating grid in front of the x-ray film. The different contrast in both images is very clear. Additional causes of "sharpness" loss due to intrinsic reasons of the detection methods are discussed later in Section 8.2.4.

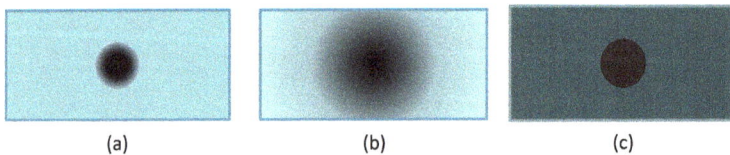

(a)   (b)   (c)

Fig. 8.5: Three types of contrast produced by an extended x-ray source: (a) penumbra; (b) Compton scattering; (c) insufficient difference of attenuation coefficients for object and surrounding.

! X-ray radiography utilizes shadow contrast by bones and organs for image formation. Contrast is achieved by laterally varying attenuation. Penumbra and Compton scattering are detrimental to sharp contours and contrast.

8.3 X-ray attenuation and contrast

8.3.1 Contrast

We consider an x-ray beam of intensity I_0 that hits a medium with varying lateral mass attenuation coefficients and thicknesses, as indicated in Fig. 8.7. A suitable detection system, such as an x-ray film[3] or a flat panel camera, records the transmitted beam intensity. Obviously, the thinner the material and the lower the mass attenuation coefficient, the stronger the film's blackening. Ideally, we will observe a sharp boundary on the film between stronger and weaker absorbing materials with a parallel incident beam of constant intensity. The difference in gray scale on the recording film is called contrast. Contrast is the degree to which light and dark areas of an image differ in brightness or *optical density* (OD).

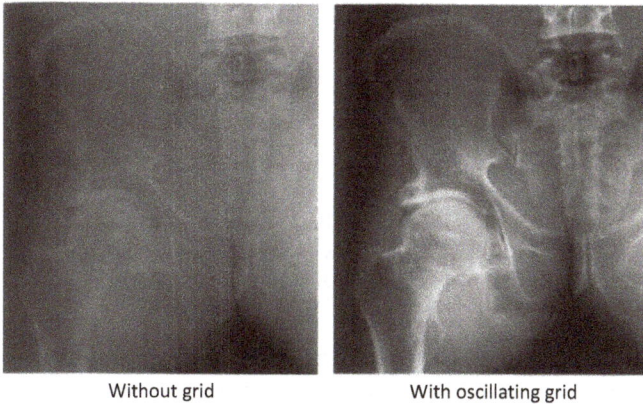

Without grid   With oscillating grid

Fig. 8.6: Radiograph of the pelvic space taken with and without an oscillating grid in front of an *x*-ray film (from Mikael Häggström, Wikimedia, © creative commons).

One way of defining the x-ray contrast C is to take the difference of transmitted intensity in neighboring regions:

$$C = I_2 - I_1 = I_0 \left[\exp\left(-\left(\frac{\mu_2}{\rho_2}\right)\rho_2 t_2 \right) - \exp\left(-\left(\frac{\mu_1}{\rho_1}\right)\rho_1 t_1 \right) \right]. \tag{8.1}$$

Here we have assumed that the mass attenuation coefficients μ_1/ρ_1 and μ_2/ρ_2 are constant within their respective thicknesses t_1 and t_2. Alternative definitions of

---

3 For didactic reasons, we sometimes show x-ray film recordings in which the grayscale is logarithmically proportional to the exposure intensity. X-ray films are no longer used and replaced by DRs.

contrast, which can be found in the literature, are the Michelson contrast[4] defined by the normalized intensity difference:

$$C_M = \frac{I_2 - I_1}{I_2 + I_1},$$ (8.2)

and the Weber contrast,[5] defined by intensity difference normalized by the higher of the two intensities:

$$C_W = \frac{I_2 - I_1}{I_2}; \quad I_2 \geq I_1.$$ (8.3)

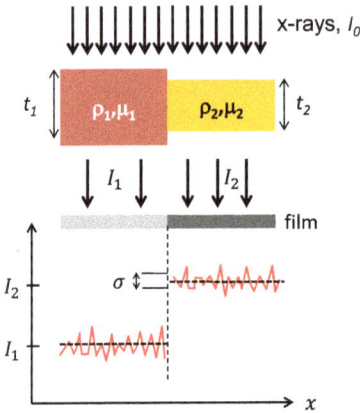

Fig. 8.7: X-ray attenuation by a laterally inhomogeneous medium recorded with an x-ray film that forms a grayscale contrast. The lower panel shows the transmitted intensity recorded with an x-ray detector. σ is the mean square amplitude of the noise.

X-ray contrast between body parts is often difficult to obtain, particularly between soft tissues like muscles and fat. However, the contrast between soft tissues and bones is pronounced because of the high mineral content of bones. The contrast between tissues and air is also significant, which is important for imaging the lung. A few examples for x-ray contrast values are provided in Tab. 8.1.

Often contrast is determined as the difference between the signal intensity and a noisy background $\Delta I = I_1 - I_{\text{noise}}$. If the noisy background signal has a mean square amplitude of σ, then the signal-to-noise ratio is defined as

$$\text{SNR} = \frac{\Delta I}{\sigma}.$$ (8.4)

More to contrast and SNR is presented in Section 8.4.

---

4 Albert A. Michelson (1852–1931), American physicist and Nobel laureate 1907.
5 Ernst Heinrich Weber (1795–1878), German physiologist and anatomist.

Tab. 8.1: X-ray attenuation contrast between different body parts of 1 cm thickness and for an x-ray photon energy of 50 keV.

C(tissue-bone)	+ 0.26
C(tissue-air)	− 0.2
C(muscle-blood)	− 0.003

Note: The contrast is defined here as the intensity difference, according to eq. (8.1).

8.3.2 Attenuation profile

For a more realistic scenario, we consider an x-ray beam parallel to the x-direction and homogeneously extended in the y,z-plane, as depicted in Fig. 8.8. The beam has a broad spectral distribution typical for a bremsstrahlung spectrum. An object with a spatially varying attenuation coefficient $\mu(x, y, z, E)$ intersects the beam.

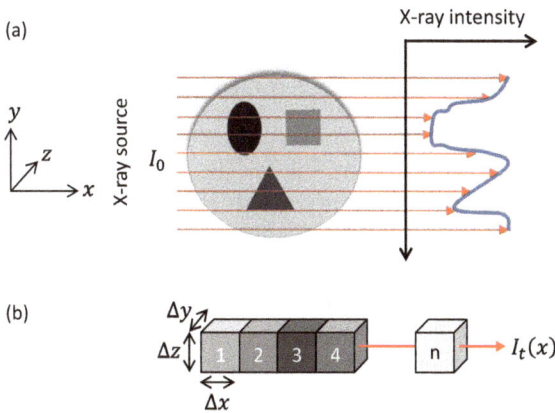

Fig. 8.8: Geometry of x-ray attenuation via an inhomogeneous object. (a) The blue line shows an attenuation profile recorded by an x-ray film or a flat panel detector. The length of the red arrows symbolizes the local x-ray intensity behind the object. Many line scans combine to a two-dimensional picture in the y,z-plane. (b) In a first approximation, the object is subdivided into a sequence of voxels of equal volume and constant attenuation coefficient.

The easiest way to calculate the attenuation of such an inhomogeneous object is to subdivide it into small equal volume elements, called voxels (Fig. 8.8(b)). Then we sum over the attenuation coefficients of all voxels in the beam path, assuming an average and homogeneous attenuation coefficient in each voxel v of size $v = \Delta x \cdot \Delta y \cdot \Delta z$. The voxel size is adapted to the x-ray beam resolution. If the voxels are smaller than the resolution, we cannot distinguish between different voxel rows. If the voxels are bigger than the resolution, the assumption of a homogenous attenuation coefficient inside may be violated. Typical x-ray voxels have an edge size of about 0.5 mm.

The transmitted intensity $I(x)$ after passing the x-ray beam through all voxels in the x-direction is

$$I(x) = I_0 \exp(-(\mu_1 + \mu_2 + \cdots + \mu_n)\Delta x) = I_0 \exp\left(-\left(\sum_{i=1}^{n} \mu_i\right)\Delta x\right). \qquad (8.5)$$

The intensity evaluation has to be repeated for all voxels in the (x,y)-plane, which constitutes one attenuation slice. Slices in the z-direction (head to toe) are later added on, similar to taking slices in magnetic resonance imaging.

In the integral form, the transmitted intensity in the x-direction for fixed y- and z-positions is

$$I(y,z,E) = I_0(y,z,E) \exp\left(-\int_0^t \mu(y,z,E)dx\right). \qquad (8.6)$$

Here the line integral is taken over a varying absorption coefficient along the x-direction and t is the thickness in the x-direction. The line integral is repeated for all points within the illuminated area of the object. Taking into account the energy-dependent quantum efficiency $\eta(E)$ of the detector, the recorded intensity in the detector is

$$I_{det}(y,z) = \int_0^{E_{max}} I_0(y,z,E)\eta(E) \exp\left(-\int_0^t \mu(y,z,E)dx\right) dE. \qquad (8.7)$$

Here the energy integral goes over the bremsstrahlung spectrum. The intensity recorded by the detector $I_{det}(y,z)$ is called a *projection image* of $\mu(y,z,E)$ [1], meaning that the voxels in the object are projected into pixels on the detector. Line integrals over longer distances t cause lower output intensity at the detector and vice versa. X-ray detectors, including x-ray films, collect the transmitted intensity over the exposure time T_{exp}. The number of photons per area integrated over the exposure time T_{exp} is then:

$$N_{film}(y,z) = \int_0^{T_{exp}} dT \int_0^{E_{max}} I_0(y,z,E)\eta(E) \exp\left(-\int_0^t \mu(y,z,E)dx\right) dE. \qquad (8.8)$$

$N_{film}(y,z)$ is also the integrated dose to the film, which causes darkening and increases the OD of the film. Repeating the line integral for all y-positions generates an attenuation profile of one particular slice in the z-direction. Figure 8.8(a) illustrates an attenuation profile detected by an x-ray film or a flat detector camera by illuminating an object with a parallel beam. Repeating the line profiles for successive slices in the z-direction will produce a projection image of the body.

8.4 X-ray recording

8.4.1 Film radiography

Originally, x-ray radiographs were recorded with x-ray films.[6] As film recording is almost entirely replaced by digital recording (DR, see next section), film recording remains only of historical interest. Therefore, a brief summary of some physical properties of x-ray films is justified but may be skipped if not incidentally required. Figure 8.9(a) shows the layering in a standard x-ray film. In the emulsion layer, transparent AgCl/AgBr particles are embedded in a water soluble gelatin like we know it from traditional camera rolls. Upon x-ray exposure, the silver halogenides become reduced and convert to metallic Ag. If at any specific spot at least five or more Ag^+ ions are reduced, they cluster up to a metal particle storing a latent image. The silver particles grow in size to amplify the latent image after developing in a watery solution containing a reducing agent. Subsequently, the unexposed halogenides and all other residuals are washed out in an oxidizing solution called fixer. After drying, the film is inspected with the help of a viewing lightbox.

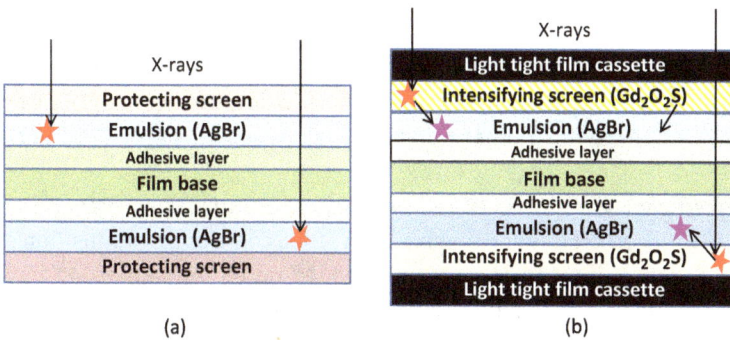

Fig. 8.9: Left: layer arrangement of an x-ray film; right: layering in an x-ray film with intensifying fluorescence screen.

When watched against the light source, optically dark areas correspond to a high density of grainy silver particles. The high particle density is due to a high x-ray exposure implying low x-ray attenuation in the exposed body part. Vice versa, lighter areas on the film are less exposed to x-rays, meaning a higher attenuation in body parts. Bones appear bright, soft tissues like the lung appears dark on the film. The x-ray radiograph is a negative image: higher x-ray exposure of the film produces darker images. An example of a chest radiograph is shown in Fig. 8.10.

---

6 This part can be skipped as film recordings are mostly replaced by digital recording

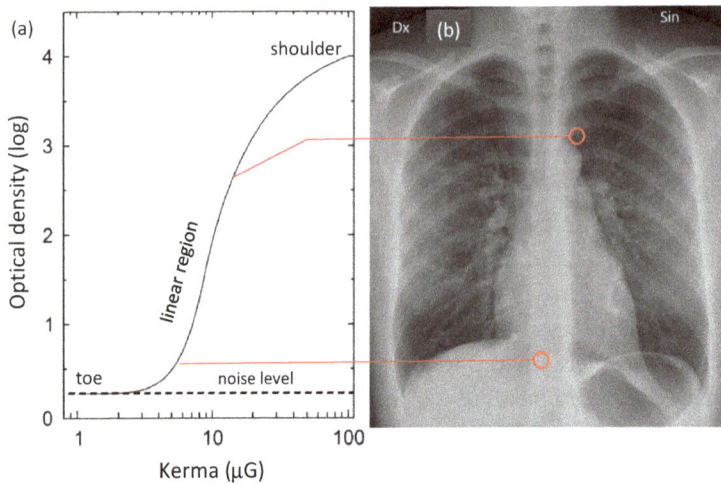

Fig. 8.10: (a) Optical density of an x-ray film plotted versus the x-ray exposure dose to the film. Note the logarithmic scale of the x- and y-axes. (b) Chest radiograph with differences in gray shades. The lungs appear dark while the bones are bright. Dx and Sin stand for "right" and "left," respectively (from Mikael Häggström, Wikimedia, © creative commons).

The lateral resolution of x-ray films depends on the density of AgCl/AgBr particles in the emulsion and the grain size of silver particles after reduction. Both can be controlled to a certain extent. The resolution is usually quoted as 0.1–0.06 mm, which is the "pixel" size of radiographic films.

The quantum efficiency of the described film process is rather low. Only 5% of the x-ray intensity is absorbed in the emulsion layer, thus contributing to the film's darkening. An additional, intensifying screen is added on either side of the film containing rare earth metal ions in more advanced versions. In most cases, gadolinium oxysulfide (Gd_2O_2S) is used for this purpose, as sketched in Fig. 8.9(b). X-rays are more effectively absorbed by the high Z material of Gd_2O_2S and sequentially emit fluorescence light in the visible green region. Thus, x-rays are effectively converted into visible light, which, in turn, exposes the silver halide emulsion layer. The resulting conversion efficiency is increased by a factor of 3 to about 15%. The screen is placed inside of a light-tight cassette and tightly pressed against the film during exposure to avoid any blurring.

Although the conversion of x-ray intensity into visible light intensity I_{vis} is linear, the conversion of visible light into darkening of the x-ray film is not a linear function but logarithmic. The OD of the x-ray films D_{opt} is defined as the negative logarithm to the base of 10 of the film transparency T, when viewed against a visible light source:

$$D_{opt} = -\log_{10} T; \ T = \frac{I_{vis}}{I_{0,vis}}. \tag{8.9}$$

An optical density of 1 corresponds to a transparency of 10%, requiring an x-ray exposure of 7 µGy. About 20% transparency is reached at an exposure of 10 µGy. The optical density as a function of exposure is plotted in Fig. 8.10(a). The relevant linear region in this double logarithmic plot lies in between the toe and the shoulder of the curve. The lower toe part is affected by noise; the film goes into saturation for high exposure.

Now we have to relate the optical density D_{opt} to the x-ray exposure of the film expressed in kerma [1, 2]. First, we realize that the transmitted light depends exponentially on the number of grains g per unit area sensitized by x-rays:

$$I_{vis} = I_{0, vis} e^{-g\sigma}, \tag{8.10}$$

where σ is the cross section of a typical Ag-grain. The number of sensitized grains per unit area increases with the dose (kerma):

$$g = G(1 - e^{-kX}). \tag{8.11}$$

Here G is the total number of grains (sum of unexposed and exposed grains), X is the kerma, and k is an appropriate coefficient. Using eq. (8.11), we find for the optical density:

$$D_{opt} = -\log T = \sigma g \log_{10}(e) = \log_{10}(e) \cdot \sigma \cdot G(1 - e^{-kX}) = D_{opt, max}(1 - e^{-kX}). \tag{8.12}$$

The last equation is intuitively clear: $D_{opt, max}$ relates to a completely black film, which is approached in saturation. In the linear region of the optical density (see 8.10(a)), the dose is linearly proportional to the logarithm of the kerma:

$$D_{opt} = \Gamma \cdot \log_{10}\left(\frac{X}{X_0}\right). \tag{8.13}$$

X_0 is the kerma without attenuation in the body, the so-called exposure dose. The factor Γ is known as the film Γ-value and corresponds to the slope in the linear region of Fig. 8.10(a). The linear region is also the region of highest contrast. Therefore, x-ray exposure and dose to the patient should preferably lie in the linear region.

We adapt the contrast definition in eq. (8.3) to the current situation and take the kerma difference divided by the exposure kerma X_0:

$$C = \frac{\Delta X}{X_0}. \tag{8.14}$$

Using eq. (8.11), we can now relate the kerma contrast to the film dose contrast, which is written for small differences:

$$\Delta D_{opt} = \Gamma \cdot \left(\log_{10}\left(\frac{X_1}{X_0}\right) - \log_{10}\left(\frac{X_2}{X_0}\right)\right) \cong \log(e) \cdot \Gamma \cdot \frac{\Delta X}{X}. \tag{8.15}$$

Values for the kerma contrast are given in Tab. 8.1. The prefactors are about 1; so the last eq. (8.15) tells us that the optical density difference ΔD_{opt} is linearly related to the *kerma contrast* $\Delta X/X_0$.

Finally, we consider the noise of the film blackening, which may limit the contrast. The noise in the film has three main contributions: (1) fluctuations in the number of absorbed x-ray photons per unit area; (2) fluctuations in the absorbed x-ray photon energies; and (3) fluctuations in the number of AgCl/AgBr particles per unit area. These fluctuations lead to a random background noise in the film, which can be estimated as follows [2]:

$$\frac{\Delta X}{X_0} = \frac{1}{\sqrt{A\varepsilon N}}. \tag{8.16}$$

Here, A is the exposed area, ε is the interaction efficiency, and N is the time-integrated number of x-ray photons. With some more manipulation, the last equation can be shown to be proportional to the root of the grain size. Noise is an important issue in x-ray radiography. We will come back to this point at several other places. A comprehensive treatment of the signal-to-noise optimization of medical imaging systems can be found in [5].

8.4.2 Fluoroscopy

In contrast to film radiography, the fluoroscope is the video camera in medicine [1, 6]. Fluoroscopes provide *real-time radiographs* (RTR) that allow monitoring surgical procedures in vivo. A schematic layout of a traditional x-ray fluoroscope is shown in Fig. 8.11. In a fluoroscope, high-energy x-ray photons are first converted to low-energy visible light. The visible photons are then converted to photoelectrons. The photoelectrons, in turn, are accelerated from the cathode to the anode When the photoelectrons hit the anode, they are converted back to photons, and made visible after analog-digital conversion on a video screen.

In some more detail, fluoroscopes work as follows. X-ray photons that have penetrated the human body pass through a glass window of the fluoroscope housing. Behind the glass window, the x-rays are absorbed by a fluorescent converter to visible light (see Fig. 8.12). For this first conversion, CsI is preferred because of the high conversion efficiency and the columnar crystal growth that guides the converted light to the photocathode. Both cesium ($Z = 55$) and iodine ($Z = 53$) are high Z ions that can readily absorb hard x-rays. The fluorescent light then arrives at the photocathode. At the photocathode, the compound $SbCs_3$ converts the photons in photoelectrons. The photocathode is negatively charged. Therefore, the freshly generated photoelectrons are immediately accelerated to the positively charged anode. Focusing electrodes guide the photoelectrons to the anode at the other end of the tube. Simultaneously, the photoelectrons are accelerated in a high potential field (25–35 kV).

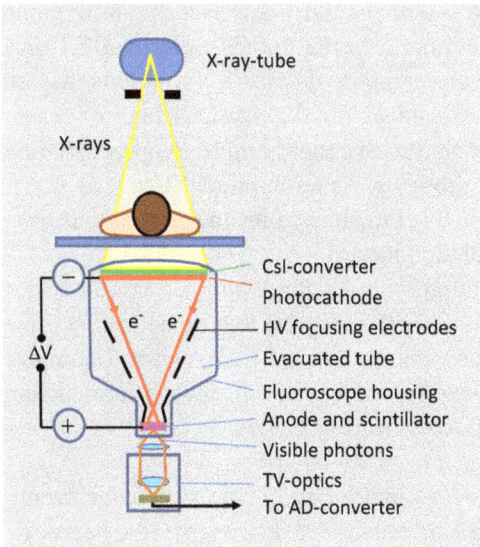

Fig. 8.11: Schematic layout of a fluoroscope. X-rays penetrating the glass housing of the fluoroscope are converted into electrons in a CsI/ZnCdS:Ag converter layer. The electrons are then accelerated to the anode, where they are converted back to light in a scintillator crystal. The visible photons go through an optical microscope and the image is displayed on a TV screen.

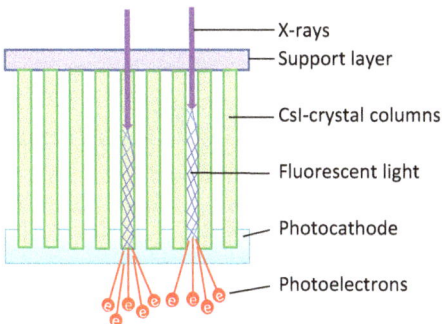

Fig. 8.12: CsI converter and photocathode in a fluoroscope. The fluoroscope first converts x-ray photons into visible photons. The visible photons are then converted into photoelectrons in the photocathode. Electrons emanating from the photocathode are accelerated to the anode.

The fast electrons are converted back to visible light by an anode coating that consists of an ZnCdS:Ag compound. Each electron arriving generates about 10^3 photons. This high photon intensification is achieved through two effects: for one, through the acceleration of electrons from the cathode to the anode, and second, through the reduction in the image size between cathode and anode. Finally, the light goes through an optical system and an analog-digital converter for display on a video screen.

Fluoroscopes are often mounted on a semicircular C-arm (see Fig. 8.16c): one side holds the x-ray generator, and the other side the fluoroscope. Together they can spin around the patient to oversee and support surgical interventions. A dose of a few nGy is sufficient for fluoroscopic imaging. The fluoroscope's high x-ray sensitivity helps keep the dose level low. Nevertheless, the C-arm is an open radiation system; therefore, the staff's radiation protection is an important issue. As previously mentioned, fluoroscopes were also used in shoe stores in the past, but have now been phased out because of the radiation hazard.

An important advantage of digital fluoroscopy is that it allows a pulsed operation. Here the x-ray tube is switched on and off according to a predefined scheme that sets the x-ray pulse frequency, pulse amplitude, and pulse width. The image recording is then phase-locked to the x-ray exposure. The advantage is a reduction of radiation dose to the patient and clinicians during the video-controlled surgery with a reasonable video quality at a frame rate of 20–30 images/s.

In the meantime, fluoroscopes are being replaced by TFT or CMOS (complementary metal-oxide-semiconductor) flat screens because of their higher x-ray sensitivity. In the future, it can be expected that the radiation hazard to the staff will be overcome by using robotics.

8.4.3 Flat panel radiography

There have been many efforts to use digital sensors for x-ray imaging. However, because of the large FOV, CCD chips are too small, and stacking them up into an array is too costly. The easiest way to get a digital image is by laser scanning an x-ray film. While this was often practiced in the past, it was not a step forward toward digital image processing and life x-ray videos that is expected from a complete DR. In recent years, significant DR progress has been achieved [7]. Two types of digital detectors are distinguished: direct and indirect. Their principal design features are shown in Fig. 8.13. By direct detectors, we understand the conversion of x-ray photons into electron–hole pairs in photodiodes. The charge is collected in a high-voltage potential at a suitable pixel electrode. Indirect detectors convert the x-ray energy first into visible photons, which are then detected by photodiodes in a second step. Light-sensitive scintillators consist either of Gd_2O_2S or CsI. Gd_2O_2S has a higher conversion efficiency, whereas the columnar growth of CsI has a better photon guiding ability to the photodiodes underneath. This means that the generated visible photons travel straight to the photon detector without being scattered in the scintillator material nor detected by a neighboring diode. Suppression of scattering in scintillator materials is important to avoid blurring and to increase contrast. Amorphous silicon (a-Si) on glass substrates is commonly used as a photodiode. The *thin-film transistors* (TFT) have a typical size of 0.2×0.2 mm^2 or less, each acting as a photodiode and representing a pixel in a large detector array. The electrical signal generated by the photodiode

(a)

x-rays

HV bias
electrode

amorphous
selenium

SCANNING CONTROL

MULTIPLEXER

storage
capacitor

TFT

pixel
electrode

(b)

x-rays

phosphor
screen

SCANNING CONTROL

MULTIPLEXER

TFT

Photodiode

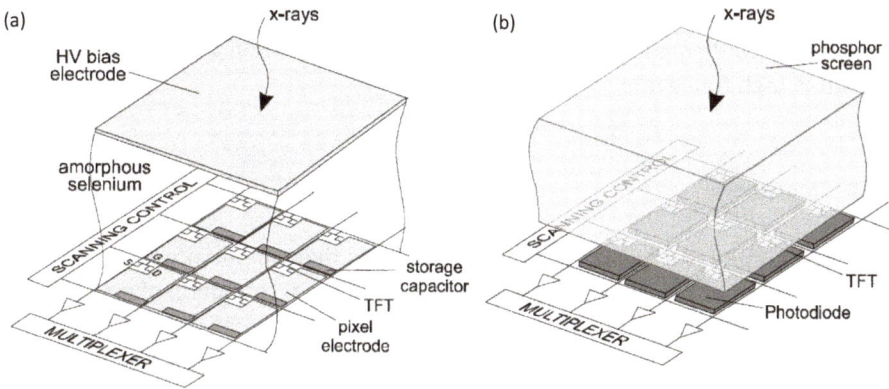

Fig. 8.13: (a) Direct and (b) indirect flat panel x-ray detectors (reproduced from http://hepwww.rl.
ac.uk/Vertex03/Talks/Row/Rowlands.pdf).

is then amplified and encoded by electronics into a gray or color scale, generating sensitive and accurate x-ray images.

Flat-panel x-ray detectors are costly. On the other hand, they are more sensitive than x-ray films and fluoroscopes. The higher quantum efficiency allows a lower dose of radiation for any picture taken. Presently, commercial TFT panels have sizes up to 41×41 cm^2, with an array of 2048×2048 diodes. This yields a spatial resolution of 0.2 mm, considering only the physical size of the diodes. Compared to traditional x-ray films, the spatial resolution of TFT panels is lower by a factor of 2–3, which is tolerable in most cases.

Upon conversion of x-rays to visible light and from light to electron–hole pairs, the information about the original x-ray photon energy gets lost in indirect detectors. Therefore, indirect detectors are energy integrating devices. However, the energy information is retained in direct detectors if the energy resolution is sufficiently high. Now imagine that each pixel would record not only the number of x-ray photons but also their energy, then the flat panel detector has an added energy dimension. Each 2D image is then a multiple energy image, and contrast can be varied by tuning through the energy landscape. Such detectors have indeed been developed for CT scanners. The new detector which enables this quantum leap is equipped with a layer of cadmium telluride (CdTe) single crystal. The principal design features are also applicable to flat panel TFTs [8].

For smaller flat panel displays, x-ray imagers based on CMOS gain increasing attention and importance, despite their high cost [9]. CMOS sensors are similar in design to TFT. However, the a-Si diodes are replaced by crystalline Si (c-Si). Advantages of the CMOS sensor include: (1) higher readout speed; (2) lower noise; (3) wider dynamic range; and (4) and higher quantum efficiency. These favorable properties are due to the much higher electron-hole mobility in c-Si than in a-Si. The lower noise, higher speed, and reduced pixel size (<100 μm) are of benefit for radiographs in demand of high resolution, such as mammography, dental applications, and cone-beam CT imaging. Compared to TFT, the wider dynamic range and higher sensitivity render a lower dose to the patient while maintaining high-resolution images and life videos with a high frame rate.

8.4.4 Comparison

Film radiography is a low-cost option in hospitals that cannot afford more costly digital equipment. Film radiography provides the highest resolution of steady-state images. The handling of the film cassettes before and after x-ray exposure requires, though, a fair amount of experience. Digital post-image processing is barely possible. Images obtained and examined with the help of a lightbox must be interpreted by an experienced physician.

Fluoroscopes allow DR of x-ray videos. The tube, sensors, cathode, and anode have been optimized over many years. Although an analog technique, advanced optics and analog-digital converter provide digital videos with an adequate frame rate and image quality. Nevertheless, in recent years, fluoroscopes have been replaced by much more costly TFT flat panels. TFT panels provide higher sensitivity and a wider dynamic range, only surpassed by even faster CMOS panels used for special applications. Common to all digital radiography recordings is the possibility of post-processing the images and enhancing the regions of interest.

The following specifications are important for the assessment of DRs:

1. The absolute sensitivity threshold is the smallest number of x-ray photons the sensor recognizes as a signal instead of noise. The absolute sensitivity threshold is determined from the value where the SNR is equal to 1, meaning that the signal is as large as the noise. The lower the threshold value, the higher the sensor sensitivity.
2. Noise is composed of dark noise and photon noise. The dark noise is due to the hardware electronics of the sensor; photon noise is caused by blackbody radiation from the environment and may eventually be reduced by cooling.
3. Dynamical range of a detective system is the ratio of the maximum number of electrons detected to the minimum number given by the noise level.
4. Quantum efficiency is the ratio of electrons generated per x-ray photons absorbed.[7]
5. Pixel size is given by the physical size of individual photodiodes. The pixel size determines the resolution of the image. The size of the pixel matrix is a question of preference and the aimed FOV or region of interest (ROI).
6. The K-factor is an amplification factor that scales the number of electrons collected to an incremental change of the gray scale by one unit. The K-factor is determined by the sensitivity and dynamic range of the sensor.

---

7 Quantum efficiency can be defined in several other ways. For example, as the number of photons detected relative to the number of photons incident on a detector pixel.

X-ray films, fluoroscopes, and flat panel detectors are the main recording media for x-ray radiography. The latter two devices can be used for life video monitoring of surgical procedures. !

8.5 Counting statistics, noise, quantum efficiency

8.5.1 Counting statistics

The following two sections focus on counting statistics and noise, which concerns
all imaging modalities not just x-ray radiography. We give a brief overview of these
topics and refer back to them in the next chapters on SPECT and PET. The outline
below is not more than a guide to more comprehensive treatments in textbooks, for
instance, the one by Huges et al. listed under "Further reading."

DR is a question of photon counting and is therefore governed by counting statistics. Each pixel is a photon detector, and the counting is incremental in each
pixel. Assuming that each pixel is independent and no correlation or proximity effects exist to neighboring pixels, an arbitrary pixel is representative of all others.
We briefly outline the counting statistics of such a pixel.

The counting statistics of photons is described by the binomial distribution that
approximates the Poisson distribution for a large number of photons. This can be
seen as follows. Suppose there are N identical photons hitting the pixel during the
time period Δt, and the probability of being detected is p. In that case, the number
of counted photons in time interval Δt is on the average $n = pN$. More generally, we
may ask what is the probability of detecting n photons out of N incident photons, if
the probability of each single photon detection is p. If one arbitrarily picks n indistinguishable photons out of N, then p^n is the probability that these n photons are
detected, and $(1-p)^{N-n}$ is the probability that they are not detected. Since we can
randomly arrange the sequence of detected and missed photons, the probability is
described by the binomial function:

$$P(n) = \frac{N!}{(N-n)!n!} p^n (1-p)^{N-n}. \tag{8.17}$$

In the limit of $n, N \to \infty$, while the ratio n/N remains finite, we call $pN = \mu$ the expectation value of the binomial distribution and rewrite:

$$P(n) = \frac{N!}{(N-n)!n!} \left(\frac{\mu}{N}\right)^n \left(1 - \frac{\mu}{N}\right)^{N-n}. \tag{8.18}$$

In the limit $N \to \infty$, the binomial distribution yields the Poisson[8] distribution, written as

---

[8] Siméon Denis Poisson(1781–1840), French mathematician and physicist.

$$\lim_{N \to \infty} P(n) = \frac{\mu^n}{n!} e^{-\mu}. \tag{8.19}$$

The Poisson distribution is described by a single parameter μ, which determines both the expectation value and the variance. The expectation value or mean number follows from

$$\langle n \rangle = \sum_{n=0}^{\infty} nP(n) = \mu, \tag{8.20}$$

and the variance is defined by

$$\sigma^2 = \left\langle (n - \langle n \rangle)^2 \right\rangle = \mu. \tag{8.21}$$

For large μ one can show that the Poisson distribution is approximated by the Gauss[9] or normal distribution of a variable x with the expectation value μ and variance μ:

$$P(x) = \frac{1}{\sqrt{2\pi\mu}} \exp\left(-\frac{(x-\mu)^2}{2\mu} \right). \tag{8.22}$$

Both distributions are compared in Fig. 8.14. With increasing μ, both distributions become indistinguishable.[10] With these distributions and expectation values, we can now determine the signal-to-noise ratio, defined by

$$\text{SNR} = \frac{\text{mean number of photons}}{\text{standard deviation (noise)}} = \frac{\mu}{\sigma} = \frac{\mu}{\sqrt{\mu}} = \sqrt{\mu}. \tag{8.23}$$

The last equation shows that the SNR increases with the square root of the mean number of photons counted. Counting longer increases the mean number and improves the SNR. This is the normal procedure in a laboratory. However, in a clinic, counting over a longer time implies a higher dose to the patient. Therefore, there is a trade-off between acceptable image quality and tolerable radiation dose to the patient.

8.5.2 Noise and quantum efficiency

For judging an x-ray image quality, we also need to investigate the contrast. We have already introduced the concept of contrast in Section 8.2. Here, based on statistics, the *contrast-to-noise ratio* (CNR) can be redefined by considering the SNR of two different pixels A and B:

---

9 Johann Carl Friedrich Gauss (1777–1855), German mathematician, astronomer, and physicist.
10 The properties of the Gaussian distribution are further discussed in Chapter 8 of Volume 3.

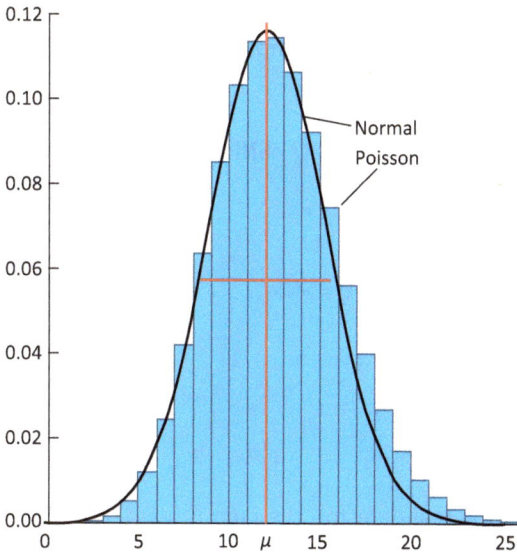

Fig. 8.14: Comparison of a Poisson distribution and a Gaussian or normal distribution for an expectation value $\mu = 12$.

$$\text{CNR} = \frac{\text{mean number difference in } A \text{ and } B}{\text{noise}} = \left| \frac{\mu_A - \mu_B}{\sigma} \right| = |\text{SNR}_A - \text{SNR}_B|. \qquad (8.24)$$

Instead of comparing two different pixels, comparing two different areas of an image is better for evaluating the image quality. In general, it turns out that two different areas can just be distinguished if the CNR ≥ 2.

When considering *noise*, we have to distinguish between background noise of the detection system and noise introduced as a result of photon counting. The background noise is the noise of the detecting system (flat panel), when all photon sources are turned off. This noise is also called dark noise. It is immanent to the system (electronics, environment, etc.) and can even be reduced by special measures but not by improving the counting statistics. The stochastic noise is added to the background noise when taking an image.

Now we consider an ideal detector with optimal SNR that converts all incoming x-rays in a useful image. A real detector also converts x-rays into a useful image, however, with a lower SNR. The *noise equivalent quanta* (NEQ) is the number of photons per pixel required by an ideal detector to create the same image quality as that of the non-ideal detector. Non-ideal detectors need more photons Q for the same image quality, because the quantum efficiency is lower. The ratio is called *detective quantum efficiency* (DQE):

$$\text{DQE} = \frac{\text{NEQ}}{Q}. \qquad (8.25)$$

We want to express this ratio in terms of SNRs of the real and ideal detectors. To this end, we rewrite NEQ and Q in terms of output and input SNR ratios. The output SNR of the ideal detector is

$$\mathrm{SNR_{out}} = \frac{\mathrm{NEQ}}{\sqrt{\mathrm{NEQ}}} = \sqrt{\mathrm{NEQ}}; \quad \text{and} \quad \mathrm{NEQ} = \mathrm{SNR_{out}^2}. \tag{8.26}$$

The input SNR for the nonideal detector is

$$\mathrm{SNR_{in}} = \frac{Q}{\sqrt{Q}} = \sqrt{Q}; \quad \text{and} \quad Q = \mathrm{SNR_{in}^2}. \tag{8.27}$$

Therefore, we find for the DQE:

$$\mathrm{DQE} = \frac{\mathrm{NEQ}}{Q} = \frac{\mathrm{SNR_{out}^2}}{\mathrm{SNR_{in}^2}}. \tag{8.28}$$

For an ideal detector, this ratio is 1; for a real detector, the ratio is less than 1. This is the quantity usually quoted for comparing detectors. The mathematics that leads to the assessment of detectors via the DQE concept is quite extensive. For further information and a more detailed discussion, we refer to [5]. A simple approach can be found in [10].

8.6 System integration

8.6.1 Projection radiography

Figure 8.15 depicts all parts required for taking an x-ray projection radiograph: x-ray source, filter, support table for body and parts to be imaged, oscillating grid, and detector.

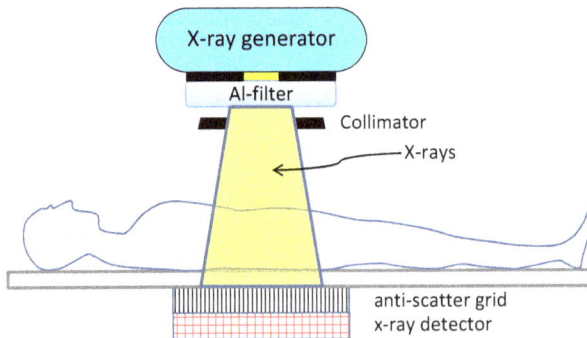

Fig. 8.15: Schematics of the arrangement for taking x-ray projection radiography images.

Depending on which part of the body is to be imaged, three main types of x-ray machines are used for projection radiography (Fig. 8.16): vertical panel, horizontal table with tilt option, and C-frame. For reasons of comparison, the standard distance between x-ray tube and detector (source-to-image distance, SID) is for vertical panels 183 cm, and for horizontal tables 100 cm.

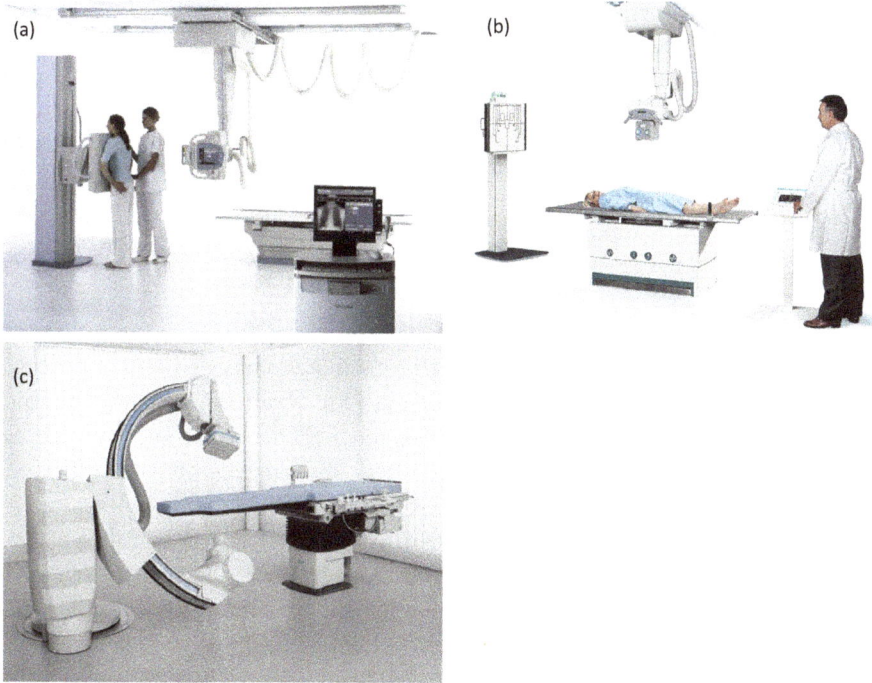

Fig. 8.16: X-ray machines for projection radiography: (a) vertical panel for chest radiographs (http://medicaloutfitter.net), (b) horizontal table with tilt option (http://www.medicalexpo.com), and (c) fluoroscopy with a C-arm for real-time x-ray imaging during surgery (reproduced from http://usa.healthcare.siemens.com).

The C-frame has the smallest detector field but the largest flexibility and is often used in surgery rooms for monitoring. In recent years with the use of flat-panel detectors, the sensitivity has dramatically increased, such that RTR with high time and spatial resolution is possible during surgical procedures. An example of monitoring the proper placement of stents during cardiac surgery is shown in Fig. 8.17.

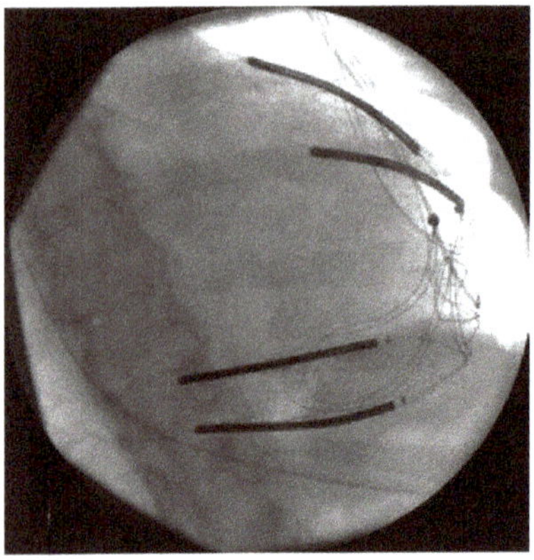

Fig. 8.17: X-ray image recorded with a C-arm mobile imaging machine during cardiac surgery and placement of stents (reproduced from https://www.healthcare.siemens.de/surgical-c-arms-and-navigation/mobile-c-arms/arcadis-avantic# publication Arcadis Avantis).

8.6.2 Mammography

A special kind of radiography is the mammography of the female breast for tumor recognition at an early stage. To secure high sensitivity to any changes in the soft tissue, the voltage of the x-ray generator is reduced to about 35 kVp, and often the Mo-Kα characteristic radiation from a molybdenum target is used as an x-ray source. During exposure, the breast is squeezed between adjustable compressor plates in order to produce a homogeneous absorption thickness, as indicated in Fig. 8.18(a). A collimated fan beam delivers a magnified image. Traditionally highly sensitive black/white films were used for mammography. But recently, *digital mammography*, also called full-field digital mammography (FFDM), replaced x-ray films. The x-ray photons are converted by a phosphorous screen to visible light and funneled off via fiberglass to a recording CCD detector. The advantage of DM is higher sensitivity and lower radiation dose per screening. Furthermore, post-processing of the digitally recorded image can enhance spots of interest for further inspection. A variant of the FFDM is digital *breast tomosynthesis* (DBT), which is a kind of 3D CT imaging of the breast [11]. In DBT, images of the breast are taken under a few different angles, usually three.

! Vertical, horizontal, and C-frame are the main designs for x-ray projection and video radiography. For mammography, special arrangements are necessary to recognize suspicious tissues in the breast: soft x-rays, squeezing, and magnification.

8.7 Attenuation contrast enhancement

When imaging the upper chest, the gastrointestinal (GI) tract, or blood vessels, the contrast to the surrounding tissue is often insufficient to recognize fine details, meaning that the attenuation profile is too flat. Table 8.1 gives a list of typical contrast

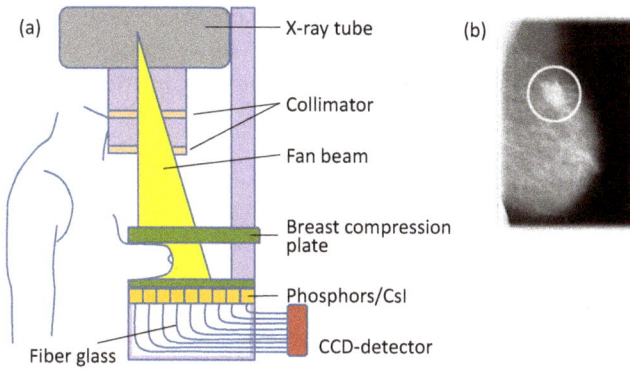

Fig. 8.18: (a) Fan beam (yellow) for image magnification of the breast squeezed between plates. The recording is performed with a photon converter, fiber glass, and CCD detector. (b) Mammography of a female breast for tumor recognition. In the encircled area, a higher absorption is observed, which could indicate the presence of tumorous tissue.

values. In this situation, it is desirable to enhance the contrast artificially. There are several means for achieving higher attenuation contrast:
1. Injection of contrast agents
2. Digital subtraction angiography (DSA)
3. Dual x-ray energy (DxE) imaging

8.7.1 Contrast agents

Any contrast-enhancing agent (CA) should fulfill the following conditions:
a. Stronger attenuation than the surrounding.
b. Easy take-up by the body parts to be imaged without altering its shape or function.
c. Short biological lifetime.
d. Pharmacological and physiological inertness.

Two examples are given in the following concerning the imaging of the GI tract and the imaging of blood vessels.

8.7.1.1 Gastrointestinal tract
For imaging, the GI tract, barium with a high atomic number ($Z = 56$) is used. Recalling from Chapter 4 that the absorption coefficient scales with the third power of the atomic number $\mu \sim Z^3$, Ba is clearly a good candidate for high x-ray absorption. For administering to the patient, barium is bound in barium sulfate ($BaSO_4$), a dry, white, chalky powder mixed with water to make a barium sulfate liquid. About 30 min after filling the liquid into the GI tract, it coats the inside walls of the esophagus, stomach, and

intestines. Hence, the inside wall coatings, size, shape, and contour become visible on x-ray radiographs by a white (highly absorbing) contrast. An example is shown in Fig. 8.19. The $BaSO_4$ slurry is used solely for diagnostic studies of the GI tract.

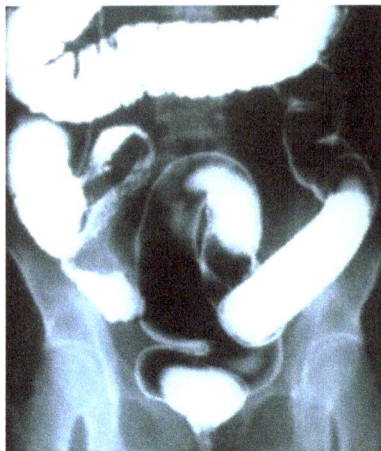

Fig. 8.19: Contrast enhancement of the gastrointestinal tract with the help of barium sulfate powder in solution (reproduced from https://openi.nlm.nih.gov/, © creative commons).

8.7.1.2 Angiography

Imaging of blood vessels with projection radiography is called *angiography*. The name comes from the Greek word angeion for "vessels." For contrast enhancement, a contrast agent is injected into the bloodstream. The liquid "dye" contains the high Z-element Iodine ($Z = 53$). Iodine may be bound in an organic (nonionic) or ionic compound. The chemical structure of iopromide is shown in Fig. 8.20. It is one of the most frequently used CAs for x-ray radiography, including CT imaging of the brain. Since these small molecules in the bloodstream are cleared out quickly by the kidneys, renal tolerance has to be considered before administering. If tolerance is not an issue, the renal clearance of iopromide can also be used for intravenous urograms, i.e., the radiography of kidneys, ureters, and bladder. Nevertheless, there is considerable interest in replacing iodine contrast agents with nanoparticle contrast media with less nephrotoxicity.

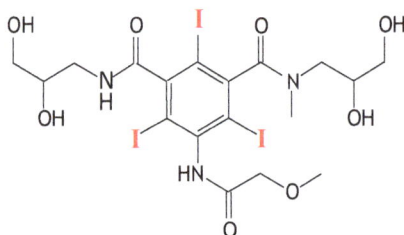

Fig. 8.20: Chemical structure of the iodinated contrast agent iopromide. Three iodine atoms, marked in red, are attached per molecule of iopromide.

8.7.2 Digital subtraction angiography (DSA)

Often angiography is combined with digital subtraction radiography, also known as *digital subtraction angiography*. For this method, two digitally recorded radiographs are taken before and after administering the dye. Obviously the object to be imaged should not move in between. The DSA at a position (x, y) is then an image of the absorption coefficient difference $\Delta\mu$ before (mask) and after injection (dye):

$$\mathrm{DSA}(\Delta\mu) = \ln\left(I_{\mathrm{dye}}(x, y)\right) - \ln\left(I_{\mathrm{mask}}(x, y)\right) \qquad (8.29)$$

The image taken before injection is called mask, as it masks off the field of interest. A DSA radiograph of blood vessels in a hand is reproduced in Fig. 8.21. The fine branching of the blood vessels would not be visible in a normal radiograph. A sequence of images is taken of the same area with a rate of 1–6 frames per second. Then it can be decided during image processing how many of them need to be subtracted from the mask in order to gain an optimized contrast.

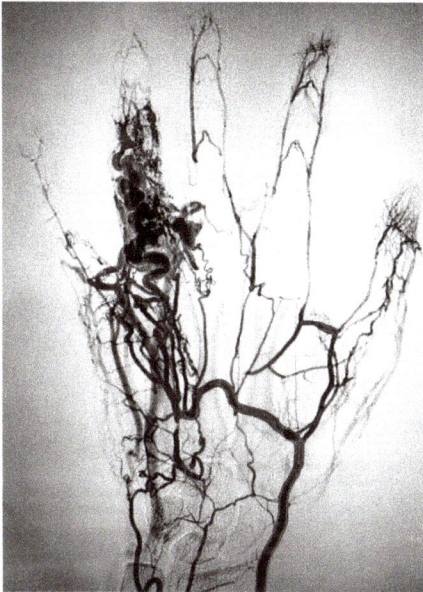

Fig. 8.21: Angiography of the blood vessels in the hand (reproduced from Michel Royon, Wikimedia, © creative commons).

8.7.3 Dual-energy x-ray absorptiometry

Even without administering any contrast agent and without using absorption edges, contrast between bones and tissue can be enhanced by taking two radiographs using different x-ray energies. This allows visualizing otherwise obstructed structures, such as tissue behind bones [12], for instance, the lung behind the rips. Contrast can be achieved by referring to the mass attenuation coefficient for bones in tissue plotted in Fig. 8.22. At low x-ray energies, the absorption in bones is higher than in tissue, while at high energies the attenuation is overall weaker and similar in bones and tissue.

As an example for the application of dual-energy x-ray absorptiometry imaging, let us assume that we take two radiographs with energies indicated by the green and yellow areas in Fig. 8.22. In the green area, the total attenuation is the sum of attenuations in bone (b) and tissue (t) times their respective thicknesses t_b and t_t:

$$(-\ln I/I_0)^{\text{green}} = (\mu_b t_b + \mu_t t_t)^{\text{green}} = b_g + t_g. \tag{8.30}$$

The last equation is a simplified notation, sufficient for the following calculations. For the yellow region, we have the respective relation:

$$(-\ln I/I_0)^{\text{yellow}} = (\mu_b t_b + \mu_t t_t)^{\text{yellow}} = b_y + t_y. \tag{8.31}$$

Now we take a weighted difference of both attenuations (radiographs) with the weighting factor w_1

$$(b_g + t_g) - w_1(b_y + t_y) = (b_g - w_1 b_y) + (t_g - w_1 t_y). \tag{8.32}$$

In order to make the bones vanish, we demand that $(b_g - w_1 b_y) = 0$, or

$$w_1 = \frac{b_g}{b_y}. \tag{8.33}$$

Then for the tissue-weighted contrast, we have

$$(t_g - w_1 t_y) = \left(t_g - t_y \frac{b_g}{b_y}\right). \tag{8.34}$$

Vice versa, for the bone-weighted contrast with $(t_g - w_2 t_y) = 0$, we find

$$(b_g - w_2 b_y) = \left(b_g - b_y \frac{t_g}{t_y}\right). \tag{8.35}$$

An example is calculated in Exercise E8.3, and dual-energy radiographs are shown in Fig. 8.23.

How can two different x-ray energies be selected with a standard medical x-ray tube? There are two possibilities. One may change the accelerating voltage and take a picture at a lower peak voltage (60 kVp) and another at a higher peak voltage (120 kVp). Although the spectrum is broad, the mean x-ray energy is shifted from low to high by increasing the voltage. Alternatively, one may use different absorber sheets, such as Al und Cu. A thin Cu sheet in front of the x-ray tube removes the soft energy spectrum and hardens the radiation compared to Al-filtered radiation. Therefore,

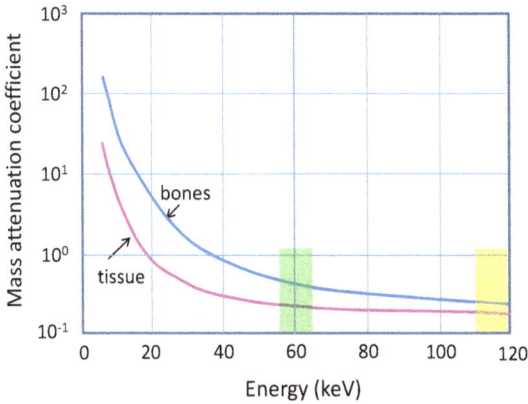

Fig. 8.22: Mass attenuation coefficient of bones and tissues as a function of x-ray photon energy.

taking pictures with an Al filter and another one with a Cu filter provides weighted changes of the x-ray spectrum and a proper contrast. Another question is whether the two images should be taken simultaneously or sequentially. There are advantages and disadvantages to both methods. When taking radiographs simultaneously, two

Fig. 8.23: Dual-energy imaging of the original radiographs and after taking weighted differences to visualize either the chest's soft tissue (bones subtracted) or bones (tissue subtracted) (reproduced from http://en.wikibooks.org/wiki/Basic_Physics_of_Nuclear_Medicine/Dual-Energy_Absorptiometry, © creative commons).

film cassettes are stacked on top of each other, sandwiching a Cu absorber sheet (or any other metal sheet) in between. Then both recordings are spatially congruent, but the lower cassette receives much less intensity than the top one. If taken sequentially, the intensities are similar, but the patient should not move between both exposures to avoid a blurry difference image. An example for dual-energy imaging is shown in Fig. 8.23. The original radiographs taken at two different average x-ray energies are shown in the top panels, and the weighted difference images in the lower panels emphasize either tissue (left) or bones (right).

The technical difficulties are overcome when using a CT unit of the newest generation equipped with two or more x-ray generators. Then the same ROI is scanned simultaneously at two different energies, and the respective tomographic recordings are digitally processed for optimal contrast.

> **!** Attenuation contrast serves the purpose to emphasize specific organs in the background of other organs and tissues. Three methods for contrast enhancement are successfully applied: contrast agents, DSA, and dual energy imaging.

8.8 Phase contrast imaging (PCI)

8.8.1 Physical background

Conventional x-ray imaging uses contrast via attenuation of x-rays. But x-ray contrast can also be generated by exploiting subtle effects of x-ray refraction that change the phase of x-ray radiation. This method is termed *phase contrast imaging* (PCI). PCI permits visualization of soft-tissue structures that would not be detectable by conventional x-ray radiography. In addition, PCI requires a much smaller radiation dose to the body for comparable image quality. Thus, PCI has great potential for special applications; its general clinical use is still hampered by specific requirements that can easily be fulfilled in laboratories but less so in clinical settings.

The physics of x-ray phase contrast lies in the refractive index for x-rays:

$$n = 1 - \delta + i\beta. \tag{8.36}$$

The real part of the refractive index is given by (see eqs. (6.8)–(6.11) and [13, 14])

$$\delta = \frac{\rho_N}{2k} \sigma_P \tag{8.37}$$

and the imaginary part is

$$\beta = \frac{\rho_N}{2k} \sigma_a, \tag{8.38}$$

where the cross sections are

$$\sigma_P = \frac{4\pi r_0 Z}{k};$$

(8.39)

and

$$\sigma_a = \frac{\mu}{\rho_N}.$$

(8.40)

Here $k = 2\pi/\lambda$ is the x-ray wave number and λ is the x-ray wavelength in vacuum, ρ_N is the atomic number density of the absorbing material, r_0 is the Thomson scattering length, also known as the classical electron radius (2.82×10^{-15} m), and Z is the atomic number. σ_P and σ_a are the phase shift cross section and the absorption cross section, respectively.

The refractive index n for x-rays penetrating matter is slightly smaller than 1 by about one part in 10^{-6}. The consequence of this is an x-ray wave with a phase velocity and a wavelength slightly bigger in matter than in vacuum, in contrast to visible light. Hence, x-rays are not only being attenuated in matter, but also phase shifted. To illustrate this effect, we compare a plane electromagnetic wave in a vacuum and one in matter, both propagating in the z-direction, described by the following equation:

$$E(x) = E_0 e^{inkzz} = E_0 e^{i(1-\delta)k_z z} e^{-\beta k_z z}.$$

(8.41)

The wave traveling in vacuum has a constant amplitude and phase. However, the wave penetrating an object is exponentially attenuated and phase shifted by half a wavelength in the example sketched in Fig. 8.24.

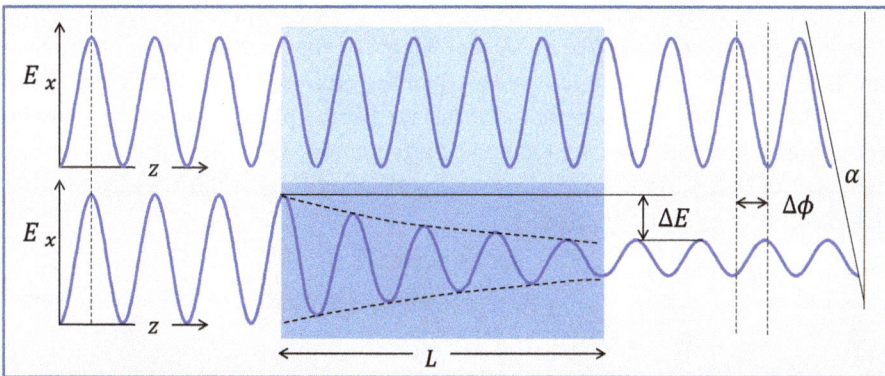

Fig. 8.24: Two EM waves are compared: the top one traveling in vacuum, and the bottom one propagating through some matter in the dark blue area of length L, causing attenuation and phase shift of the EM wave. Indicated are the attenuation of the amplitude ΔE, the phase change $\Delta\phi$, and the angular change α of the propagation direction.

In the object, the E-field is attenuated by the amount:

$$\Delta E(x) = E_0\left(1 - e^{-\beta k_z z}\right),\tag{8.42}$$

and the phase shift is

$$\Delta\phi = \delta k_z L.\tag{8.43}$$

The last equation is expressed in more general terms by a line integral of the thickness L:

$$\Delta\phi = k_z \int_0^L \delta(x,z)\,dz.\tag{8.44}$$

Furthermore, like in visible optics, the phase change also has an effect on the propagation direction. The angular change α is given by

$$\alpha = \frac{1}{k_z}\frac{\Delta\phi}{\Delta x},\tag{8.45}$$

and is shown in Fig. 8.24. This simple example already shows all important facts: x-rays that penetrate matter become attenuated and propagate faster, and consequentially change phase and direction compared to a plane wave propagating in vacuum.

The ratio of the phase shift cross section σ_P and absorption cross section σ_A can be estimated as follows:

$$\frac{\sigma_P}{\sigma_A} = \frac{\delta}{\beta}.\tag{8.46}$$

This ratio is on the order of 10^3 to 10^4, implying that the cross section for the phase shift is 3–4 orders of magnitude greater than the cross section for the absorption. This, in turn, implies that over the same distance L, the phase shift of x-rays can be much larger than the intensity change by attenuation. While this is a convincing argument for PCI, it turns out that in practice, detecting phase shifts is much harder than measuring intensity changes.

8.8.2 Detection of phase contrast

Several methods have been proposed to observe x-ray phase shifts; only one of them is briefly presented here, and further methods are discussed in [13]. Figure 8.25(a) illustrates the so-called *in-line holographic imaging* approach [16]. This method requires the wave front of the incoming beam to be parallel but not necessarily monochromatic. When the beam strikes an object, the wave front becomes distorted due to different phase velocities of the x-ray beam inside and outside of the object and due to refraction at surfaces, which are not perpendicular to the wave front. Altogether, this leads to edge effects, i.e., edges are emphasized. To allow the effect to develop, a certain distance between object and film is required (air gap), just in opposite to conventional x-ray radiography, where all unnecessary air gaps should be avoided. The phase contrast depends on the distance of the x-ray source to the object and from the object to the detector. If the source is too close, the incident beam will not be sufficiently parallel. If the detector is too close, distorted and undistorted rays will overlap and the contrast is lost. In a more advanced stage, a diffraction grating G is placed between object and detector to enhance the contrast, as shown in Fig. 8.25(b). When the grid is moved in the direction perpendicular to the beam, the contrast changes and different parts of the object can be emphasized. An early review of the PCXI method is presented in [17].

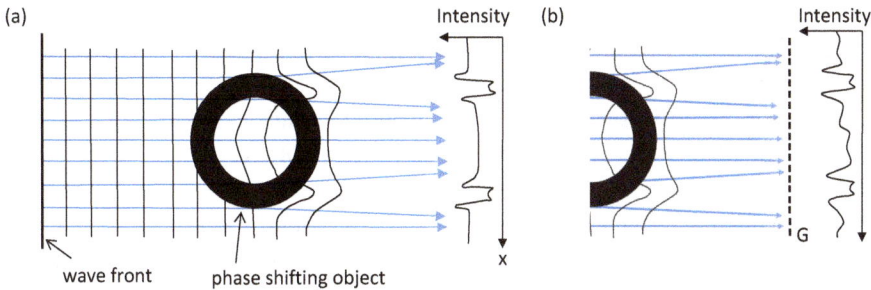

Fig. 8.25: Distortion of a planar wave front due to refraction effects in an object without additional grating (a) and with an enhancement diffraction grating G (b) (adapted from Ref. [16] by permission of IOP Publishing).

The phase analysis works even better with two grids, where the second grid acts as a Fourier analyzer of the diffraction pattern generated behind the first grid. The first grid G1 in Fig. 8.26 has a periodicity g_1, the second grid G2 has a periodicity $g_2 = g_1/2$. In the near field behind the first grating, a diffraction pattern occurs with maxima at the distances [18, 19]:

$$d_T^n = n \frac{(g_1)^2}{8\lambda},$$

(8.47)

where $n = (1, 3, 5, \ldots)$ is an odd number. These distances are known as the *Talbot*[11]-*distances*. If the second grating is positioned at one of those Talbot distances marked as red dashed lines, the diffraction pattern of G1 can be analyzed with different phases and contrast that leads to phase contrast images in the detector.

---

11 Henry Fox Talbot (1800–1877), English scientist, inventor, photographer, and mathematician.

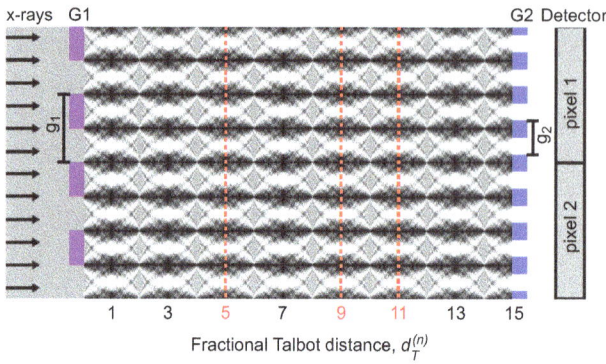

x-rays G1   G2 Detector

Fractional Talbot distance, $d_T^{(n)}$

Fig. 8.26: Talbot diffraction pattern behind a diffraction grating. The Talbot pattern can be analyzed with the help of a second grating that is positioned at the maxima indicated by the dashed red lines (reproduced from [18]).

PCI by the use of additional diffraction gratings have produced remarkably contrast-rich images, as shown in Fig. 8.27 from [18]. All three images from the same object (mouse) are intrinsically perfectly registered as they are extracted from the same data recorded with a grating interferometer. Examples of regions of enhanced contrast as compared to a conventional attenuation image in (a) are marked with arrows, showing the refraction effects at the trachea (b) and scattering effects at the lungs (c). The dark field image in (c) is taken in scattering geometry in contrast to the image (b) in transmission geometry. PCI can provide 3D tomographic images [19], and even the fourth dimension, time, has recently been explored [20]. A useful introduction to the PCI method is given in [3].

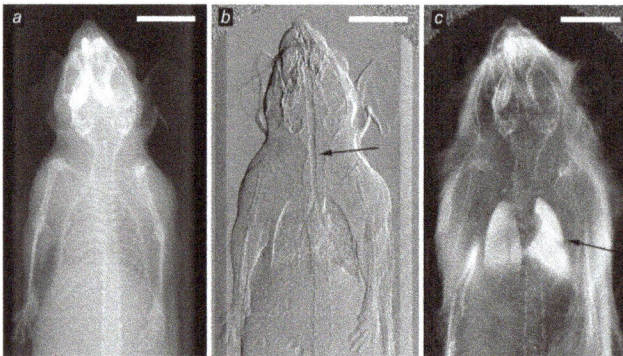

Fig. 8.27: In vivo multicontrast x-ray images of a mouse: (a) conventional x-ray attenuation image; (b) differential phase contrast image based on x-ray refraction; (c) dark-field image based on x-ray scattering. The white bars correspond to 1 cm (courtesy of F. Pfeiffer and reproduced from [18] by permission of Nature Publishing Group).

X-ray imaging can be performed conventionally via attenuation or alternatively via phase con-　!
trast to highlight edges in soft tissues.

8.9 Computed tomography (CT)

8.9.1 Overview

Projection radiography is very valuable for surveying large body parts, revealing bone fractures, foreign objects, infections, signs of cancer, and much more. The projected image is typically in the yz-plane (coronal plane) with incident x-rays parallel to the x-direction, as indicated in Fig. 8.28. However, any spatial variation of the absorption coefficient in the x-direction is integrated and remains unresolved. Dual-energy x-ray imaging (see Section 8.7.3) is a crude attempt to gain depth resolution in the x-direction. X-ray CT aims at a high-resolution 3D representation of inner organs.

The term *tomography* is derived from the Greek word *tomos*, meaning slice or section, and *grapho*, meaning to write. Tomography is used in many fields of science and technology, including radiology. In the medical field, CT is a diagnostic tool that combines usual x-ray attenuation contrast with computation to generate cross-sectional radiographs of the body in the xy-plane (horizontal, axial, or transverse plane).

Figure 8.28 shows the general layout of a *CT scanner*. The x-ray generator and a detector bank are positioned on opposite sides of a circle that rotates about a stationary patient at the center. The detector bank sits on a ring with twice the diameter of the generator ring. Unlike planar radiography, in a CT scanner, the patient is exposed to a fan beam, whose opening angle is limited by collimators. An oscillating grid optionally covers the detectors. During rotation, many frames are taken at different angles by DR. Upon one full rotation, one slice of the body is completely projected in M directions with N detectors, filling an $M \times N$ matrix (up to 1024×1024). Upon backprojection of the attenuation profile, a picture of the cross section is generated.

The basic idea of CT scans is shown in Fig. 8.29. The attenuation profile will always be different when x-ray images of an inhomogeneous body containing different objects with different mass absorption coefficients are taken from different orientations. From these different profiles, it should be possible to reconstruct the spatial position and orientation of the objects inside the body. The procedure is as follows. First, a cross-sectional slice is divided into voxels. It is assumed that each voxel at position (x_i, y_i, z_i) has a specific mass absorption coefficient $\mu(x_i, y_i, z_i)$ on average. The coefficient is projected onto the (x,y) plane and each pixel of a voxel is given a

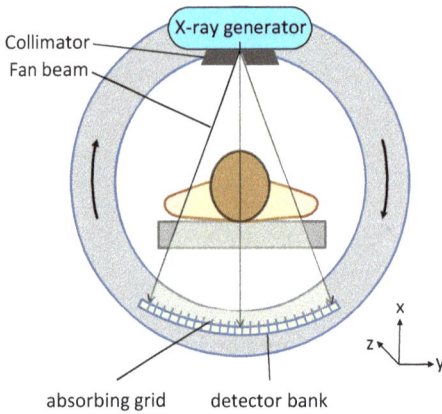

Fig. 8.28: CT geometry: x-ray generator and detector bank rotate simultaneous about a stationary patient. The ring supporting the x-ray generator and the detector bank is in the x,y-plane. CT-scans can be either taken by rotation to generate a single attenuation slice or by a combination of rotation and translation in the z-direction, generating a sinogram.

CT number representative of the x-ray absorption in that voxel. Upon backprojection of the *attenuation profile*, an image of the objects inside is retrieved.

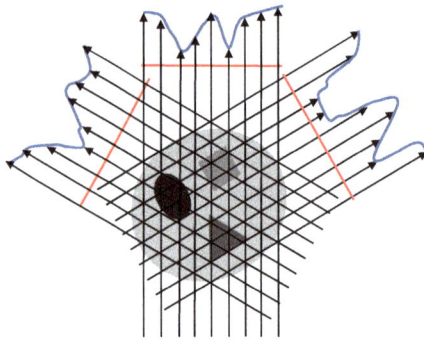

Fig. 8.29: Attenuation profiles (solid lines connecting arrows) of three objects from three different orientations. Here the x-ray trajectories are shown as parallel rays. In a CT scanner, the x-rays from the source to the detector are fan-like.

8.9.2 The Hounsfield scale

CT numbers are assigned to each pixel according to the *Hounsfield scale*. In the Hounsfield[12] scale, the x-ray attenuation of water is taken as a reference. Anything which has a higher linear attenuation than water is assigned a positive number, in

---

12 Godfey Hounsfield (1919–2004), British electrical engineer and Nobel laureate in medicine 1979, pioneered CT scanners.

the opposite case, a negative number. CT numbers on the Hounsfield scale are defined as normalized differences:

$$CT(x_i, y_i, z_i) = 1000 \times \frac{\mu_{object}(x_i, y_i, z_i) - \mu_{water}}{\mu_{water} - \mu_{air}}. \tag{8.48}$$

Although the result is just a number, it is given an artificial unit, the *Hounsfield unit*
(HU).

According to the definition, water with a linear attenuation of 0.19 cm^{-1} has 0
HU. For hard x-rays, air has an absorption coefficient that is nearly zero. Thus the CT
number of air is -1000 HU. Bones with a linear attenuation coefficient of roughly
0.39 cm^{-1} have a CT number of +1000 HU. Each CT number is associated with a grayscale. Bones and air are the two extremes of the Hounsfield scale spanning 2000
shades from white for bone to black for air. Since the eye cannot distinguish between
2000 shades of gray, a Hounsfield window is usually set: a range of ±500 HU for a
rough overview and ±100 HU for a detailed inspection. The Hounsfield scale is graphically illustrated in Fig. 8.30.

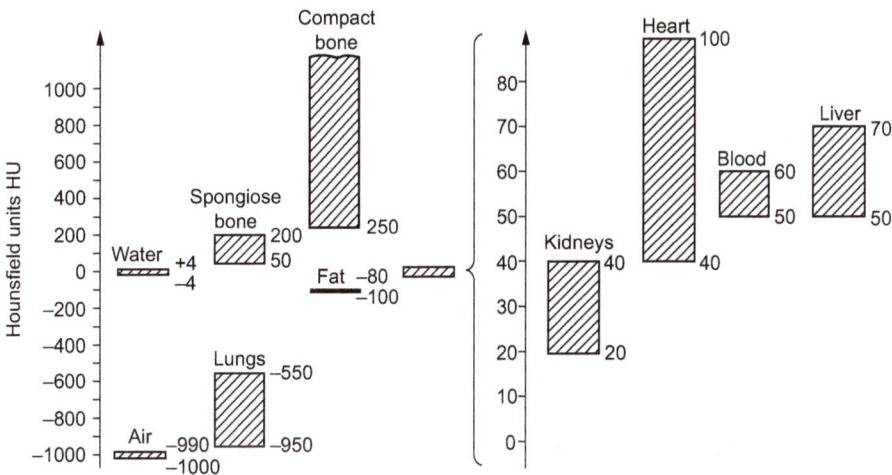

Fig. 8.30: Hounsfield scale for x-ray absorbing tissues, bones, and organs.

In CT terminology, tissues are hypodense if they have a negative HU; vice versa, tissues are hyperdense if they have a positive HU. Isodense are two pixels with the
same HU. The task of CT is to reconstruct the 3D distribution of $\mu_{object}(x_i, y_i, z_i)$ in a
slice of thickness Δz from 2D projections at angles 0–360°. For the next slice, the
scanner is translated in the z-direction by Δz.

8.9.3 Specifications of CT scanners

In practice, many technical issues must be considered for engineering CT equipment, and mathematical methods must be developed to analyze projections taken, such as those developed by Cormack.[13] First to be considered is a beam from the x-ray tube that diverges as indicated in Fig. 8.28 instead of being parallel, as suggested in Fig. 8.29. This is not a problem, as long as x-rays do not cross before detection. Grids are used in front of the detector bank to avoid the detection of crossed or scattered x-rays, like for flat projection radiography. X-ray generator and detector bank are mounted rigidly on opposite sides of a rotating gantry, and the gantry spinning about a stationary patient in the center. The rotation and recording speed can be as high as 200–300 rpm in order to take quasi-static pictures of moving parts like the heart. During rotation, the supporting table can be translated in the z-direction to continuously record slices of the body forming a spiral, as illustrated in Fig. 8.31. The pitch p of the slice is given by the ratio of the table translation per rotation ΔL to the slice thickness Δz:

$$p = \Delta L / \Delta z. \tag{8.49}$$

For $p < 1$, the slices overlap; for $p = 1$, the slices match; for $p > 1$, gaps occur between the scanned slices.

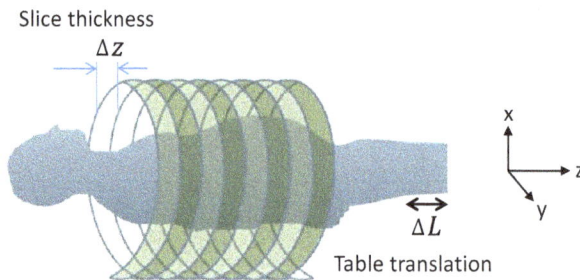

Fig. 8.31: Geometry for taking slices by rotating source and detectors while simultaneously translating the patient in the z-direction (reproduced from www.siemens.com/press).

The detector sampling rate is typically 2 kHz. If the gantry rotates with a speed of 120 rpm, i.e., 500 ms per rotation, then 1000 projections can be taken during one full rotation. The detector bank sits on a circle with a radius of about 500 mm, which amounts to 3140 mm for the complete circumference. If 1000 projections are taken, then the sampling distance between frames is about 3 mm for a single detector.

---

13 Allan M. Cormack (1924–1998), South African-American physicist and Nobel laureate in Medicine 1979, developed mathematical methods for the CT analysis.

However, the detector bank covers a sector containing many detectors, increasing the resolution to about 0.3 mm. The detector sector in the front view is schematically seen in the left panel of Fig. 8.32(a).

Furthermore, the detector bank contains a multi-slice arrangement of detectors, seen by the side view in the middle panel of Fig. 8.32(b). The size of the detectors determines the thickness of a slice. For high resolution, the slice should be as thin as 2 mm. The divergence of the x-ray beam in the z-direction has to be confined by collimators to the detector size. If an organ spanning 200 mm needs to be scanned, 100 slices are required for a full coverage. The recording time of each slice depends on the rotational speed, which we assume to be 0.5 s. Thus, a full scan takes about 50 s. This long scan time is drastically reduced by using multidetector arrays.

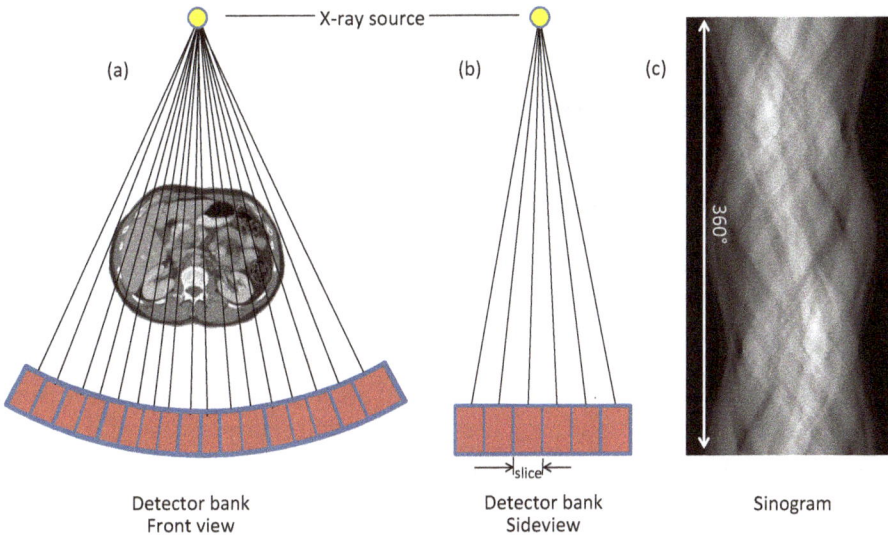

Fig. 8.32: Schematic view of a sector detector bank seen from the front and the side. Only six detectors are shown in parallel defining $M = 6$ slices; up to $M = 256$ detectors in parallel are used nowadays. The sinogram in the right panel is a sequence of pictures taken during one 360° rotation by all M detectors simultaneously. Reconstruction of the sinogram yields the image of a slice shown in the left panel (sinogram courtesy of Dr. Siegel, University of Maryland, School of Medicine).

Presently, solid-state detectors with a width of 0.5 mm are available. In detector banks, up to 256 detectors are mounted in parallel to each other in the z-direction, covering a total area of up to 120 mm. For exploiting the multidetector array over its total width of 120 mm, the x-ray beam divergence has to be relaxed accordingly. Now only two rotations are needed for scanning 200 mm, requiring only about 1 s. The radiation exposure of the patient is thus drastically reduced.

The present state of the art is a CT voxel size of about $0.5 \times 0.5 \times 0.5$ mm^3 and a gantry rotation of 214 rpm, yielding an exposure time of only 280 ms for one revolution. For special applications such as the inner ear, the voxel size can be reduced to $0.25 \times 0.25 \times 0.25$ mm^3. Furthermore, ECG synchronization with a time resolution of better than 75 ms has been reached, and a dose of less than 2.8 mSv is required for a full exposure. A third-generation CT scanner is shown in Fig. 8.33 with open and closed gantry.

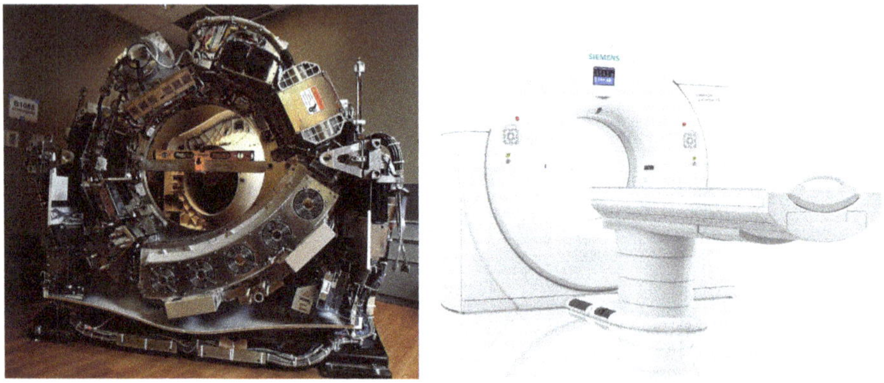

Fig. 8.33: CT unit without cover and with cover. X-ray generator and detector bank are housed inside behind the cover (http://www.healthcare.siemens.com/computed-tomography/).

The CT scanners described here belong to the so-called third-generation scanner manufactured by GE, Siemens, and Toshiba and are presently in operation at most clinics. The fourth-generation CT scanner employs a full stationary ring of detectors and a rotating generator inside the ring. Only exposed detectors within the fan area are activated one at a time.

The raw data taken by the detectors during one full rotation is called a sinogram. An example is shown in the right panel of Fig. 8.32(c). The sinogram requires reconstruction with appropriate algorithms to obtain real space pictures of slices like the one shown in Fig. 8.32(a) [21]. The reconstruction is performed by a 2D Radon[14] backtransformation. The Radon transformation resembles a Fourier transformation generating an approximate reconstruction of the object [22]. The Radon transformation is described further in the next section.

> **!** CT is an imaging modality that takes attenuation projection images as a function of rotation angle, generating sinograms. Sinograms must be backprojected in order to gain images of body parts.

---

14 Johann Radon (1887–1956), Austrian mathematician.

8.9.4 Contrast enhancement

CT scans can be combined with contrast enhancement procedures, as discussed earlier, including blood vessel angiography. The x-ray generator may be tuned to higher or lower accelerating voltages as needed, and the x-ray beam can be hardened with the help of different absorbers.

Figure 8.34 reproduces a state-of-the-art CT-radiographic image of the head and the upper chest. In order to obtain contrast of the blood vessels, a CA was injected. The dose delivered to the person was as low as 1.1 mSv for taking a high-contrast radiograph with a remarkable resolution and richness of detail. This CT radiograph clearly demonstrates the improvements concerning spatial resolution and dose reduction that have been achieved during recent years.

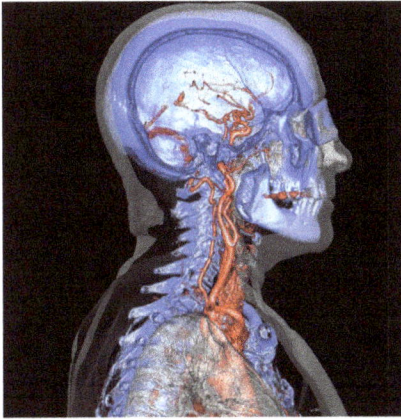

Fig. 8.34: Three-dimensional CT radiograph of the head and upper chest, including part of the vascular system (from http://www.medicalradia tion.com/ provided by www.healthcare.siemens. com).

An alternative approach to enhance contrast is the dual x-ray energy imaging, as already discussed in Section 8.7.3 for projection radiography. The usefulness of this technique is unquestionable [21–23], but the technical difficulties are serious. A particular problem with conventional DEX imaging is the time interval between the first and second exposure due to organ or patient movements. Dual-energy CT (DECT) offers new technical solutions to solve these problems of traditional DEX. There are at least three approaches that can be followed [21].

The first and most expensive approach is sketched in Fig. 8.35(a). A second x-ray generator is mounted on the gantry at an angle of 90° to the first one and operated at a different kVp. This approach also requires the installation of a second detector bank. The second, more cost-effective solution sketched in panel (b) is to rapidly alternate the operating voltage of a single x-ray generator during rotation. The third and most elegant solution from a technical point of view is as follows. Each pixelated sensor in the detector bank acts as an energy-dispersive detector. This means that the sensors count the x-ray photons and simultaneously register their energy, as already

mentioned in Section 8.4.3. The x-ray generator emits a broad spectrum of x-ray photon energies. Each pixel sensor in the detector bank represents this broad energy distribution. Therefore, the necessary information on the body's energy-dependent x-ray
attenuation is already stored in the detectors. Instead of reading the counts integrated
over the entire energy spectrum, the energy-dispersive count rate distinguishes counts
recorded in one particular energy window compared to counts in another energy window of the same detector, as indicated in panel (c). This method is well known as
energy-dispersive x-ray detection and is applied here to DECT. Clinical applications
are discussed in [22, 23]. Beyond diagnostic applications, dual energy recording is
also important for determining the local electron density [24], an information that is
needed to calculate the proton stopping power and therefore the range of proton
beams in the body.

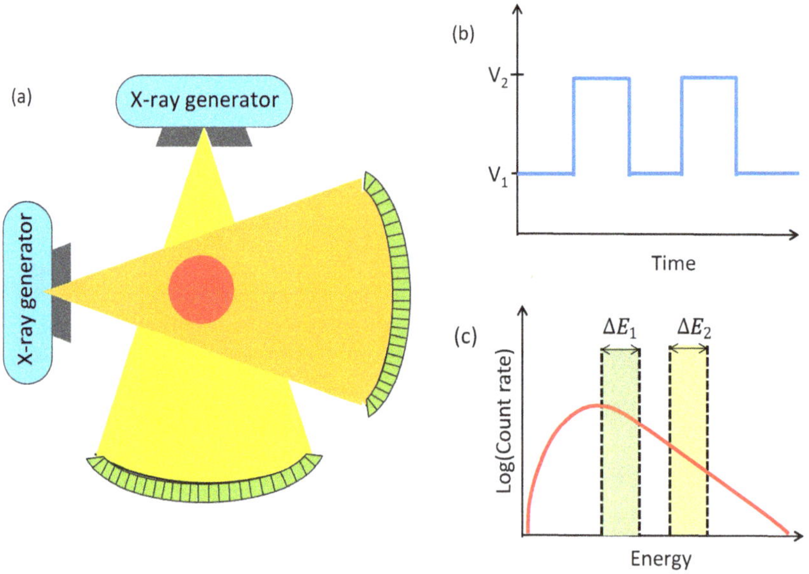

Fig. 8.35: Dual-energy CT scans can be performed in three different modes: (a) use of two x-ray
generators operating at different kVp; (b) fast voltage switching during rotation of a single
generator; and (c) use of energy-dispersive pixel detectors and setting two energy windows.

8.9.5 Radon transformation

CT imaging is an inverse Fourier transformation process that requires some significant mathematical tools developed over the past 30 years. Here we give a simplified
account on the main concept of the Radon transformation. The Radon transformation is a special kind of Fourier transformation. We recommend consulting several
excellent biomedical imaging texts for further information, such as [3, 4, 25, 26].

The mathbox briefly introduces Fourier transforms (FT) and convolutions; more on FTs can be found in [27].

We recall that the intensity measured in a detector pixel (i,j) is the result of a *line integral* over attenuation coefficients along the path of the x-ray beam from the x-ray generator to the detector:

$$I_{det}(x,y,z) = I_0(x,y,z) \exp\left(-\int_0^t \mu(x,y,z)dx\right). \tag{8.50}$$

Here and in the following, we neglect the energy integration of the detector and the detector efficiency. We focus on the first slice in the (x,y)-plane perpendicular to the y,z-detector plane and fix the slice index z. The attenuation along one path in the x-direction is then for each y_i-position:

$$p_z(x,y_i) = -\ln\left(\frac{I_{det,z}(x,y_i)}{I_{0,z}(x,y)}\right) = \int_0^t \mu_z(x,y_i)dx. \tag{8.51}$$

Referring to Fig. 8.36, we may select the first ray in one slice of fixed z that gives rise to a signal $p_1 = p_z(y_1)$ in the detector bank. Next, we add all other beams in the y-direction (parallel or fan), $p_1, p_2, \ldots, p_n$ that are recorded simultaneously by the detector bank. In the end, we have a 1D attenuation projection profile for one fixed source-detector orientation as shown in Fig. 8.36. The 1D attenuation profile is the projection of one slice in the x,y-plane. The other slices through the 3D object with different z-values can be added later.

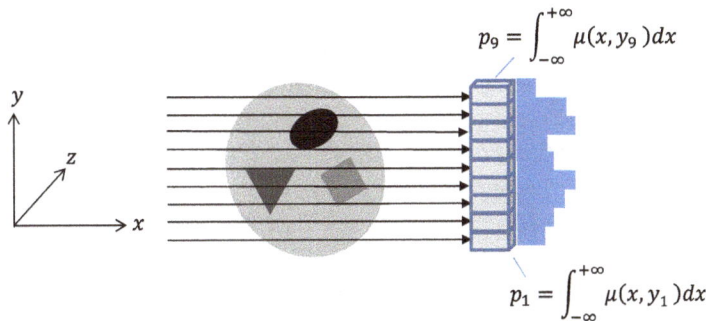

$$p_9 = \int_{-\infty}^{+\infty} \mu(x,y_9)dx$$

$$p_1 = \int_{-\infty}^{+\infty} \mu(x,y_1)dx$$

Fig. 8.36: Attenuation profile of one slice.

Next we want to take an attenuation profile of the same slice but at a different angle. Then it is preferable to change the Cartesian coordinate system fixed to the laboratory to a polar one that rotates with the object or the gantry:

$$r = x \cos\theta - y \sin\theta,$$
$$s = x \sin\theta + y \cos\theta. \tag{8.52}$$

In the polar system, the attenuation projection is now denoted as

$$p_z(r,\theta) = \int_0^t \mu_z(r,\theta)ds. \tag{8.53}$$

Here the r-axis is oriented parallel to the detector bank and s is directed parallel to the x-rays at orthogonal angle to r, while θ is the angle between r and x, as shown in Fig. 8.37.

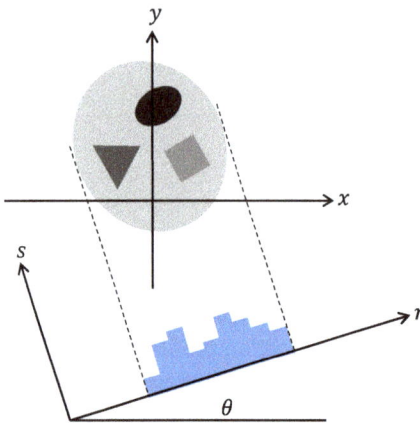

Fig. 8.37: Definition of the polar coordinate system.

Recording the attenuation profile as a function of the rotation angle θ will generate a map of sinusoidal curves. Therefore, such recordings are called *sinograms*. A schematic example is shown in Fig. 8.38, demonstrating how a sinogram is generated. In this example, the x-ray source is fixed, and the object attenuating the beam is rotated. In reality, the object (body and organs) is fixed, and the x-ray source is rotated. In Figure 8.38, the attenuation profile of only a single object is shown. In real sinograms, we observe the superposition of many profiles. Note that the same object gives rise to different, angular-dependent attenuations. Here, the rectangular bar exerts a higher attenuation on the x-ray beam at 90° and 270° than at other angles.

More generally speaking, any 2D function $f_z(r,s)$ projected by a line integral into a 1D profile constitutes a *Radon transformation*. Thus, $p_z(r,\theta)$ is a Radon transformation of $f_z(r,s)$:

$$p_z(r,\theta) = \mathcal{R}f_z \equiv \hat{f}_z = \int f_z(r,\theta)ds. \tag{8.54}$$

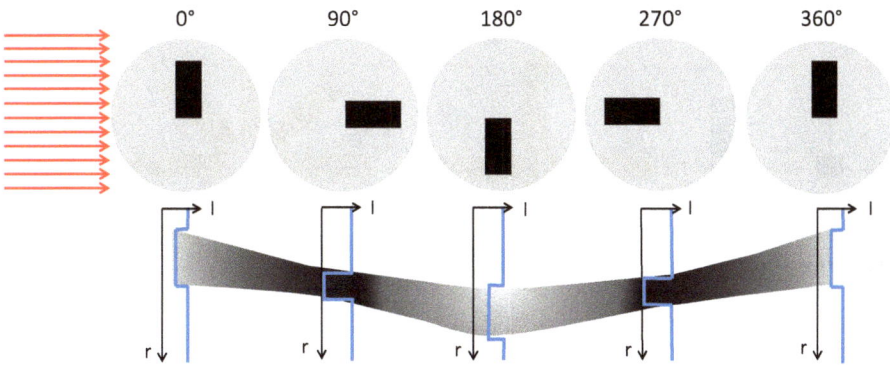

Fig. 8.38: Generation of a sinogram with a parallel incident x-ray beam. The x-rays are absorbed by an excentric object, which changes position upon rotation. If the long side of the object is oriented parallel to the x-ray beam, the absorption of x-rays is higher than in the perpendicular orientation. Therefore, the attenuation profile, shown in the lower part of the panel, changes continuously with the rotation angle. After a rotation by 360°, a sinogram of one slice is taken.

8.9.6 Backprojection

The main challenge is how to perform the back transformation to obtain information about an object in the x,y-plane. In general, the backprojection operation is the *inverse Radon transform*:

$$f_z = \mathcal{R}^{-1} p_z(r, \theta). \tag{8.55}$$

We go step by step and take the backprojection for a couple of discrete angles θ_i ($i\_1 \ldots n$) and then add them up:

$$\hat{f}_z(x,y) = \sum_{i=1}^{n} p_z(r, \theta_i) d\theta_i. \tag{8.56}$$

Figure 8.39 shows a simple example: a high-density opaque square in the center of an object in the x,y-plane. The dashed lines surrounding the objects mark the field of view. When we take x-ray radiographs from such an object at two orthogonal rotation angles θ, we will obtain the attenuation profiles labeled P_1 and P_2 in Fig. 8.39(a). If these projections were taken with a film, the film would be black everywhere but transparent and unexposed in the shadow region of the object. For backtransformation we take a flashlight with a parallel beam and shine through the transparent holes. Projections P_1 and P_2 in panel (b) will generate two perpendicular light bars from the two holes in panel (a). The center is brighter and indicates the position and orientation of the object in space. We may add more projections $P_1 \ldots P_4$ to refine the positional information. Then the center of the projections will eventually become over-exposed,

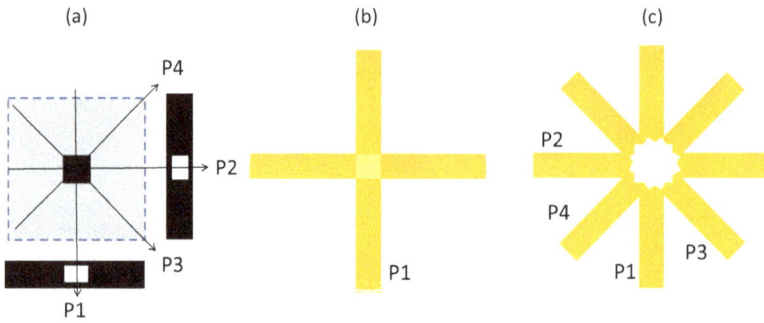

Fig. 8.39: (a) Attenuation by an objection at the center; (b) two backprojections P_1 and P_2; and (c) four backprojections $P_1 \ldots P_4$ from different sides.

and artifacts appear close to the center indicated in panel (c). These artifacts have to be corrected for by filters, as described in the next section.

8.9.7 Filter

The backprojection procedure suggested so far has two fundamental problems: first, it generates edge effects, and second, the center is emphasized disproportionately. In order to overcome these problems, weighting filters $h(r)$ are applied before backprojection:

$$p_z'(r, \theta) = p_z(r, \theta) * h(r). \tag{8.57}$$

Here, the symbol "$*$" indicates a convolution integral.

A popular filter is the Ramachandran[15]-Lakshminarayanan (Ram–Lak) filter, expressed by

$$h_{RL}(r) = \frac{1}{2(\Delta r)^2} \left[\mathrm{sinc}\left(\frac{r}{\Delta r}\right) \right] - \frac{1}{4(\Delta r)^2} \left[\mathrm{sinc}^2\left(\frac{r}{\Delta r}\right) \right], \tag{8.58}$$

where Δr is the sampling interval along the r-axis, and sinc $= \sin(x)/x$.

Figure 8.40 gives an example of how the Ram–Lak filter acts on the rectangular attenuation profile, sketched in panel (a). The $h_{RL}(r)$ function is plotted in panel (b), and the result of the convolution is displayed in panel (c). The effect of the convolution is to decrease the "star artifact," which occurs by simple backprojection.

The computational effort to perform the convolution in real space is significant. Therefore, the convolution is performed in Fourier space, using fast FT methods [27].

---

15 G. N. Ramachandran (1922–2001), Indian biophysicist.

(a)   (b)   (c)

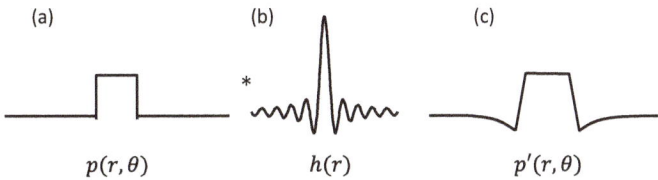

$p(r,\theta)$ $h(r)$ $p'(r,\theta)$

Fig. 8.40: Ram–Lak filter applied to a rectangular attenuation profile.

It can be shown that convolution in the spatial domain is equivalent to multiplication in the Fourier k-domain. This has the decisive advantage that the multiplication process can be performed much faster than computing a convolution. Each projection $p_z(r,\theta)$ is first Fourier transformed along the r-direction to yield $P_z(k,\theta)$. Then $P_z(k,\theta)$ is multiplied with the FT of $h_{RL}(r)$, which is $H(k)$, to give the filtered projection in Fourier k-space:

$$P_z'(k,\theta) = P_z(k,\theta)\cdot H(k). \tag{8.59}$$

The Fourier transformed filter $H(k)$ is a simple ramp-function, as shown in Fig. 8.41. This is the most simple filter one can apply. Other more realistic filters are discussed in the pertaining literature.

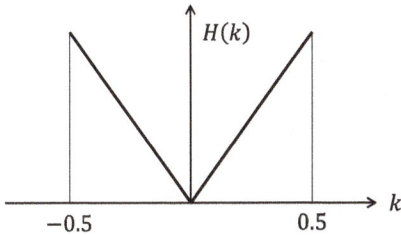

-0.5 0.5 Fig. 8.41: Fourier transform of the Ram-Lak filter.

Now the inverse FT $\mathcal{F}^{-1}$ is applied to transform $P_z'(k,\theta)$ back to the spatial domain. Back Fourier transformation and backprojection will then give the final filtered image in real space called the *filtered backprojection* (FBP):

$$\hat{f}(x,y) = \sum_{i=1}^{n} \mathcal{F}^{-1}\{P_z'(k,\theta_i)\}d\theta. \tag{8.60}$$

The final step is done by calculating the CT number for each pixel according to eq. (8.44):

$$CT(x_i, y_i, z_i) = 1000 \times \frac{\hat{f}(x_i, y_i, z_i) - \mu_{water}}{\mu_{water}} \tag{8.61}$$

and plot the result as a grayscale image.

In the biomedical imaging literature, it is customary to test the Radon transformation and filtered back transformation using a phantom representing roughly the contrast expected from the skull and brain. The phantom shown in Fig. 8.42(a) is known as the Shepp–Logan phantom [28]. The coordinates are posted in the internet [29]. The computed sinogram of the Shepp–Logan phantom is reproduced in Fig. 8.42(b). Panel (c) shows an FBP of the phantom, reproducing well the features of the original image. In contrast, panel (d) shows an example containing a reconstruction error. Often the programming of the transformations is performed with the python code [30]. The python code for programming the Radon transformation and filtered back transformation using different filters can be found in [31].

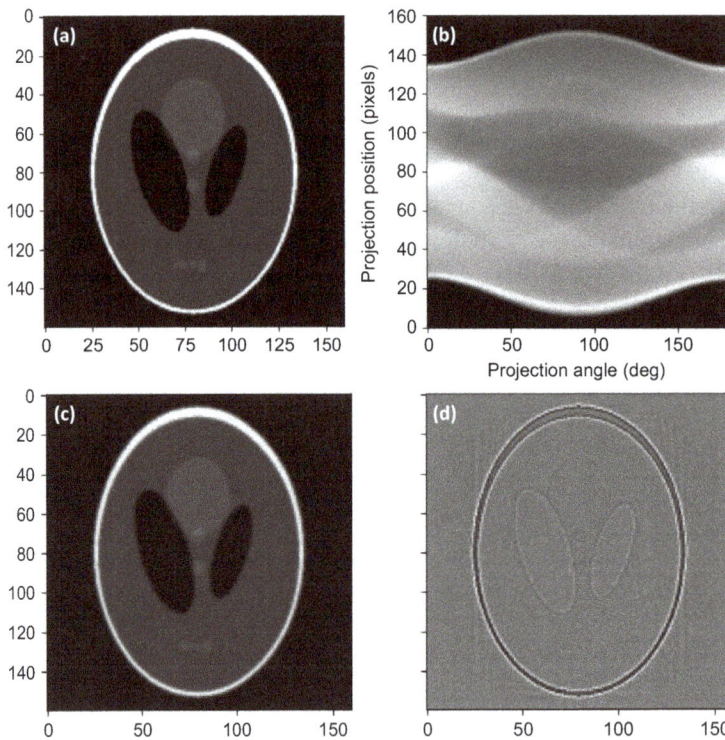

Fig. 8.42: (a) Original Shepp-Logan phantom; (b) sinogram or Radon transformation of the Shepp-Logan phantom; (c) filtered backprojection of the sinogram; and (d) reconstruction error in the filtered backprojection (adapted from [30]).

> ! Recording a CT is equivalent to performing a Radon transformation. A Radon transformation is a line integral projected into 1D as a function of a rotation angle. FBP yields the real space image.

Mathbox: Fourier transformation and convolution

An FT of a signal $f(x)$ in spatial domain is expressed as follows:

$$F(k) = \frac{1}{\sqrt{2\pi}} \int_{-\infty}^{+\infty} f(x) \exp(-ikx) dx.$$

Here x is a spatial variable and k is the wave number. Fourier transformation is a filtering process. It tells us which wave numbers are needed to describe any function $f(x)$ in terms of cos- and sin-functions. FTs can also be expressed in the time-frequency domain. Then the FT of a function $f(t)$ in the time domain is

$$F(\omega) = \frac{1}{\sqrt{2\pi}} \int_{-\infty}^{+\infty} f(t) \exp(-i\omega t) dt,$$

where $F(\omega)$ represents all frequencies relevant for describing $f(t)$. The FT can be back-transformed via

$$FT^{-1}(F(k)) = f(x) = \frac{1}{\sqrt{2\pi}} \int_{-\infty}^{+\infty} F(k) \exp(ikx) dx.$$

Three examples for FTs are shown below for illustration. (a) A Gaussian curve in real space is also a Gauss curve in k-space. (b) The FT of an exponentially decaying function is a Lorentzian curve in k-space. (c) A step function in real space becomes a sinc-function in k-space.

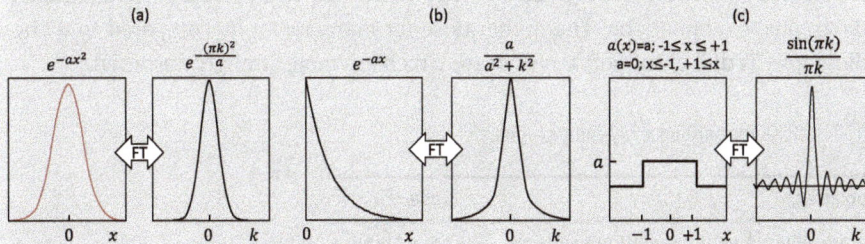

The FTs can be expanded to two and three dimensions. For instance, the FT of a function $f(x, y)$ in the x, y-plane is

$$F(k_x, k_y) = \frac{1}{\sqrt{2\pi}} \int_{-\infty}^{+\infty} f(x, y) \exp(-i(k_x x + k_y y)) dx dy.$$

FTs have important properties. One useful property concerns multiplication and convolution. If two signals are multiplied in the spatial domain, then in the wave number domain the signals are the convolution of the two individual FTs:

$$f_1(x) \times f_2(x) \Leftrightarrow FT \Leftrightarrow F_1(k) * F_2(k).$$

The convolution of two functions is not a simple multiplication but defined by the following integral:

$$F_1(k) * F_2(k) = \int\limits_{t=-\infty}^{+\infty} F_1(k-t)F_2(t)dt.$$

In practice, the convolution integral is time consuming to perform. Therefore, the inverse FT helps speeding up the computation. Vice versa, two functions, which are convoluted in the spatial domain are simply multiplied in the Fourier domain:

$$f_1(x) * f_2(x) \Leftrightarrow FT \Leftrightarrow F_1(k) \times F_2(k).$$

Hence, a convolution in the spatial domain is equivalent to a multiplication in Fourier domain and vice versa. This theorem helps performing the filtering and the back Fourier transformation.

8.10 Risks and comparisons

State-of-the-art x-ray radiographs not only provide an unprecedented imaging quality with respect to spatial resolution and contrast. The radiation exposure time has also decreased dramatically due to the use of highly sensitive detectors, fast scanning, and enormous computer power for image processing. Reduced exposure time translates into less x-ray radiation dose. Some average values are given in Tab. 8.2 for standard radiographic examinations. The higher dose for mammography compared to a chest radiography is due to the soft x-rays used, which are more strongly absorbed.

Tab. 8.2: Dose for various x-ray radiographs.

Exposure	Dose
Average environmental radiation exposure	$3.5 \pm 1\,mSv/a$
Chest projection radiograph	0.1 mSv
GI tract with $BaSO_4$ contrast	15 mSv
Mammography	3 mSv
CT skull	1.5–2.3 mSv
CT chest	5.8 mSv
CT cardio-angiogram	7–13 mSv

These low-dose levels still compete with a zero dose in MRI. It is therefore appropriate to make a first comparison between these two advanced imaging modalities, which have reached a very high technological standard. What are the pros and cons? When should one or the other imaging methods be used? A comparison is given in Tab. 8.3. In general, it can be said that imaging the lungs is challenging for both methods, but x-rays will always show more contrast if given a choice. In most cases, these two methods are complementary and images captured using one method can be confirmed by using the other method.

Tab. 8.3: Comparison of x-ray radiography and MRI scanning techniques for various body parts and diagnostics.

Body part	X-ray radiography	MRI
Tendons and ligaments	Not seen in CT scans	Shows up in MRI
Fracture of bones and vertebrae	Better seen in CT	
Brain tumor		Better recognized by MRI
Tears and organ injuries, bleeding in brain	CT scans are quicker	
Spinal cord injuries		MRI is preferred
Pneumonia	CT is preferred	
Early cancer recognition	CT is preferred	

8.11 Summary

S8.1 X-ray attenuation contrast is defined as the difference of transmitted intensity in neighboring regions normalized by the sum.

S8.2 Contrast on x-ray films or flat panels is achieved by attenuation of the penetrating radiation in tissues of different densities and thicknesses.

S8.3 The transmitted intensity in projection radiography follows from the Lambert-Beer equation. In the exponent a line integral is taken over the attenuation path with spatially varying attenuation coefficients.

S8.4 Extended x-ray sources and Compton scattering cause a penumbra at the border of shadow projections.

S8.5 Oscillating grids in front of the recording detector improve contrast by suppressing stray x-rays.

S8.6 Beam hardening is required to eliminate soft x-rays, which cause high dose without contributing to the image quality.

S8.7 Most x-ray radiographs are recorded digitally with a flat panel detector containing TFTs.

S8.8 X-ray machines for projection radiography are either vertical, horizontal, or have a C frame.

S8.9 Mammography is taken with lower x-ray kVp (35 kVp) and often with Mo Kα characteristic radiation in contrast to normal radiographs taken with 80–140 kVp.

S8.10 Contrast can be enhanced by injection of contrast agents, by DSA, by DxE imaging, or by PCI.

S8.11 CT is a diagnostic tool that combines usual x-ray attenuation with rotating x-ray source, and computation to generate cross-sectional radiographic images of the body in the xy-plane.

S8.12 The gray scale of CT images is determined by the Hounsfield scale.

S8.13 The Hounsfield scale compares the attenuation of water with attenuation of organs in the body.

S8.14 High-speed CT scans of the thorax take only 3 s and the dose is in the order of 6 mSv.

S8.15 Contrast enhancement of DECT scans can be achieved by using two x-ray generators mounted on the rotating gantry, by switching periodically between two energies, or by energy-dispersive detector recording.

S8.16 Sinograms can be considered as FTs of real space objects. Back transformation yields the real space image.

S8.17 The Fourier transformation of line integrals is called Radon transformation.

S8.18 Radiography and CT compares favorably to MRI for scanning of fractures, lung, and for early cancer recognition.

? Questions

Q8.1 How is x-ray attenuation contrast defined?

Q8.2 What is an attenuation profile?

Q8.3 What are the two reasons for a blurry x-ray radiograph?

Q8.4 How can the contrast in a blurry radiograph be sharpened?

Q8.5 How can the x-ray spectrum be hardened, meaning that the average x-ray energy is shifted to higher energies?

Q8.6 What are the two effects of x-ray filters?

Q8.7 For recording x-ray radiographs, what has completely replaced the exposure of x-ray films?

Q8.8 Does the DR of x-ray intensity have the same spatial resolution as x-ray films?

Q8.9 What is the main advantage of DR x-ray radiographs?

Q8.10 Why does mammography use different x-ray sources than in standard x-ray radiography?

Q8.11 Contrast enhancement modalities are frequently used in x-ray radiography. Which ones are available?

Q8.12 Does PCI utilize the real or the imaginary part of the x-ray refractive index?

Q8.13 What is the main advantage of PCI?

Q8.14 Give a brief motivation for CT imaging.

Q8.15 How is the Hounsfield scale defined?

Q8.16 In Hounsfield-weighted images, what appears black and what appears white?

Q8.17 In today's CT scanners, a number of detectors work in parallel defining a slice in the z-direction. How many detectors are mounted in parallel for scanning which slice thickness at once?

Q8.18 What is the voxel size of a standard CT scan?

Q8.19 What is the dose delivered during a standard CT scan?

Q8.20 When is x-ray CT imaging preferred over MRI?

Q8.21 What are the three main reasons for loss of contrast?

Q8.22 What are the three mechanisms for obtaining x-ray contrast?

Q8.23 What are the three mechanisms for obtaining DECT contrast enhancement?
Q8.24 What is a sinogram?
Q8.25 What is a Radon transformation?

Attained competence	+	0	-	⚡
I know what the difference is between x-rays and γ -rays.				
I know how contrast is achieved with x-ray radiography.				
I know what moving grids in front of detectors are used for.				
I am aware of different x-ray detection systems.				
I appreciate the usefulness of fluoroscopes.				
I am aware of different contrast enhancement techniques.				
I know what dual-energy radiography is good for.				
I can draw the main parts of a CT scanner.				
I recognize the importance of the Radon transformation.				
I know what the term "filtered backprojection" implies.				

Exercises

E8.1 **Magnification and penumbra:** For the geometric arrangement shown in Fig. 8.3 and assuming a distance of 1 m from the anode to the film and 10 cm from the bone to the film, calculate:
   a. the magnification factor and
   b. the penumbra extension, if the source is extended to 20 mm.

E8.2 **Mass attenuation coefficient and contrast:** Confirm the contrast numbers in Tab. 8.1 using the values in the following table. In the table you find mass attenuation coefficients for 60 keV photons, provided by NIST https://physics.nist.gov/PhysRefData/XrayMassCoef/tab4.html

	$\mu/\rho(\text{cm}^2/\text{g})$ @ 60 keV	$\rho\ (\text{g/cm}^3)$	$(\mu/\rho)\rho\ (\text{cm}^{-1})$ (@ 60 keV)	$\dfrac{I(t=1\ \text{cm})}{I_0}$ @ 60 keV
Air	0.187	0.0012	0.00022	1
Bone	0.314	1.85	0.58	0.55
Muscle	0.208	1.04	0.216	0.805
Whole blood	0.2	1.06	0.212	0.808
Tissue	0.22	1.04	0.22	0.802
Brain	0.206	1.04	0.214	0.807

E8.3 **Dual-energy radiography:** We have taken x-ray radiographs of body parts at two energies: 60 keV and 100 keV. The x-rays penetrate 7 cm of tissue and 3 cm of bone material. What are the weighting factors of the images taken that must be applied to emphasize bones and tissues separately.

E8.4 **Bone contrast:** At low x-ray energies, the absorption in bones is higher than in tissue, while at high energies, the attenuation is overall weaker and similar in bones and tissue. Why is this so? Try to explain the difference.

E8.5 **Attenuation profile:**

  a. An object is illuminated with a homogeneous intensity I_0 along the x-direction. Determine the attenuated intensity profile behind the object. Each voxel has an attenuation coefficient of 1 cm^{-1}, and an edge length of 0.1 cm.

  b. Indicate by gray shade the local intensity in each voxel. The lower the intensity, the darker is the voxel.

  c. Determine the Weber contrast between pixels 1 and 13.

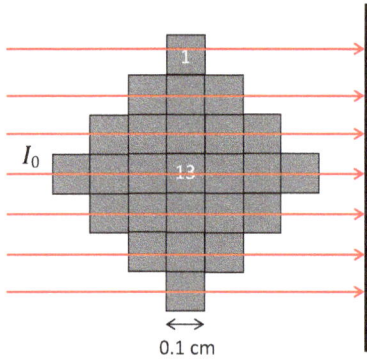

E8.6 **Backprojection:** We consider a very simple example of a CT with four projection at 0°, 45°, 90°, and 135°. The object is a square containing 9 voxels. Using the expression

$$-\ln\left(\frac{I}{I_0}\right) = (\mu_1 + \mu_2 + \cdots)\Delta x = A,$$

the recorded A_p value for each projection is given in the following graph:

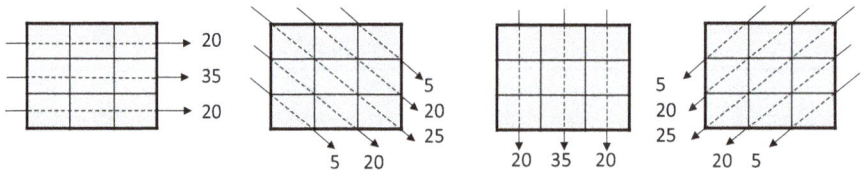

The task is to find the A_i values in all nine individual voxels by backprojection.

References

[1]   Dance DR, Christofides S, Maidment ADA, McLean ID, Ng KH. Diagnostic radiology physics –
      A handbook for teachers and students. Vienna: International Atomic Energy Agency; 2014.
      IAEAL 14–00898.
[2]   Bushberg JT, Seibert JA, Leidholdt EM Jr, Boone JM. The essential physics of medical imaging.
      3rd edition. Philadelphia, Baltimore, New York, London: Lippincott Williams & Wilkins,
      Wolters Kluwer; 2012.
[3]   Salditt T, Aspelmeier T, Aeffner S. Biomedical imaging. Principles of radiography,
      tomography, and medical physics. Berlin, Boston: De Gruyter. Graduate Text; 2017.
[4]   Maier A, Steidel S, Christlein V, Hornegger J. Medical imaging system. An introductory guide.
      Springer Open.
[5]   Cunningham IA, Shaw R. Signal-to-noise optimization of medical imaging systems. J Opt Soc
      Am A. 1999; 16: 621–632.
[6]   Wang J, Blackburn TJ. The AAPM/RSNA physics tutorial for residents: X-ray image intensifiers
      for fluoroscopy. RadioGraphics. 2000; 20: 1471–1477.
[7]   Lanc L, Silva A. Digital imaging systems for plain radiography. New York: Springer Science +
      Business Media; 2013.
[8]   Flohr T, Petersilka M, Henning A, Ulzheimer S, Ferda J, Schmidt B. Photon-counting CT review.
      Physica Medica. 2020; 79: 126–136.
[9]   Russ M, Shankar A, Setlur Nagesh SV, Ionita CN, Bednarek DR, Rudin S. A CMOS-based high
      resolution fluoroscope (HRF) detector prototype with 49.5 μm pixels for use in endovascular
      image guided interventions (EIGI). Proc SPIE Int Soc Opt Eng. 2017; 10132: 101323W.
[10] Rashidian B. DQE, A simplified view. A simple description of Detective Quantum Efficiency.
      Teledyne DALSA; 2013. available from: https://de.scribd.com/document/325963130/DQE-
      Simplified
[11] Ciatto S, Houssami N, Bernardi D, Caumo F, Pellegrini M, Brunelli S, Tuttobene P, Bricolo P,
      Fantò C, Valentini M, Montemezzi S, Macaskill P. Integration of 3D digital mammography with
      tomosynthesis for population breast-cancer screening (STORM): A prospective comparison
      study. Lancet Oncol. 2013; 14: 583–589.
[12] Menten MJ, Fast MF, Nill S, Oelfke U. Using dual-energy x-ray imaging to enhance automated lung
      tumor tracking during real-time adaptive radiotherapy. Med Phys. 2015; 42: 6987–6998.
[13] Als-Nielsen J, McMorrow D. Elements of modern X-ray physics. 2nd edition. New York, London,
      Sydney, Toronto: Wiley & Sons; 2011.
[14] Jensen TH. Refraction and scattering based x-ray imaging. Dissertation. Niels Bohr Institute,
      Copenhagen, Denmark; 2010.
[15] http://en.wikipedia.org/wiki/Phase-contrast_x-ray_imaging
[16] Lewis RA. Medical phase contrast x-ray imaging: Current status and future prospects. Phys
      Med Biol. 2004; 49: 3573–3583.
[17] Zhou SA, Brahme A. Development of phase-contrast x-ray imaging techniques and potential
      medical applications. Physica Medica. 2008; 24: 129–148.
[18] Bech M, Tapfer A, Velroyen A, Yaroshenko A, Pauwels B, Hostens J, Bruyndonckx P, Sasov A,
      Pfeiffer F. In-vivo dark-field and phase-contrast x-ray imaging. Nat Sci Rep. 2013; 3: 3209.
[19] Bech M, Jensen TH, Bunk O, Donath T, David C, Weitkamp T, Le Duc G, Bravin A, Cloetens P,
      Pfeiffer F. Advanced contrast modalities for x-ray radiology: Phase contrast and dark-field
      imaging using a grating interferometer. Z Med Phys. 2010; 20: 7–16.
[20] Hoshino M, Uesugi K, Yagi N. 4D x-ray phase contrast tomography for repeatable motion of
      biological samples. Rev Sci Instrum. 2016; 87: 093705, p. 1–8.

[21] Johnson TRC, Fink C, Schönberg SO, Reiser MF. Dual energy CT in clinical practice. Berlin, Heidelberg, New York: Springer Verlag, Springer; 2011.

[22] Coursey CA, Nelson RC, Boll DT, Paulson EK, Ho LM, Neville AM, Marin D, Gupta RT, Schindera ST. Dual-Energy Multidetector CT: How Does it Work, What Can it Tell Us, and When Can We Use It in Abdominopelvic Imaging? RadioGraphics. 2010; 30: 1037–1055.

[23] Grajo JR, Patino M, Prochowski A, Sahani DV. Dual energy CT in practice: Basic principles and applications. App Radiol. 2016; 45: 6–12.

[24] Landry G, Reniers B, Granton PV, van Rooijen B, Beaulieu L, Wildberger JE, Verhaegen F. Extracting atomic numbers and electron densities from a dual source dual energy CT scanner: Experiments and a simulation model. Radiother Oncol. 2011; 100: 375–379.

[25] Kevin L, La Rivière PJ. Sinogram restoration in computed tomography with an edge-preserving penalty. J Med Phys. 2015; 42: 1307–1320.

[26] Gabor HT. Fundamentals of computerized tomography: Image reconstruction from projections. 2nd edition. Berlin, Heidelberg, New York: Springer Verlag; 2009.

[27] Brown A. Guide to the Fourier transform and the fast Fourier transform. London, New York: Cambridge paperbacks; 2019.

[28] Shepp LA, Logan BF. The Fourier reconstruction of a head section. IEEE Trans Nucl Sci. June 1974; NS-21: 21–43.

[29] https://en.wikipedia.org/wiki/Shepp%E2%80%93Logan_phantom

[30] https://www.python.org/

[31] https://scikit-image.org/

Further reading

Bushberg JT, Seibert JA, Leidholdt EM Jr, Boone JM. The essential physics of medical imaging. 3rd edition. Philadelphia, Baltimore, New York, London: Lippincott Williams & Wilkins, Wolters Kluwer; 2012.

Smith NB, Webb A. Introduction to medical imaging: Physics, engineering and clinical applications. Cambridge, New York, Melbourne: Cambridge Texts in Biomedical Engineering; 2010.

Salditt T, Aspelmeier T, Aeffner S. Biomedical imaging. Principles of radiography, tomography, and medical physics. Berlin, Boston: De Gruyter. Graduate Text; 2017.

Zeng GL. Medical image reconstruction. A conceptual tutorial. Berlin, Heidelberg, New York: Springer Verlag; 2010.

Useful website

www.upstate.edu/radiology/education/rsna/radiography/index.php

9 Scintigraphy (SPE and SPECT)

Physical properties of scintigraphy

Detector	Scintillator, CTZ
Spatial resolution	5–10 mm
Dead-time	200 ns
Sensitivity	0.01% (scintillation), 0.12% (CTZ)
Most important isotopes for SPECT	^{99m}Tc, ^{201}Tl
^{99m}Tc – half-life, emission energy	$\tau = 6$ h, $E_y = 140$ keV
^{99m}Tc – production	fission product of ^{235}U
^{99m}Tc – use	brain, bone, liver, spleen, etc.
^{201}Tl – half-life, emission energy	$\tau = 73$ h, $E_y = 135$ keV, 167 keV
^{201}Tl – production	proton activation
^{201}Tl – use	cardiac stress test

9.1 Introduction

In x-ray radiography, the radiation source is outside the body. In scintigraphy, the radiation source is inside. The radioisotopes are administered to the patient by intravenous injecting of a small dose of radiotracers. The isotopes, bonded to pharmaceuticals, accumulate in targeted tissues. The emitted γ-radiation reveals their location. X-ray films or specialized flat-panel cameras detect the radiation and form images.

Scintigraphy is not a general imaging technique like x-ray radiography. The latter is used for examining healthy and diseased body parts alike. In contrast, scintigraphy is used to confirm the tumor's presence, location, and distribution in the body. The difference between these two imaging methods is highlighted in Fig. 9.1. Additionally, scintigraphy is used for several nontumor-related examinations, such as bone metabolism, thyroid disorder, kidney clearance, lung ventilation, and myocardial infarction perfusion. Aside from the similarities, there are fundamental differences between x-ray radiography and scintigraphy. X-rays reveal anatomical structures of body parts; scintigraphy highlights healthy versus diseased organs.

Moreover, there are essential differences in the imaging technique. In x-ray radiography, a straight line of radiation connects the source with the pixel on the detector. This line can be traced back to reconstruct an image. In scintigraphy, the location of radiation is not known a priori. Furthermore, each radioisotope emits γ-rays in 4π. It is the task of the detection system to locate the γ-source and to generate meaningful images for subsequent therapeutics. Because of this difference, scintigraphy and PET (Chapter 10) are known as *emission radiography* in contrast to *transmission radiography* using x-rays.

https://doi.org/10.1515/9783110757095-009

a. X-ray
radiography

X-ray
source

X-ray
image

b. Scintigraphy

Radioactive
Isotopes
inside body

Scintigram

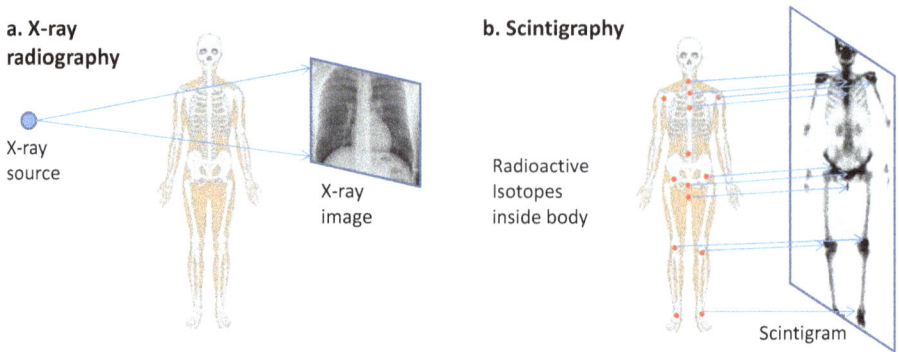

Fig. 9.1: The graphic compares X-ray radiography and scintigraphy. Note that the radiation source is outside the body in X-ray radiography, while in scintigraphy, it is inside.

In the following sections, we discuss the essential aspects of scintigraphic imaging. We distinguish between *single-photon emission* (SPE) scintigrams and *single-photon emission computed tomography* (SPECT). The difference is like in the x-ray analog: SPE uses a single flat-panel detector (called camera), which can be translated for a full-body scan. In the case of SPECT, two or three cameras rotate about the body at the center and take tomographic images of specific organs, such as the heart or the lungs.

This chapter will first discuss topics that concern both imaging modalities, including the detection system, the most important isotopes, and some of the main clinical applications performed with SPE. In the last part, we present some aspects of SPECT and discuss a few clinical applications. For basic information on radioisotopes, radioactivity, and radiation safety, we refer to Chapters 5 and 7. In the last part on detector techniques and image reconstruction, there will be some overlap with Chapter 8 on x-ray CT.

9.2 Collimators for scintigraphy

Scintigraphic imaging is most frequently performed with the radioisotope ^{99m}Tc. This isotope is a γ-emitter with a photon energy of 140 keV and a physical half-life of 6 h. After injection, the γ-emitting isotopes must accumulate in the tissue above equal distribution to generate contrast. Further details on the generation of the isotopes and their targeting of specific organs are discussed below. For now, we use this information to describe the scintigraphic imaging components consisting of collimators, photon detectors, and computer power, schematically illustrated in Fig. 9.2.

In x-ray radiography, the optional grid in front of the detector improves the contrast. In scintigraphy, the collimator grid is mandatory for generating contrast. Without a grid, there is no contrast and no image. This is because the γ-rays are emitted

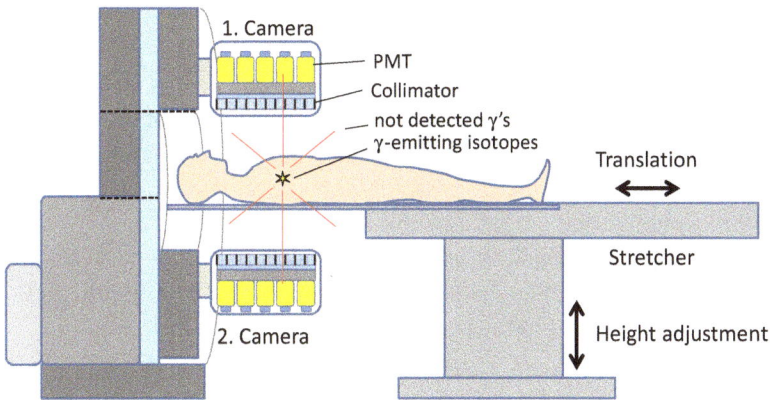

Fig. 9.2: Overview of components used for SPE and SPECT imaging. SPE imaging uses only one camera, while SPECT imaging requires both cameras, which rotate about the patient.

isotropically in all directions. Only those rays are useful and detected for imaging whose trajectories run perpendicular to the detector bank. All other beams with oblique incidence should be eliminated. Rays parallel to a collimator can be traced back to the origin and form an image of the isotope location [1].

Collimators can be made tall and narrow or short and wide (Fig. 9.3). The walls (septa) must be thick enough to absorb the high-energy γ-rays (see Exercise E9.1). Tall and narrow collimators promise high spatial resolution but low transparency. On the contrary, short and wide collimators have higher transparency at the expense of a lower spatial resolution. Therefore, the trade-off between transparency and resolution has to be tailored to the specific application. With a given collimator width d and height L, a field of view (FOV) with the diameter R at a distance z in the body is projected into the detector below. R, L, d, and z are interrelated by geometry (see Fig. 9.3(a)):

$$R = \frac{d(2z + L)}{L} \qquad (9.1)$$

This relation shows that the area projected by a single slit increases with the distance z. If two objects are laterally separated by a distance less than R, they cannot be distinguished by the detector. The diameter d defines the *resolution* of the collimator according to eq. (9.1) and the *transparency* scales with d/L. In practice, the spatial resolution of γ-cameras is about 5–10 mm.

The *sensitivity* of collimators is defined as the ratio of photons passing through the slits by geometric means to the total number of emitted photons. For hexagonal collimators with parallel walls, the sensitivity is expressed by [2, 3]:

$$S = \frac{\sqrt{3}}{8\pi} \left(\frac{d}{L}\right)^2 \left(\frac{d}{d + t_{\text{eff}}}\right)^2 \qquad (9.2)$$

Here, t_{eff} is the effective septal thickness, which is the combination of the physical septal thickness times a transparency factor depending on the attenuation. In general, the sensitivity of conventional collimator systems is rather low. On average, only 1 out of 5000 γ-photons pass the collimation system (0.02%).

Collimators are usually arranged on top of the detectors in a close-packed fashion, like the honeycomb grid in Fig. 9.3(b) [3]. The walls consist of a high Z material, either lead or tungsten, to keep the penetration as low as possible.

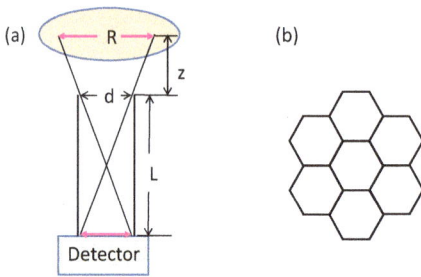

Fig. 9.3: (a) Collimator design characterized by length L and width d. z is the distance to the area of interest above the collimator, and R is the diameter of the FOV. (b) Hexagonal arrangement of collimators above the detectors.

The collimator walls may be straight (Fig. 9.4(a)), tilted toward the edges for some mimification (b) or tilted in the opposite direction for magnification (d). A pinhole camera can be used for special applications with high magnification, such as sketched in panel (c).

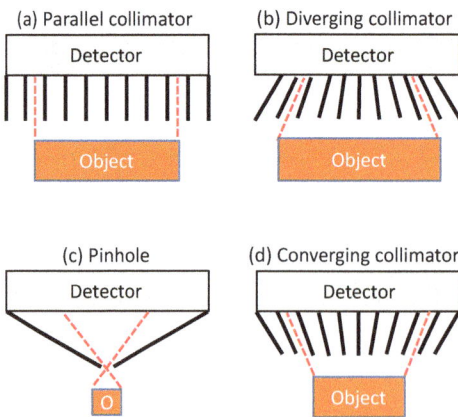

Fig. 9.4: Collimator arrangements for scintigraphic imaging: (a) collimators with straight walls; (b)–(d) collimators with inclined walls; (c) pinhole camera.

9.3 Detectors, counting, and artifacts

9.3.1 Photomultiplier tube

In X-ray radiography, digital images are recorded using 2D flat-panel TFT detectors that deliver high-resolution images. Scintigraphy has an inherently lower resolution, enabling the use of less expensive 2D scintillation detectors common in nuclear physics. Images recorded with scintillation counters are referred to as SPE scintigrams since each γ-photon is counted individually.

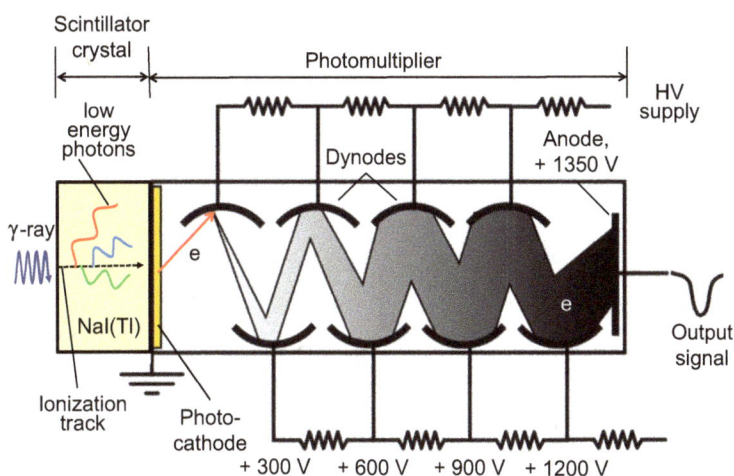

Fig. 9.5: Schematics of a scintillation counter including a scintillation crystal, photocathode, dynodes, and anode, forming the photomultiplier.

The components of a scintillation detector can be seen in Fig. 9.5. A scintillation crystal is placed in front of a photon counter cathode. The scintillation crystal converts γ-rays into visible photons. The most common scintillator material is a single crystal made of sodium iodide (NaI) that is drifted with thallium (Tl) ions. The high atomic number of Tl ($Z = 81$) is added to NaI in order to increase the quantum efficiency of the γ-ray detection and to enhance the linear attenuation coefficient of the NaI scintillator. However, the Tl concentration should not be too high to maintain the crystallinity and transparency of the NaI–photon conductor.

One γ-quant generates in NaI(Tl) about 5000 photons in the visible wavelength spectrum. The photons are directed to a photocathode, which generates around 400 photoelectrons per γ-quant. These electrons are then accelerated in a cascade of cathodes, called dynodes, toward the anode. The potential difference between the dynodes is about 100–200 V. Each time the electrons hit the next dynode, more electrons are produced until all electrons arrive at the anode and deliver a sizeable current pulse of about 10^5 to 10^6 electrons per detected γ-quant. The pulse height at the anode is proportional to the incident photon energy. The current pulse is then converted into a voltage pulse and further amplified by the electronics. The entire arrangement is referred to as a photomultiplier tube (PMT) [4].

Since the output signal of the PMT depends not only on the energy of the γ-rays but also on the voltage applied to the dynodes, the PMT array must be fine-tuned before use. Without calibration, artifacts can occur as some tubes are more or less efficient than others and produce higher or lower signals than average. The calibration is performed with a radioactive source of known activity and photon energy. To eliminate spatial variations between detectors, all should show the same count rate.

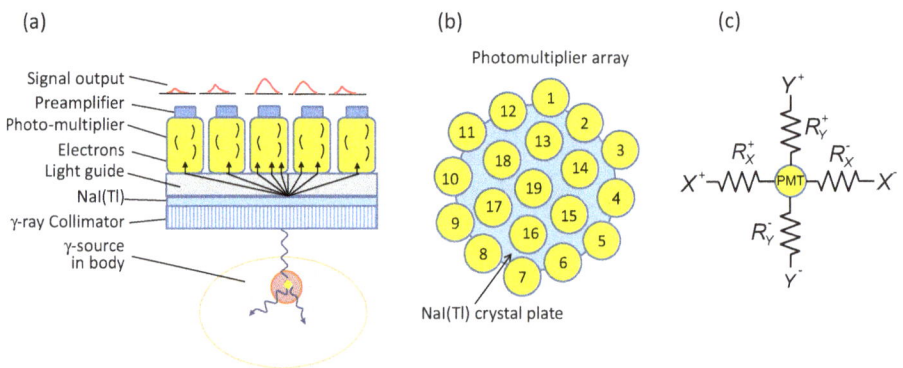

Fig. 9.6: Detector system for scintigraphy consisting of collimator, scintillation crystal, and an array of photomultipliers.

Because of the proportionality between the energy of the incident γ-ray and the pulse height at the anode, scintillation counters are also used for γ-spectroscopy in nuclear physics [5], and in some cases, also in medicine [6]. An example is discussed in the Infobox.

Scintillation detectors are much faster than ionization chambers or Geiger-Müller detectors (presented in Chapter 7.4), and their respective dead-times are much shorter, on the order of 200 ns. Therefore count rates from 10^4 to 10^5 are possible without saturating the detector. They are also much more sensitive than detectors used for X-ray radiography. Hence, SPE detectors deliver a γ-ray image with a lower spatial resolution but higher sensitivity than x-ray images.

Because of the higher sensitivity, the dose required for a scintigram is 10^3 to 10^4 times lower than for an x-ray radiograph. However, when an x-ray tube is switched off, the x-ray dose is also ended simultaneously. In contrast, the radioisotopes administered for taking a scintigram remain in the body and contribute to the accumulated dose to the patient over the biological half-life of the isotope.

9.3.2 Anger counting

The PMTs are usually arranged in a hexagonal array covered by a scintillation crystal and a light-guiding material, as illustrated in Fig. 9.6(a,b). In the end, the analog signal output is converted to a digital voltage by a high-speed A/D converter.

In order to localize the PMT with the highest output signal, one could simply compare the count rates of all PMTs individually after A/D conversion. However, it is faster to work with the analog signal. For this purpose, all PMTs are embedded in a resistor matrix that connects each PMT with horizontal and vertical resistors. An equivalent circuit is shown in Fig. 9.6(c), which is representative of all PMTs in the matrix. Now, like a potentiometer, the horizontal resistance difference between the left ($R_{i,x}^+$) and right resistance ($R_{j,x}^-$) is determined and normalized to the sum of all horizontal resistance values:

$$X_{PMT} = \frac{R_{1,x}^+ + R_{2,x}^+ + \ldots - \left(R_{n-1,x}^- + R_{n,x}^-\right)}{R_{1,x}^+ + R_{2,x}^+ + \ldots + R_{n-1,x}^- + R_{n,x}^-} = \frac{X^+ - X^-}{X^+ + X^-} \tag{9.3}$$

The same calculation is performed for the vertical direction, yielding:

$$Y_{PMT} = \frac{R_{1,y}^+ + R_{2,y}^+ + \ldots - \left(R_{n-1,y}^- + R_{n,y}^-\right)}{R_{1,y}^+ + R_{2,y}^+ + \ldots + R_{n-1,y}^- + R_{n,y}^-} = \frac{Y^+ - Y^-}{Y^+ + Y^-} \tag{9.4}$$

These two equations provide the (X,Y)-coordinates of the PMT with the highest signal output, indicating the position of the emitting isotope at a straight line (within the spatial resolution) below the detector. According to its inventor Anger,[1] who developed this detection scheme in the 1950s, the analog position logic is referred to as *Anger logic* [7].

9.3.3 CZT detectors

Commercial SPE/SPECT scanners have been equipped with cadmium-zinc-telluride (CdZnTe or simply CZT) detectors for about a decade. CZT is a solid-state p-type semiconductor material with a wide indirect bandgap of 1.68 eV and a high density of 5.78 g/cm^3 [8–10]. The high density is responsible for a significant photon absorption coefficient of 95% for photon energies <50 keV, but still high also for photon energies

---

1 Hal Anger (1920–2005), American engineer and biophysicist.

>50 keV. The wide energy gap between the valence and conduction band implies that CZT detectors have low dark current and hence can be operated at room temperature without cooling.

Figure 9.7 shows a schematic energy band structure of a p-type semiconductor. An energy gap of 1.5 eV in the case of CZT separates the full valence band from the empty conduction band. The p-type conductivity implies that electric transport is supported by positively charged holes in the valence band rather than by electrons in the conduction band. Thermal excitation transfers electrons to local acceptor energy levels from the full valence band. This creates positively charged holes in the valence band, which are mobile in an electric field.

By γ-ray absorption, the photon energy is converted in electron-hole pairs, electrons in the conduction band, and holes in the valence band. For detection of the γ-rays, the electron-hole pairs are separated in a high electric field. The positive charges move to the cathode (front surface) and negative charges to the pixelated anode. Both cathode and anode are metal contacts. Each anode acts as a detector.

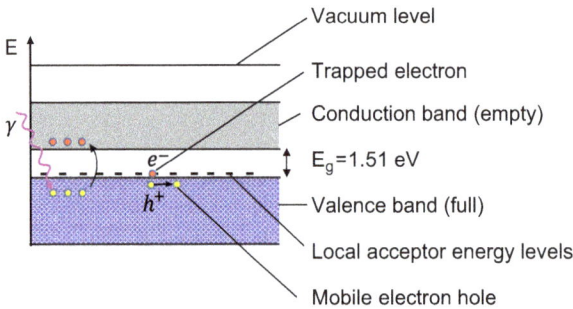

Fig. 9.7: Schematic energy band diagram of a p-type semiconductor. During absorption, the γ-photon energy is converted in many electron-hole pairs.

Two effects distinguish scintillators from CZT detectors. First, there is a chance of signal loss in CZT due to electron-hole recombination before detection. And second, the cloud of electron-hole pairs created will diffuse laterally and may affect neighboring pixels, which reduces the spatial resolution. Overall, the advantages of CZT detectors outweigh the disadvantages, such that CZTs increasingly replace PMTs.

Tab. 9.1: Comparison of CZT and scintillation detectors concerning spatial and energy resolution and sensitivity [9].

	CZT	Anger camera
Spatial resolution	6.7–7.7 mm	7.7 mm
Energy resolution for 140 keV photons	5.5% (7.7 eV)	9.2% (12.9 eV)
Sensitivity (detected photons per 1 MBq source activity)	1150 photons	144 photons

A comparison of Anger PMTs with CZT detectors regarding spatial resolution, energy resolution, and sensitivity has been carried out using a whole-body phantom [8]. This study shows that the CZT camera has an eightfold higher sensitivity, a higher energy resolution, and better image contrast, but similar spatial resolution compared to conventional SPECT cameras. The higher sensitivity of the CZT detectors allows a decrease of the acquisition time and a lower activity injected to patients. Moreover, the pixilated CZT detectors allow different and adaptive designs of the detector system, as we will see later.

9.3.4 Artifacts: Compton scattering

So far we have assumed that all isotopes emit γ-rays isotropically and along a straight line from the origin to the detector. This assumption should be questioned and justified. The straight-line assumption is sketched in Fig. 9.8 (2) and compared with other potential scenarios. The tilted beam (1) is absorbed in the collimator, while the tilted beam (3) penetrates the collimator wall and reaches the detector. Beam (4) is Compton-scattered by the electrons in the collimator and reaches the detector instead of being absorbed. Ray (5) is Compton-scattered in the detector and escapes before it is

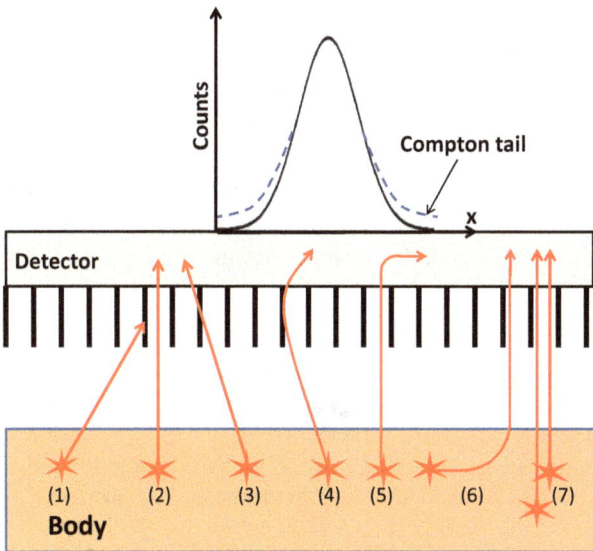

Fig. 9.8: Seven possible scenarios for the emission and detection of γ-rays. In the presence of Compton scattering, the Gaussian-shaped tails of the intensity peak are intensified: (1) absorption in the collimator; (2) straight through; (3) penetration through the collimator wall; (4) Compton scattering at the collimator wall; (5) Compton scattering in the detector crystal; (6) Compton scattering in the body; (7) obscured depth information by superposition of two γ-emitters.

detected. Ray (6) is Compton-scattered in the body and gives incorrect information about the location of the isotope. Finally, overlapping rays (7) obscure the depth information about the position of the isotopes. There is nothing that one can do about these processes than to take them into account by critically evaluating the recorded images. However, the tail of the recorded intensity distribution can assess how significant the contribution of Compton-scattered photons is. Compton-scattering increases the tails of an otherwise Gaussian distribution, as indicated in Fig. 9.8.

9.3.5 SNR and CNR

The signal-to-noise ratio is defined as (see eq. (8.23)):

$$\text{SNR} = \frac{\mu}{\sigma} \tag{9.5}$$

Here μ is the mean number of photons detected, and σ is the noise, i.e., the standard deviation of a Poisson or normal distribution. The contrast-to-noise ratio is defined as:

$$\text{CNR} = \left| \frac{\mu_A - \mu_B}{\sigma} \right| \tag{9.6}$$

where μ_A and μ_B are the mean numbers of photons detected in area A and B. More details about the evaluation of SNR and CNR can be found in Section 8.5.1–2. CNR applied to scintigraphy implies that there is no intrinsic noise in an area where no radiopharmaceuticals have been distributed. Therefore, if $\mu_B \approx 0$, SNR and CNR are essentially equal. Only Compton scattering and incomplete accumulation of radiopharmaca may reduce the CNR. The CNR in SPE scintigrams is generally better than for x-ray radiographs. But the spatial resolution is much lower compared to x-ray radiography.

Infobox: Nuclear spectroscopy

Because of the proportionality between the γ-photon energy and the PMT signal height, PMTs are suitable as energy analyzer of γ-emission spectra. An example of an γ-emission spectrum is shown below. Such a spectrum is recorded by first setting an energy window ΔE, which is equivalent to the energy resolution. In fact, what we analyze is not the energy but the pulse height. Therefore this method is also called pulse height analysis (PHA). Once we have set the resolution, we move the window from the lowest voltage output up to the highest value and record for each window setting the accumulated counts for a specified time interval. The results of such a measurement are shown below. It consists of a photopeak at high energies (pulse height) and a long tail stretching to lower energies. The main photopeak is due to the γ-emission, for instance, from the ^{99m}Tc isotope. The width of the photopeak is a good measure of the detector's energy resolution since the intrinsic width of the γ-emission line is rather narrow. The long tail is caused by Compton scattering of γ-quants at electrons in the NaI(Tl) crystal. During the inelastic scattering process, the photon loses energy, the amount depending on the scattering angle (see Section 6.2.4). The red-dashed line indicates count rates originating from various scattering processes.

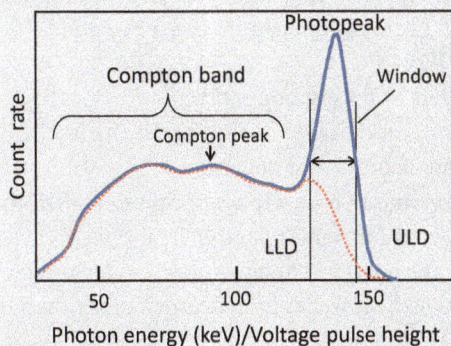

Photon energy (keV)/Voltage pulse height

Once we have determined the energy spectrum, we can set an energy window that limits the detection of counts to those energies between the upper-level discrimination (ULD) and lower-level discrimination (LLD). This window is usually wider than the energy window used for the spectroscopy analysis. The new and broader window serves the purpose to execute a γ-ray detection with a high count rate that simultaneously surpresses background noise from Compton-scattered photons. For this purpose, it is customary to set a 20% acceptance window centered on the photopeak.

An advanced version of this detection scheme would use a multichannel analyzer that allows to detect either large parts or the entire spectrum at once. Energy windows of interest can then be selected electronically. [5]

9.4 Isotopes for scintigraphy

9.4.1 Radioisotopes and radiopharmaceuticals

All radioisotopes used for scintigraphy are produced artificially. The most common ones in nuclear medicine are listed in Tab. 9.1 [11]. The ideal radioisotope is a high-energy, pure γ-emitter akin to the body's metabolism. Unfortunately, isotopes of light elements common in organic molecules, such as H, C, N, O, and P, do not have emission lines in the 80–300 keV energy range favored for scintigraphy. Only iodine with the isotopes ^{123}I and ^{131}I is part of the thyroid metabolism and can be used for diagnostics (^{123}I) and for therapy (^{131}I). Other useful γ-emitters for scintigraphy such as ^{99m}Tc, ^{111}In, and ^{211}Tl have to be bound to ligands that promise a similar and analogous biochemical behavior as the natural ones.

The isotope ^{99m}Tc is the most common γ-emitting radioisotope for scintigraphy. It is used for imaging of tumors of the skeleton, in the brain, kidneys, liver, gallbladder, lungs, etc. and for perfusion studies of blood in the heart and liquids in the kidneys. Table 9.1 lists the main applications of ^{99m}Tc and other γ-emitting isotopes for scintigraphy. Before administering radioisotopes, they have to be tagged onto radiopharmaceuticals with specific functions that transport them to the targeted destination. Some frequently used radiopharmaceuticals for tagging ^{99m}Tc and their targets are listed here (Fig. 9.9) [12]:

–   The isotope ^{99m}Tc is chemically tagged to an organic molecule $C_{13}H_{25}N_4O_3Tc$, called exametazime (HMPAO). HMPAO can cross the blood–brain barrier (BBB) and is used to study cerebral perfusion disorder and dementia.
–   A radiopharmaceutical ^{99m}Tc-hydroxy methylene diphosphonate (HMDP) or methylene diphosphonate (MDP) is used for bone imaging. The phosphonate concentrates in the mineral phase of the bone by chemisorption. Accumulation in the bone depends on blood flow, which is higher in cancerous areas than in healthy parts.
–   Imaging the renal glomerular filtration is done by tagging ^{99m}Tc to mercaptoacetyl triglycine known as MAG_3 scan. The molecule is small enough to pass the glomerular filter (see Vol. 1/Chapter 10).
– ^{99m}Tc tagged to methoxy isobutyl isonitrile (MIBI) ligands is used for myocardial perfusion imaging. Scans performed with the use of MIBI are commonly referred to as "MIBI scans".

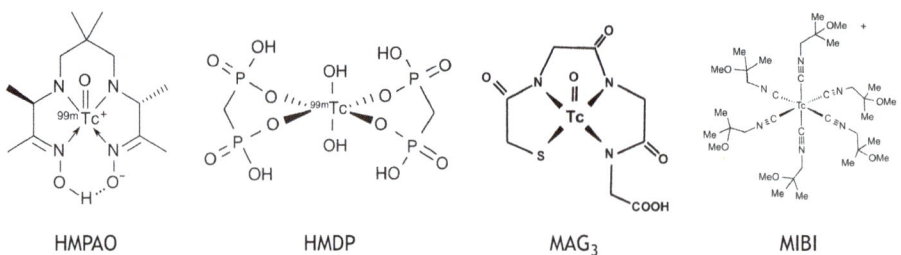

Fig. 9.9: Common radiopharmaceuticals used for tagging ^{99m}Tc.

In the past, radiation clinics used different isotopes listed in Tab. 9.2. However, more recently, most scintigrams are performed with ^{99m}Tc in general and ^{123}I specifically for thyroid examinations.

Tab. 9.2: Radiosotopes and their main use in scintigraphy. Adapted from: http://www.nucmedtuto rials.com/dwclinical/in111.html.

Isotope	Energy of γ-photons (keV)	Physical half-life	Application
^{99m}Tc	140	6.005 h	Most widely applied radioisotope for diagnostics. Different radiopharmaceuticals are used for brain, bone, liver, spleen, kidney imaging, and for blood flow studies.
^{67}Ga	94, 184, 296	79.2 h	Gallium citrate used for demonstrating the presence of several types of malignancies.
^{111}In	173, 247	67 h	Used for tumor recognition
^{123}I	159	13.3 h	Widely used to diagnose thyroid disorders
^{201}Tl	135, 167	73 h	Used in cardiac stress test
^{133}Xe	81	5.3 days	Used for studying lung ventilation
^{51}Cr	321	28 days	Used for radiolabeling of red blood cells for specialized in vivo/in vitro studies, not for imaging

Isotopes for scintigraphy are artificially produced γ-emitters with energies in the 100–200 keV range and lifetimes between a few hours to a few days. The most important isotopes for scintigraphy are ^{99m}Tc and ^{123}I.

9.4.2 Isotope generators

Isotope generators provide short-lived radioisotopes in clinics for diagnostics and for therapy. Short-lived radioisotopes are extracted from longer-lived radioisotopes by *elution*. In equilibrium, both nuclides, parent and daughter, appear to decay with the same half-life.

The best known example for the elution process is the ^{99}Mo/^{99m}Tc generator [13]. After ^{99}Mo is produced by fission reaction of ^{235}U (see Section 5.6.5), it is chemically purified and passed on an ion-exchange column composed of alumina (Al_2O_3). The doubly negative charged and water-soluble molybdate ^{99}MoO$_4^{2-}$–ion is firmly bound to the alumina substrate. In contrast, the single charged pertechnetate ^{99}TcO$_4^-$-ion is less bound to the substrate. Technetium unbinds from the substrate whenever ^{99}Mo decays to ^{99m}Tc. Then ^{99m}Tc goes in solution and can then be washed out. The column is placed in a lead container with tubes attached to allow column elution. The column is schematically shown in Fig. 9.10.

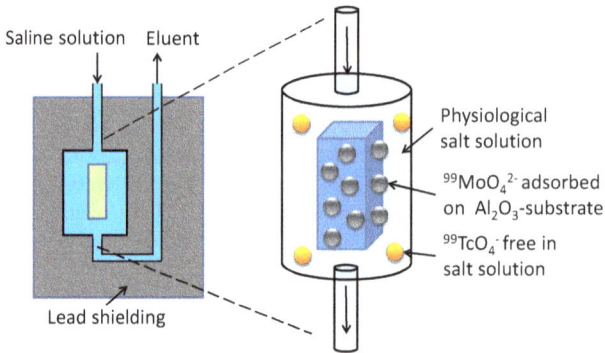

Fig. 9.10: Generator system for elution of ^{99m}Tc from a column-containing ^{99}MoO$_4^{2-}$ ion bound to alumina substrate.

The activity of ^{99m}Tc reaches a maximum after 23 h and then decays with the same effective half-life as ^{99}Mo, illustrated in Fig. 9.11(a). Elution at the peak maximum is optimal and usually executed in clinics. After elution, the activity drops dramatically but is recovered after a few hours. Therefore several elutions are possible during a week-long period, as depicted in Fig. 9.11(b). This process is sometimes also referred to as "milking" the generator. The elute-containing salt solution and ^{99m}Tc radioisotopes are safely placed in a vial, from which a patient's dose is withdrawn. The dose has to be calibrated before administering.

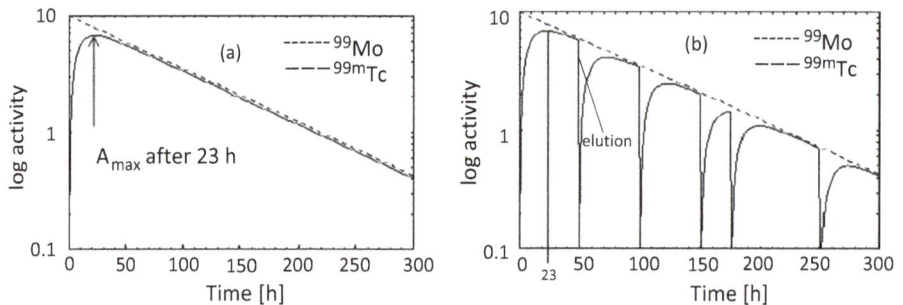

Fig. 9.11: (a) Build-up of ^{99m}Tc in a ^{99}Mo/^{99m}Tc generator. Note the maximum ^{99m}Tc activity after 23 h. (b) Multiple elutions of ^{99m}Tc are possible during one week.

> ! ^{99m}Tc is the most widely used radioisotope for scintigraphy. The isotope is made ready for use in clinics with the help of a generator.

9.5 Full body SPE scans

As described above, scintigraphy is traditionally performed with two-dimensional arrays of scintillation detectors. 2D images provide information on the location of the γ-emitting radioisotope in the body. Simultaneously they record the temporal evolution of the radioisotope distribution. Collimators in front of the detector are obligatory for localizing the emitter, enhancing contrast, and increasing the spatial resolution. The contrast is usually high since only the radioisotopes but not the rest of the body contribute to the image, in distinction to attenuation radiography via x-rays. However, the spatial resolution is much lower (~cm) compared to x-ray radiography (<mm). Figure 9.12 shows a full-body scan of the ventral and dorsal body parts after three hours of ^{99m}Tc injection. Another instructive sequence of scintigrams is depicted in Fig. 9.13. The perfusion of liquids through the renal system from the kidneys to the bladder was recorded over a time span of 12 min. The tubular filtration was probed by using a ^{99m}Tc-MAG$_3$–scan [14].

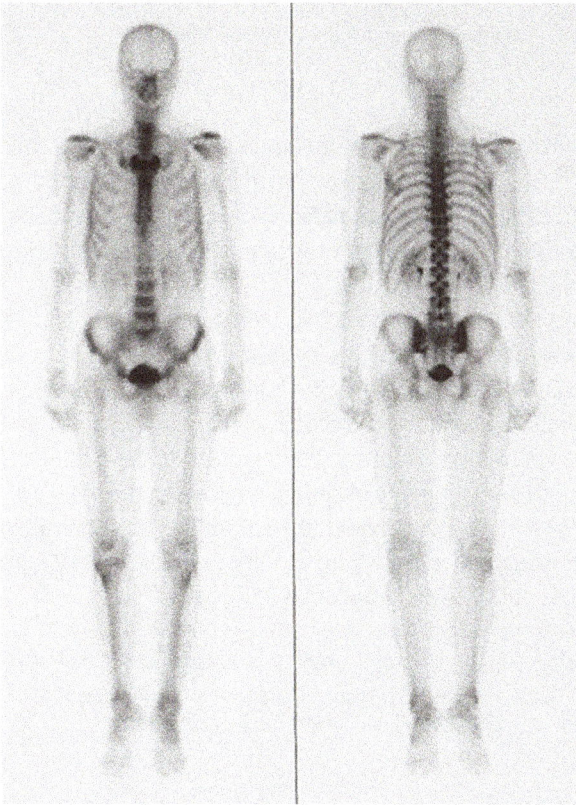

Fig. 9.12: Ventral and dorsal scintigram of the skeleton. (Adapted from https://en.wikipedia.org/wiki/Bone_scintigraphy#/, © creative commons).

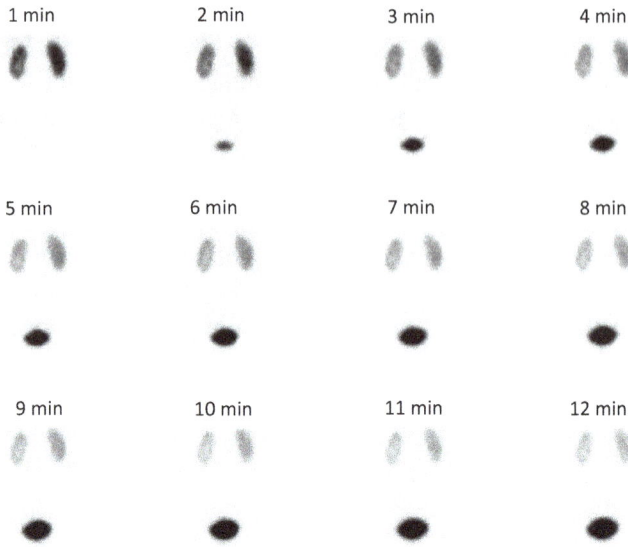

Fig. 9.13: Time sequence of the perfusion of liquids through kidneys and bladder. (Adapted from http://www.people.vcu.edu/~mhcrosthwait/).

The average activity injected intravenously is in the order of 500 MBq for a full-body bone scintigram using ^{99m}Tc. This is sufficient for recording a full-body γ-radiation image and low enough to keep the patient radiation level within reasonable margins. The corresponding dose is about 0.1–0.2 μSv. In comparison, for a full-body x-ray CT-scan, a dose of more than 5 mSv is deposited.

Planar scintigraphy or projection scintigraphy using a single detector head has the drawback of providing just a single projection. This is similar to an x-ray radiogram that lacks depth information. Another disadvantage is the high dose to patients while the spatial resolution is rather limited. Therefore, projection scintigraphy with just one detector array is applied only for confirming tumors, inflammations, etc. An additional justified application is the test of the functionality of certain organs like the renal system or the respiratory system. For scanning the thyroid, there is, in fact, no alternative to planar scintigraphy. A typical examination setting is reproduced in Fig. 9.14 [15]. As discussed in the following section, SPECT is preferred in all other cases.

> **!** Scintigrams, although similar to x-ray radiographs, are not taken for imaging body parts and organs, but for studying perfusion (renal and heart), localizing tumors, and identifying specific disorders. Scintigraphy therefore is a functional imaging modality not an anatomical.

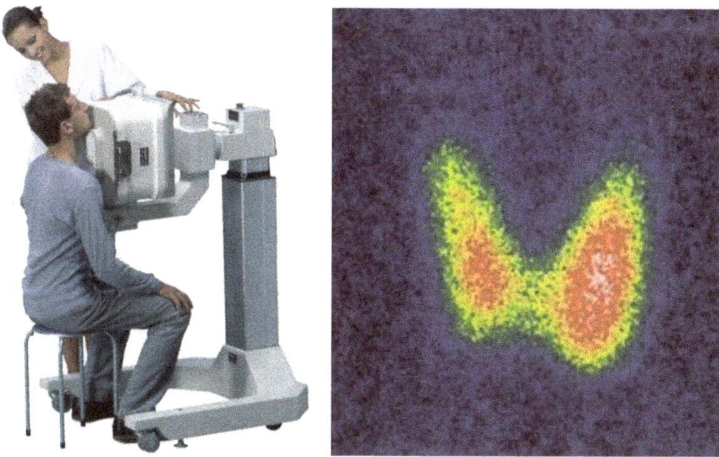

Fig. 9.14: Left: SPE scintigraphy of the thyroid with a flat panel camera. Right: Tc-99m scintigraphy of the thyroid gland (Reproduced from https://openi.nlm.nih.gov/ © creative commons).

9.6 Single-photon emission computed tomography (SPECT)

9.6.1 SPECT systems and detectors

The shortcomings of flat panel projection scintigraphy can be overcome by taking scans simultaneously under different angles. One detector can be rotated around the patient like it is done in x-ray CT scans, or several planar detector arrays may be mounted on a circular support (gantry) recording simultaneously the γ-emission of isotopes in the body. Three detector arrangements are shown in Fig. 9.15. Two detector panels may enclose an angle of 180° or 90° as seen in panels (a) and (b), or three detector panels may be arranged at angles 120° apart. Each planar detector array (camera) is similar to the one used for SPE. In *single-photon emission computed tomography* (SPECT), however, rotatability of the detector arrays and computer power are added for higher contrast and depth resolution of the scintigrams taken.

SPE uses the analog signal output from NaI (T1)/PMTs and the Anger logics for 2D mapping of the isotope distribution and image reconstruction. In contrast, with SPECT, the PMT signal is digitized by an A/D converter, and the entire calculation is carried out using digital electronics. A SPECT system with one or more cameras equipped with NaI (Tl)/PMT detectors is referred to as an Anger SPECT system [7]. In recent years solid-state detectors have been developed that use semiconducting CdZnTe (CZT) single crystals as sensors [9]. As mentioned before, CZT-based detectors have a higher sensitivity and a higher energy resolution with a spatial resolution similar to that of PMTs. With their increased sensitivity, the new SPECT cameras enable a

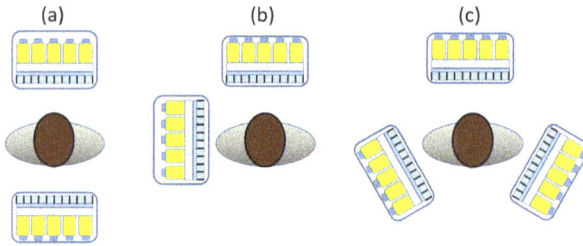

Fig. 9.15: Detector arrangement for SPECT: (a) double detector arrangement separated by 180° and in (b) by 90°. (c) Triple detector arrangement.

lower isotope dose and a significantly shorter recording time compared to the Anger cameras. The 90° arrangement sketched in Fig. 9.15(b) is mainly used for myocardial diagnostics since the heart is located close to the front on the left side of the body. Scanning on the opposite side would only increase the noise instead of contributing to a useful signal. When the detectors in panels (a) and (c) are rotated, the center of rotation must be chosen either at the gantry's symmetry center or at the organ's center, such as the heart. In any case, the speed is kept low because of the low counting rate.

The latest SPECT emission scanners offer a few more advantages [8]. The CZT pixel arrays are mounted on retractable arms that surround the patient (Fig. 9.16(a)). As soon as the patient is fixed in the scanner, the detector arms move inwards and are individually adjusted. The detectors can focus on the coronary system for a myocardial CT or form an ellipse for a whole-body bone scan. Special ergonomic and patient-friendly armchairs have been designed for myocardial and coronary stress tests that support a detector bank focusing on the heart.

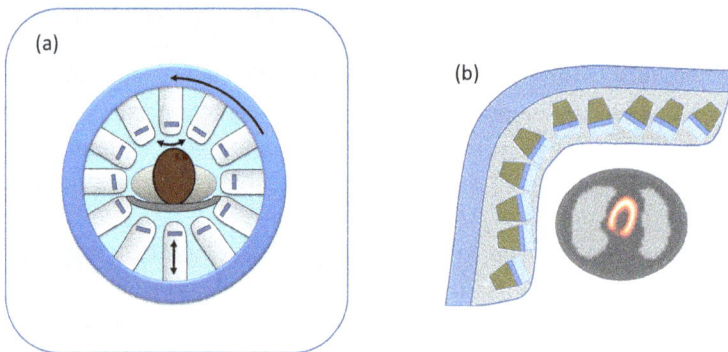

Fig. 9.16: (a) Design of a multipurpose, whole body CZT-based SPECT-camera. Twelve detectors arranged around the patient, each one move in and out and tilt to adjust to the region of interest; (b) "armchair" design for myocardial stress tests. The adjustable CZT detectors with parallel slits in front surround the region of interest.

The major manufacturers of SPECT systems offer combinations of SPECT and x-ray CT. This has the advantage that anatomical information can be correlated to functional records in one setting by a short z-translation of the patient. For image processing, the CT voxel v_{ijk} is reusable as SPECT voxel. The different spatial resolutions of CT and SPECT result in different voxel sizes. However, both voxel sizes can be adapted by binning the CT voxels. CT takes almost instantaneous body images, whereas SPECT averages over many cardiac and breathing cycles. The result is a blurring of images taken close to the heart and lung, which must be corrected for during image processing.

9.6.2 Clinical applications

In clinics, SPECT is mainly applied for the following diagnostic procedures:
- Bone-SPECT is used to localize regions with different and conspicuous metabolism.
- Brain perfusion SPECT is used for the diagnosis of normal functionality versus Parkinson's and Alzheimer's diseases and epilepsy.
- Myocardial perfusion-SPECT is used for inspecting the vitality of the myocardium.

In the following, we will discuss only the myocardial perfusion SPECT [16, 17]. After injection of a radioactive tracer, usually ^{201}Tl or ^{99m}Tc, and waiting a few minutes for homogeneous distribution of the isotopes, scintigraphic images are recorded. The SPECT images integrate over blood flow to and from the heart. Hence, the images recorded are superpositions of a complex flow pattern in the coronary artery and the cardiac chambers. The images observed are not easy to interpret and require some experience. Usually, three slices (projections) are taken of the heart along three perpendicular axes: short axis, vertical long axis, and horizontal long axis. The orientation of the axis together with characteristic images of a healthy person is shown in Fig. 9.17 panel (a). These images serve as a reference against which any heart disease affecting the blood flow can be evaluated.

The most common test is a stress test. First, myocardial perfusion images are taken of the patient at rest. Then the patient carries out some exercise like walking or cycling such that the heartbeat goes up markedly. The exercise requires an increased oxygen supply by higher blood flow through widened vessels. Subsequentially, another series of images are taken. A comparison of the images will allow the evaluation of the blood flow under different levels of exercise (stress). The images may show different shades of color that indicate which areas of the heart absorbed more of the radioactive tracer and which ones less. A normal test result indicates sufficient and unrestricted blood flow in and out of the heart, while an abnormal result means that the heart's blood flow is insufficient due to narrow or blocked arteries, as indicated in panel (b) of Fig. 9.17. Any damage due to, for instance, a heart attack can then be located and further diagnosed.

(a) Healthy | (b) Abnormal

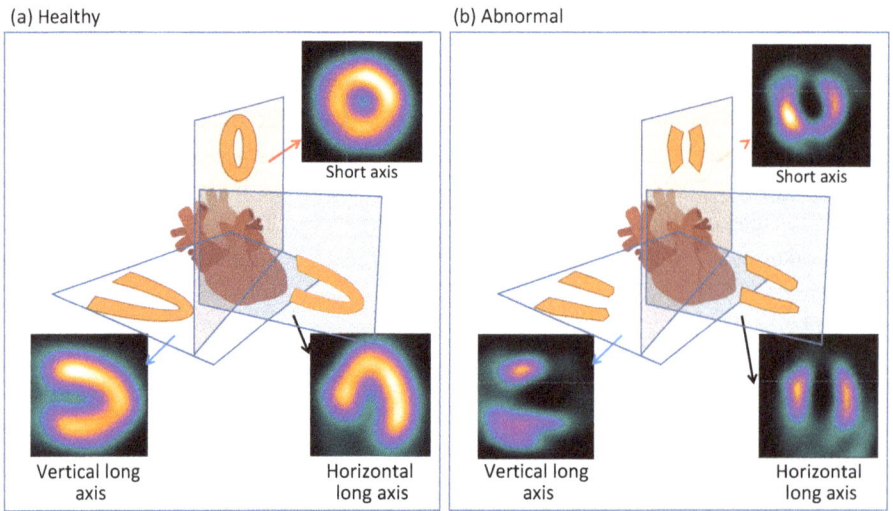

Fig. 9.17: (a) The SPECT images show the heart in a "resting" position. The tomographic orientations are indicated by plains: long vertical axis; long horizontal axis; and short axis. The SPECT images refer to a healthy person at rest; (b) same as in (a) but for a patient with a heart disease that reduces the blood flow.

Another application of myocardial SPECT is the control of artery bypass grafting. Preoperative and postoperative myocardial SPECT images are shown in Fig. 9.18 allowing to review the improvements after the intervention.

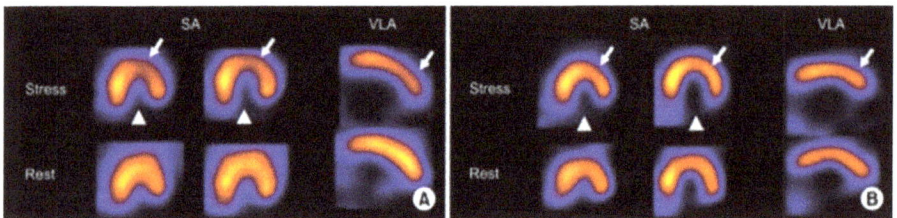

Fig. 9.18: (a) Preoperative and (b) postoperative myocardial SPECT. The patient underwent a left carotid-to-subclavian artery bypass grafting. Postoperative myocardial SPECT shows improved stress perfusion in the anterior and anterolateral walls when compared with the preoperative SPECT (arrows). SA, short axis; VLA, vertical long axis. (Reproduced from https://openi.nlm.nih.gov/© creative commons).

9.6.3 SPECT image processing

In Chapter 8.9.5 on CT, we have seen that the intensity recording in a particular detector pixel is the result of a linear x-ray attenuation integrated over the path length and expressed in polar coordinates:

$$p_z^{CT}(r,\theta) = \int_0^t \mu_z(r,\theta)ds \qquad (9.7)$$

Here the projection $p_z^{CT}(r,\theta)$ results from a 1D line integral into one particular zero-dimensional (0D) pixel. By repeating the line integrals for different (r,θ)-values, a sinogram results as a representation of all projections from one slice of width Δz perpendicular to the z-direction.

In SPECT, similar projections are performed. However, the integration is not taken over an attenuation profile but instead over the spatial distribution $f_z(r,\theta)$ of radioisotopes in a slice of the body:

$$p_z^{SPECT}(r,\theta) = k \int_0^t f_z(r,\theta)ds. \qquad (9.8)$$

k is a proportionality constant between the actual PMT signal amplitude and the isotope concentration in the body.

As stated in Section 8.9, any 2D function $f_z(r,s)$ projected by a line integral into a 1D profile constitutes a Radon transformation:

$$p_z(r,\theta) = \mathcal{R}f_z = \int f_z(r,\theta)ds. \qquad (9.9)$$

In this sense, the SPECT line integral is also a Radon transformation. Therefore, the projections and filtered back projections can be treated similarly to what is already detailed for x-ray CT. However, there is one difficulty that distinguishes x-ray CT and SPECT. In SPECT, the projections taken are made under the assumption of zero-tissue attenuation. If attenuation effects have to be considered, the line integral is modified to:

$$p_z^{SPECT}(r,\theta) = k \int_0^t f_z(r,\theta)\exp\left[-\int \mu_z(r,\theta)ds'\right]ds \qquad (9.10)$$

This projection double integral analysis is much more difficult to solve than the simpler one in eq. (9.8) and can be solved only via an iterative approach. For simplification, one may assume that the attenuation coefficient is a constant within the body tissue of known thickness t. Then

$$p_z^{\text{SPECT}}(r, \theta) = \frac{k}{\mu_z} \int_0^t f_z(r, \theta) \exp\left[-\mu_z t\right] ds \qquad (9.11)$$

In most cases, absorption corrections are applied. For further information on the data analysis of SPECT, we refer to specialized literature such as [18, 19].

9.7 Summary

S9.1   Scintigraphy records γ-emission of radioisotopes that are injected into the body.

S9.2   Radioisotopes for scintigraphy are bound in ligands that have analogous biochemical properties to those without isotopes.

S9.3 ^{99m}Tc is the most frequently used radioisotope in scintigraphy.

S9.4 ^{99m}Tc is a decay product of ^{99}Mo, which in turn is a fission product of ^{235}U after slow neutron capture.

S9.5 ^{99m}Tc is extracted in the clinic with the help of a generator separating parent (^{99}Mo) and daughter isotopes (^{99m}Tc) after the activity of the daughter isotope has reached the first maximum.

S9.6 ^{99m}Tc is used for brain, bone, liver, spleen, kidney imaging, and for blood flow studies.

S9.7   Scintigrams consist of digitally recorded projection images with an array of collimators in front of scintillation counters.

S9.8   Scintigraphy using scintillation counters is known as single-photon emission (SPE).

S9.9   SPE scintigraphy with a single flat-panel detector provides two-dimensional projection scintigrams.

S9.10 2D projection SPE scintigrams are used for full-body scanning, thyroid inspection, and perfusion studies of the myocardium and the renal system.

S9.11 Arrangement of two or three flat-panel detectors allows depth resolution (SPECT), extending perfusion studies to the brain, and myocardial investigations under various stress conditions.

? Questions

Q9.1   In scintigraphy the source of the electromagnetic radiation is in the body. How do they get there?

Q9.2   What type of radioactive emitter is used for scintigraphy: alpha, beta, or gamma emitters?

Q9.3   Why are heavy radioisotopes required?

Q9.4   How are the radioisotopes produced for use in scintigraphy?

Q9.5   Why is the gamma emission of ^{137}Cs not used for scintigraphy?

Q9.6   What are the most frequent radioisotopes used for scintigraphy?

Q9.7   What is the photon energy of the gamma emission of ^{99m}Tc?

Q9.8   Are the radiopharmaceuticals used in scintigraphy authentic or analog?

Q9.9   What is the basic working principle of a ^{99}Mo/^{99}Tc generator?

Q9.10   After what amount of time can the Tc–elution be extracted from the generator?

Q9.11   What detection system is used for recording scintigrams?

Q9.12   What is the use of collimators in front of the detector?

Q9.13   How does a scintillator counter including photomultiplier tube work?

Q9.14   What does the acronym SPECT stand for?

Q9.15   How many detector arrays are used for SPECT scanning?

Q9.16   How does SPECT compare with x-ray CT?

Q9.17   What is SPECT used for?

Q9.18   What is the typical spatial resolution of SPE/SPECT images, and how does it compare with x-ray radiography?

Attained competence   +   0   –

	+	0	–
I know SPE/SPECT uses radioisotopes that have been injected in the body			
I know which isotopes are mainly used for SPE/SPECT diagnosis			
I can name and identify the main components of a SPE γ-camera			
I am aware of the fact that the radiopharmaca are analogs			
I have an idea of how the Anger camera works			
I know the principle mechanism of a PMT			
I am aware of alternative detection systems			
I recognize the difference between SPE and SPECT			
I appreciate that SPECT requires filtered backprojection for image reconstruction			
I am aware of new designs of SPECT scanners			

Exercises

E9.1   **Wall thickness for 1% transmission:** Determine the grid wall thickness made of lead that absorbs 99% of 140 keV photons at normal incidence.

E9.2   **Effective wall thickness:** The γ-rays usually do not arrive at normal incidence when penetrating the collimator wall, as assumed in E9.1. Instead, those rays to be absorbed have an angle $\alpha' = \alpha + \Delta\alpha$, where α is the limiting penetration angle, defined by the length L and width d of the collimator.

     a.   What is the effective wall thickness t_{eff} in terms of L, d, t?

     b.   If $t_{eff} = 2$ mm, what is the wall thickness t at the ratio $d/L = 0.1$ and $\Delta\alpha = 10°$?

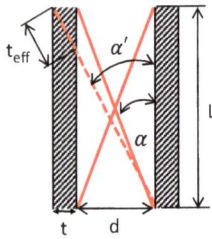

E9.3 **Detector quantum efficiency:** One γ-quant of 140 keV generates in a NaI(Tl) single crystal in the order of 5000 photons in the visible wavelength spectrum at a wavelength of about 500 nm. What is the quantum efficiency of this conversion process?

E9.4 **Activity uptake:** For a body scan, usually 500 MBq of ^{99m}Tc are administered. We assume that the isotopes homogenously distribute in the body of an area of 1.2 m^2. What is the expected count rate per pixel of size 1 cm^2 and a counting efficiency of 1%?

E9.5 **SNR:** The SNR measured is 50 per pixel. In order to improve the SNR to 100, you have two options: either increase the counting time of presently 50 s or increase the count rate of presently 50 s^{-1}. Discuss the pros and cons of both options. Which one would you prefer?

E9.6 **Compton peak:** In the infobox, the energy spectrum of the 140 keV ^{99m}Tc photopeak and the Compton band is shown. The energy spectrum shows a Compton peak at about 90 keV. What is the origin of this peak?

E9.7 **Compton peak – continued:** The Compton peak can be explained by the cutoff energy transfer. However, why is there a peak and why can scattered intensity still be detected for energies lower than 90 keV?

References

[1] Madsen MT. Recent advances in SPECT imaging. J Nucl Med. 2007; 48: 661–673.
[2] Wieczorek GA. Analytical model for SPECT detector concepts. IEEE Trans Nucl Sci. 2006; 53: 1102–1112.
[3] Van Audenhaege K, Van Holen R, Vandenberghe S, Vanhove C, Metzler SD, Moore SC. Review of SPECT collimator selection, optimization, and fabrication for clinical and preclinical imaging. Med Phys. 2015; 42: 4796–4813.
[4] Hamatsu Photonics, Inc. Photomultiplier tubes, basics and applications. 3rd. 2007.
[5] Knoll G. Radiation detection and measurement. NY: John Wiley & Sons, Inc; 2000.
[6] Jenkins D. Applications of gamma-ray detection for society, medicine and other areas of science. In: Radiation detection for nuclear physics, methods and industrial applications. 7–1 to 7–19: 2020.
[7] Peterson TE, Furenlid LR. SPECT detectors: The Anger Camera and beyond. Phys Med Biol. 2011; 56: R145–R182.
[8] Ljungberg M, Pretorius PH. SPECT/CT: An update on technological developments and clinical applications. Br J Radiol. 2018; Jan 91: 20160402.

[9]   Desmonts C, Bouthiba MA, Enilorac B. et al, Evaluation of a new multipurpose whole-body
      CzT-based camera: Comparison with a dual-head Anger camera and first clinical images.
      EJNMMI Phys. 2020; 7: 18.
[10] Shkir M, Khan MT, Ashraf IM. et al., High-performance visible light photodetectors based on
      inorganic CZT and InCZT single crystals. Sci Rep. 2019; 9: 12436.
[11] The supply of medical isotopes: An economic diagnosis and possible solutions. Paris: OECD,
      Publishing; 2019.
[12] Ogawa K. Biocomplexes in radiochemistry. Phy Sci Rev. 2016; 1: 20160005.
[13] Moore PW. Technetium-99 in Generator Systems. J Nucl Med. 1984; 25: 499–502.
[14] Ceylan Gunay E, Erdogan A. Ring sign over left kidney in 99mTc DTPA renal scintigraphy. Rev
      Esp Med Nucl. 2010; 29: 140–141.
[15] Subramanyam P, Palaniswamy SS. Pictorial essay of developmental thyroid anomalies
      identified by Technetium thyroid scintigraphy. Indian J Nucl Med. 2015; 30: 323–327.
[16] Berrington de Gonzalez A, Kim KP, Smith-Bindman R, McAreavey D. Myocardial perfusion
      scans: Projected population cancer risks from current levels of use in the United States.
      Circulation. 2010; 122: 2403–2410.
[17] Navare SM, Mather JF, Shaw LJ. et al., Comparison of risk stratification with pharmacologic and
      exercise stress myocardial perfusion imaging: A meta-analysis. J Nucl Cardiol. 2004; 11: 551–561.
[18] Maier A, Steidel S, Christlein V, Hornegger J. Medical imaging system. An introductory guide.
      Springer Open.
[19] Hosny T, Khalil MM, Elfiky AA, Elshemey WM. Image quality characteristics of myocardial
      perfusion SPECT imaging using state-of-the-art commercial software algorithms: Evaluation
      of 10 reconstruction methods. Am J Nucl Med Mol Imaging. 2020; 10: 375–386.

Further reading

Mettler FA, Guiberteau MJ. Essentials of nuclear medicine imaging. 6th. Philadelphia, London, New
      York: Elsevier-Saunders; 2012.
Bushberg JT, Seibert JA, Leidholdt EM, Jr, Boone JM. The essential physics of medical imaging. 3rd.
      Philadelphia, Baltimore, New York, London: Lippincott Williams & Wilkins, Wolters Kluwer; 2012.
Smith NB, Webb A. Introduction to medical imaging: Physics, engineering and clinical applications.
      Cambridge, New York, Melbourne: Cambridge Texts in Biomedical Engineering; 2010.
Webb A. Introduction to Biomedical Imaging. New York, London, Sydney, Toronto: Wiley & Sons;
      IEEE Press Ser Biomed Eng. 2002.
Wernick MN, Aarsvold JN editors. Emission tomography: The fundamentals of PET and SPECT. Sann
      Diego, London: Imprint Academic Press, Elsevier; 2004.
Schlegel W, Karger CP, Jäkel O. Medizinische Physik. Berlin, Heidelberg, New York: Springer
      Spektrum; 2018.

10 Positron emission tomography

Physical properties of PET

Attenuation length for 511 keV photons in water	7 cm
Spatial resolution	3–8 mm
Straggling length ^{18}F – β^+-annihilation	1 mm
Detectors	NaI(Tl), SiPM
Coincidence time resolution	5–10 ns
Most important PET isotopes	^{18}Fe, ^{68}Ga
^{18}Fe half-life	110 min
^{18}Fe production	^{18}O(p,n)^{18}F
^{68}Ga half-life	68 min
^{68}Ga production	Ge-68 Gen
Positron tracers	^{18}F-FDG, ^{18}F-FET, ^{68}Ga-PSMA

10.1 Introduction

Positron emission tomography (PET) is a nuclear imaging modality that is used in clinics for cardiological, neurological, and oncological diagnostics. The PET process is based on the mutual annihilation of electron-positron pairs and the conversion of their rest mass into two γ-photons flying opposite to each other. The emission of photon pairs reveals the location of the decay, which can be used to map the β^+-emitting isotopes in the body. The imaging technique is similar to the one of SPECT, but uses an additional coincidence counter that records events only when they coincide in space and time at detectors 180° apart. Electron–positron annihilation is the reverse effect of electron-positron pairing that occurs when photons with energies greater than 1 MeV interact with nuclei. Pairing is an important process in treating cancer with very hard x-rays (see Chapter 2, Volume 3). Pair annihilation is used in medicine to locate cancerous tissues.

The pairwise existence of particles and antiparticles was predicted by Dirac's[1] relativistic formulation of quantum mechanics in the early 1930s. The existence of electron-positron pairs was discovered shortly afterward by Anderson[2] in cosmic rays. However, it took the combined effort of physicians, physicists, radiochemists, engineers, and mathematicians to put PET to work in clinics. The first scanners were designed and built in the mid-1970s, while commercial scanners did not come into the market until the turn of the century. A brief history of PET development up to the late 1990s is provided in [1], and an update on recent developments can be found in [2].

---

1 Paul Dirac (1902–1984), British theoretical physicist and Nobel laureate in Physics 1933.
2 Carl Anderson (1905–1991), American physicist and Nobel laureate in Physics 1936.

https://doi.org/10.1515/9783110757095-010

PET requires positron-emitting isotopes that are generated artificially and administered to patients. Following pair annihilation, the γ-radiation is recorded externally with an array of counters arranged in a circle and wired for coincidence
counting. Finally, tomographic images are computed via Radon transformations,
similar to the ones generated in x-ray CT and SPECT.

For PET, low atomic number β^+-emitters are preferred, ideally those which take
part in the metabolism of the body. β^+-emitters of C, N, O, and F are indeed available
but must be produced artificially by proton bombardment of light atoms. Upon proton
uptake, these elements have a deficiency of neutrons and therefore decay over time by
converting protons back into neutrons plus positrons. The general nuclear reaction
path is expressed as $_Z^A X + p \rightarrow {}_{Z-1}^A Y + \beta^+$; in short notation:[3] $^A X(p, n)^A Y$. In most cases,
the β^+-emitters have very short lifetimes ranging from 2 to 110 min (Tab. 10.2), too
short for time-consuming long-distance shipping. Hence, the β^+-emitting isotopes are
preferably produced on-site via a cyclotron facility. In addition, a radiochemistry laboratory has to prepare the radiopharmaceuticals for injection. Hence, the application of
PET requires an extensive and expensive infrastructure with well-coordinated facilities.
In highly developed countries, one can find about one to two PET scanners per million
inhabitants on average.

10.2 Basic principle of PET

10.2.1 Energy and momentum

After uptake, the radioactive tracers perfuse the circulatory system and diffuse to
the target organ, where they preferentially accumulate and decay. Once decayed,
the high-energy β^+-particles slow down by colliding with electrons and emitting
synchrotron and Cherenkov radiation. As soon as the β^+-particles have come to
rest, antiparticles β^+ and particles (β^-) annihilate together. For this purpose, a valence electron with opposite spin is captured from the surrounding tissue and together they form a singlet state with spin $S = 0$. The combined rest mass of the
electron–positron pair is then instantaneously converted into two γ photons. The
rest mass and thus the γ-radiation energy are $2 \times m_e c^2 = 2 \times 0.511$ MeV $= 1022$ MeV.

The particle momentum is also conserved. Since the momentum of both particles
is zero at rest before annihilation, the combined momentum must also cancel after
annihilation. Because of the momentum conservation, these two γ-photons fly in opposite directions (180°). Figure 10.1 is a sketch of the positron straggling and the
prompt emission of γ-radiation. The average straggling length between β^+– decay
and γ-emission in soft tissue depends on the maximum decay energy of the positron,

---

3 For more information on the nuclear part, please refer Section 5.6.4.

which ranges from 1 to 3 MeV. Accordingly, the straggling length stretches from about 1 to 5 mm.

The main goal of PET is to map the distribution of β^+-emitting isotopes in the body, which tag the targeted tissue. But actually, PET is sensitive to the annihilation site at some distance to the positron emitter. Thus, the shorter the straggling distance, the higher is the spatial resolution for localizing the β^+-emitting isotopes. This condition translates into the energy constraints: The lower the isotope's maximum decay energy, the shorter the straggling path and the higher the spatial resolution that can be achieved with PET. According to Tab. 10.2, the highest spatial resolution is achievable with ^{18}F isotopes.

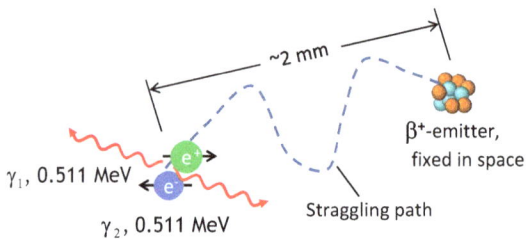

Fig. 10.1: After decay of a positron emitting isotope, the positron straggles for about 2 mm in soft matter before slowing down and annihilating together with an electron. Electron-positron annihilation takes place by the emission of two γ photons flying in opposite directions.

> **!** PET takes maps of positron-emitting radioisotopes in the body that tag via radiopharmaceuticals to malignant tissues.

10.2.2 Coincidence counting

Both γ-photons emitted simultaneously in opposite directions are used for imaging. They are detected by a coincidence detection scheme determining the angle of emission and the time of arrival. The schematics of PET coincidence electronics is shown in Fig. 10.2. Detectors are arranged in a circle around the patient in the center. If detector D1 registers a γ-photon and the outgoing detector pulse surpasses a threshold value V_{tres}, a digital (logic) pulse of height ΔV_{D1} is set off at time t_1, starting a timer for a predetermined duration τ. A coincidence event is registered if a second pulse from an opposite detector CD1 is registered at the time t_2 within the coincidence time window: $t_2 - t_1 < \tau$.

There are several scheme of how to perform the coincidence counting. Important is that photons detected are due to $\beta^+ - \beta^+$-annihilation and not due to any other unrelated γ-emission. Detection based on this principle is called *annihilation coincidence detection*. The line that connects the coinciding detector pair is called the *line*

(a)

(b)

Coincidence counting

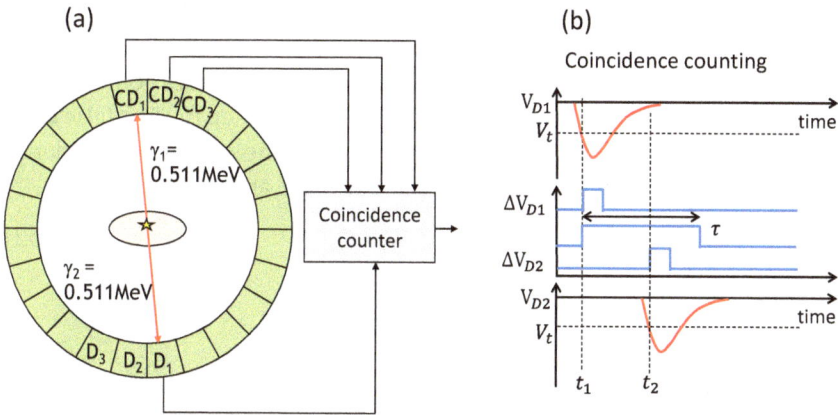

Fig. 10.2: (a) Schematics for PET coincidence counting. The yellow star marks the origin of two-photon γ-emission. Scintillation detectors are arranged in a circle around the source in the center. (b) A pulse in detector D1 sets off the coincidence counter for a time window of length τ. The signal is counted, if within the time τ a second pulse arrives in detector CD1 on the opposite side of the detector ring.

of response (LOR). The logic pulse length τ determines the temporal resolution of the PET scan and is usually set from 5 to 10 ns.

10.2.3 Artifacts

The double conditional counting with respect to angle and time provides very clean, essentially background-free spectra. Because of the coincidence counting, proximity effects between counters can be neglected. In contrast to SPECT, collimators in front of the detectors are not mandatory. Despite the "clean" detection scheme, some limitations are present nevertheless. The angular resolution in the ring-plane increases with the number of detectors mounted on the ring. However, the physical size of the (scintillation) detectors sets an upper limit, which will be discussed further. Furthermore, coincidence may not only occur by events, starting from the center of the detection ring, as illustrated in Fig. 10.3(a). In practice, the distance between the annihilation center and detector D_n will always differ from the distance to the detector CD_n, as indicated in panel (b). The question is how much difference is tolerable. In any case, the LOR should cross the center.

A PET ring usually has a diameter of 80–90 cm. Therefore, crossing the ring for γ-photons takes 2.5–3 ns in the extreme, which is well within the coincidence time limits. The different arrival times are useful for localizing the annihilation site in a high-resolution time-of-flight (TOF) mode (see below), but are not registered in normal mode. Another possibility is an event that originates from an off-center location and makes it into the counter by Compton scattering, as sketched in panel (c). This process has a longer optical path length and can be suppressed by tightening up the time discriminator. The fourth kind of coincidence shown in panel (d) is one by rare accidental events in which two unrelated annihilation processes occur simultaneously, and one photon of each pair arrives at detectors 180° apart. This false coincidence cannot be avoided but should be very rare. In the end, the coincidence counter collects counts from three events: true, scattering, and accidental [3]:

$$C(D_n; CD_n) = C_t + C_s + C_a. \tag{10.1}$$

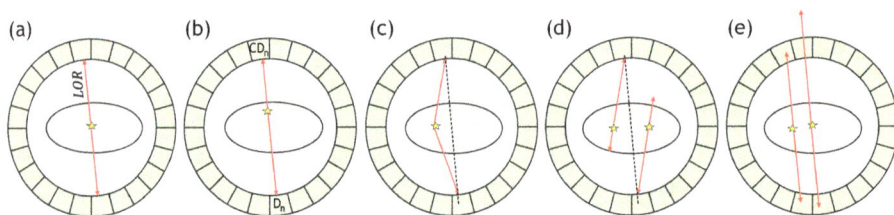

Fig. 10.3: Different coincidence events: (a) true coincidence; (b) off-center coincidence; (c) coincidence by scattering; (d) accidental coincidence; (e) missed coincidences either because the opposite detectors are not 180° apart or one of the γ-rays is not absorbed.

There are also true events that may be missed (panel (e)). For instance, when only one of the γ-rays is absorbed in the scintillation crystals and the other one escapes. Another unsuccessful event occurs when both γ-photons hit opposite detectors, which are not 180° apart, i.e., their LOR does not go through the center of the PET ring. It is a question of the tolerance-level set whether to accept those events or not and to what degree.

10.2.4 Spatial resolution

Four contributions determine the total spatial resolution of PET. Assuming that the measured spatial resolution is due to different and independent contributions ΔR_i, then the absolute error ΔR follows from the algebraic sum of the individual errors (see Chapter 8, Volume 3, eq. (8.11)):

$$\Delta R = k \sqrt{(\Delta R_{dw})^2 + (\Delta R_{ar})^2 + (\Delta R_{nc})^2 + (\Delta R_{sr})^2}. \tag{10.2}$$

Here the error ΔR_{dw} refers to the detector width, ΔR_{ar} is related to the positron acceptance range, ΔR_{nc} takes into account the non-collinearity effects of the γ–photons, and ΔR_{sr} is the uncertainty with respect to the straggling range. Finally, k is a proportionality factor related to the ring construction.

Tab. 10.1: Contributions to the spatial uncertainty of PET diagnostics. ΔR_{dw} = detectorwidth; ΔR_{ar} = acceptance range; ΔR_{nc} = noncollinearity effects; ΔR_{sr} = straggling range; k = proportionality factor.

ΔR_{dw}	2–4 mm
ΔR_{ar}	0.2–2.5 mm
ΔR_{nc}	1.8–2.0 mm
ΔR_{sr}	1–5 mm
k	1.2–1.5

The different factors entering eq. (10.2) can be estimated and are listed in Tab. 10.1. The total spatial error ranges accordingly from 3.1 to 7.3 mm. This is comparable to the SPE and SPECT spatial resolution but a factor of 10 less than the resolution achieved with x-ray radiography. If necessary, the contributions ΔR_{dw}, ΔR_{nc}, and ΔR_{nc} can be tightened up. But the main contribution to the spatial resolution is the straggling width ΔR_{sr}, which is intrinsic, depending only on the kinetic energy of the β^+-particle after emission and the local electron density. The highest spatial resolution is achievable with ^{18}F isotopes because of the lowest β^+-emission energy of 0.9 MeV (see Tab. 10.2).

Overall, PET detection systems feature a higher spatial resolution of about 3–6 mm compared to SPECT, which has a spatial resolution of about 2–4 cm. The resolution of PET is only limited by the number of detectors on the ring, by the straggling length of the positrons before they come to rest, and by Compton scattering of γ-photons in tissue, causing a deviation from 180° angular correlation. In practice, there are also limiting factors related to patient movement, mainly during respiration. All other factors set aside, the intrinsic straggling length is the most severe limitation with respect to spatial resolution.

10.2.5 TOF-PET

The development of fast detectors, as outlined further, allows using TOF technics for localizing the position of the positron emitter along the LOR [4, 5]. The opposing coincidence detectors not only register the arrival of two γ-quants within the time window τ but also the difference of the arrival times Δt. The time difference is related to the off-center location Δx of the emitter via:

$$\Delta x = \frac{1}{2} c\Delta t, \tag{10.3}$$

(a)   (b)

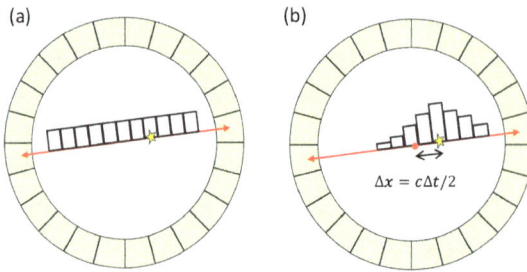

$\Delta x = c\Delta t/2$

Fig. 10.4: Probability distribution for the location of the β^+-emitter. (a) In conventional PET cameras, an equal distribution is assumed. (b) With TOF-PET, a first estimate of the β^+-emitting location can be made.

where c is the light speed. The time resolution is about 10% of the coincidence time window τ and about 20% of the standard arrival time. Therefore, TOF-PET provides not more than an estimate of the most likely positron location. In conventional PET, the positron location is assumed to have equal probability along the LOR. With TOF-PET, the probability distribution is uneven and provides a first hint toward the β^+-distribution in an organ for further tomographic analysis. The differences in the probability distributions for conventional PET and TOF-PET are illustrated in Fig. 10.4.

! Standard PET scanners determine the angular correlation of two γ-quants within a coincidence time slot τ. Advanced PET scanners determine, in addition, the arrival time of the photon along the LOR.

10.2.6 Ring designs

The pairwise γ-emission is an incoherent process, meaning that one γ-pair is uncorrelated to the next and to all other pairs, which radiate into angular space of 4π. The detector ring, however, covers only a 2π slice of all events. Other photons remain undetected. The detection efficiency can be improved by stacking up more detector rings as shown in Fig. 10.5(a). If septa separate the PET rings to suppress cross-detection, we speak of 2D PET cameras. The septa can also be retracted yielding a 3D detection capability with higher sensitivity (b). Then coincidence detectors have to correlate events crosswise on different rings to avoid losing the spatial resolution, i.e., detector D_{i+1} correlates with detector CD_{i-1}, etc. Often PET cameras feature retractable septa for selecting either higher sensitivity or higher spatial resolution. It has also been discussed to increase the number of rings to enable full-body scans (c). This would reduce the counting time during which the patient should not move. However, full-body scanners have not been realized because of the increasing investment cost for

(a)   (b)   (c)   (d)

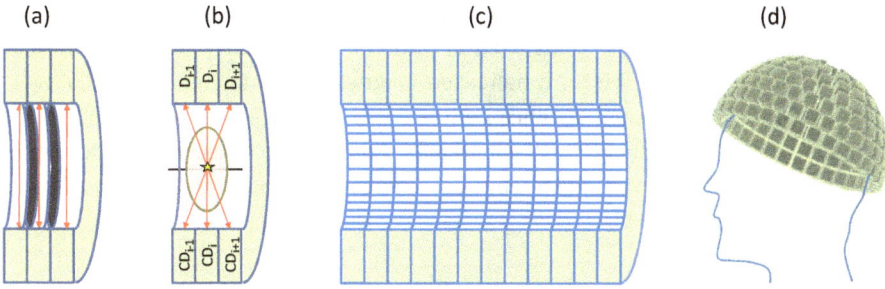

Fig. 10.5: (a) Side view of a 2D PET scanner containing three rings of detectors separated by septa; (b) 3D scanner with septa retracted, allowing cross-communication; (c) many rings for full-body scanning; and (d) sketch of a helmet PET scanner.

hardware and computing power. The current compromise is a scanner with an axial FOV of about 15–20 cm, which allows scanning whole organs at once, such as the heart, lung, liver, or the brain. A novel innovative design is a brain PET scanner shown in panel (d). First, prototypes have been built and depending on their performance, brain scanners will not only be very useful for early diagnosis of gliomas, dementia, and the like but also for a broad range of neuroscience studies [6, 7].

> PET scanners detect coincident γ-photons flying in opposite direction after electron-positron annihilation. PET scanners consist of an array of detectors arranged in one or several rings for organ or full-body screening. Helmet-shaped brain scanners are of interest to neurology.

10.2.7 PET scanner and combinations

Figure 10.7(a) shows a complete PET scanner, including a detector ring and stretcher for the patient. In (b) the cover of the detector bank is removed. During a scan of a particular organ, either the detector ring or the patient is moved across the FOV. Note that the detector ring is static in contrast to the spinning detector ring in x-ray CT.

In recent years, imaging systems have been combined which complement each other [8]. For instance, if a cancerous region is identified with PET or SPECT, then this region can be scanned with x-ray CT, providing much higher spatial resolution. System combinations of SPECT/x-ray CT, PET/CT, and PET/MRI have been discussed and constructed to improve diagnostic accuracy. In all cases, multimodal imaging aims to combine high-resolution anatomic structures with functional imaging and possibly also metabolic properties. The combination of PET and MRI has become a powerful technic, providing metabolic information via PET, high-resolution anatomic data, and functional imaging properties via fMRI. However, this combination is non-trivial as the high magnetic fields used in MRI can disturb the photon counting of photomultiplier tubes (PMTs). An increased magnetic screening can partially overcome this problem, a spatial distance between MRI and CT also helps, or the use of solid-state detectors in the PET replacing PMTs. An overview of recent advancements of scanners with improved image contrast, image resolution, and reduced image noise can be found in [8, 9].

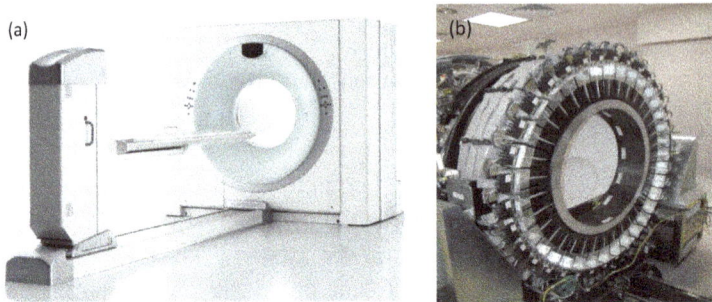

Fig. 10.6: PET setup: (a) full setup from Siemens (TruePoint PET·CT eco) and (b) cover removed from the detector ring (reproduced from www.siemens.com/press).

10.3 Data acquisition and image reconstruction

10.3.1 Detectors

An important consideration is the quantum efficiency of the detector system. Let us assume an efficiency of 0.7, i.e., 70% of all γ-photons arriving in the detector are converted into photons of the visible spectrum. Then the efficiency of detecting two photons is only 49%. It is evident that the quantum efficiency of the detectors should be as high as possible. In the past, NaI(Tl) scintillation crystals were used in combination with PMT. The higher energy γ-photons (0.511 MeV) in PET compared to the 140 keV γ-photons used for SPECT is taken into account by increasing the thickness of the scintillation crystal. Alternatively, NaI(Tl) scintillators are replaced by bismuth-germinate (BGO $=Bi_4Ge_3O_{12}$) crystals [10]. Bismuth has a slightly higher atomic number (83) than thallium (81) and higher quantum efficiency for hard γ-photons. However, the drawback is a significantly lower light output and longer light decay time than for NaI(Tl), increasing the dead time of the detector such that time coincidence detection below 30 ns becomes unrealistic. Because of these shortcomings, further detectors have been tested and are still under development. Promising materials are lutetium oxyorthosilicate, lutetium–yttrium oxyorthosilicate, and gadolinium oxyorthosilicate [11]. Detectors with these materials combine high-density, high Z values, and better timing resolution reduced to about 5–6 ns. The detectors used are so-called block detectors based on pixelated BGOs with light channeled to large PMTs shared with neighboring blocks [12]. In the newest genera-tion of PET units, Si-based complete solid-state semiconductor detectors are used, so-called Si photomultiplier (SiPM) with a time resolution of about 400 ps [13].

10.3.2 Counting statistics

Both PET and SPECT are imaging modalities that deal with low count rates, much lower than those common in x-ray CT. The low count rate results in a counting statis-tics that is always Poisson distributed. Any analysis that assumes a Gaussian distribu-tion of events is problematic.

Because of the coincidence counting, the counting statistics have a very low background, lower than SPECT. The low noise ensures high image quality even at low count rates, as can be seen, for example, in Fig. 10.12, which compares x-ray CT, SPECT, and PET images. Because of the low noise, SNR and CNR, defined in eqs. (9.4) and (9.5), are essentially the same.

The γ-photon energy of 511 keV used for PET imaging is much higher than the 140 keV γ-radiation used for SPECT or the 140 kVp bremsstrahlung radiation em-ployed for x-ray CT. The higher γ-photon energy entails a lower photon attenuation. The attenuation that takes place is mainly due to Compton scattering, which, in turn, has an effect on the artifacts, as already discussed. For a 511 keV photon beam, the calculated half-value attenuation length is about 7 cm in soft tissue, in contrast to 1 cm for 140 keV x-rays.

10.3.3 Image reconstruction

The computation of PET images follows the same principles as already sketched in Chapter 9 for SPECT. In case of PET, however, the line integral is not taken over the spatial distribution $f_z(r, \theta)$ of radioisotopes in a slice of the body. Instead, the line integral goes over all pairs disintegrating along a LOR [14]:

$$p_z^{\text{PET}}(r, \theta) = \alpha \int_0^{\text{LOR}} \lambda_z(r, \theta) dl. \tag{10.4}$$

α is an appropriate proportionality constant that takes into account detector effi-ciency, resolution, dead time, etc., and $\lambda_z(r, \theta)$ is the corresponding positron activity. The polar coordinates r, θ are defined in Fig. 10.7(a). Two events are shown, distin-guished by primed (double primed) letters and solid (dashed) lines. θ' is the detector angle and $\Delta r'$ is the orthogonal distance of the first LOR. In PET, coincidence detection events that are not separated by 180° are rejected. Therefore, the coincidence indicated in Fig. 10.7(a) would not be registered. However, to construct a sinogram, it is shown here. The registered coincidence event is projected in a 2D (r', θ') matrix, indicated by the striped bar. The black pixel records the distance $\Delta r'$, and the angular orientation of the bar encodes the θ' value. θ' versus r' is plotted in the panel (b). From the same location, another coincidence event is detected, however, with a different angle θ'' and a different distance $\Delta r''$ of LOR'' from the center. The respective $(\Delta r'', \theta'')$ pair is also plotted in panel (b). Like in x-ray CT and SPECT, all off-center events describe a sinusoidal curve in the sinogram. Finally, filtered back projections generate maps, showing where the body's positron annihilation processes have occurred. Examples are shown in Figs. 10.12 and 10.13. With increased computer power, an iterative reconstruction instead of a Radon back-transformation is the preferred data analysis approach in recent years. As this approach goes beyond the scope of this chapter, the interested reader is referred to the expert literature [15, 16].

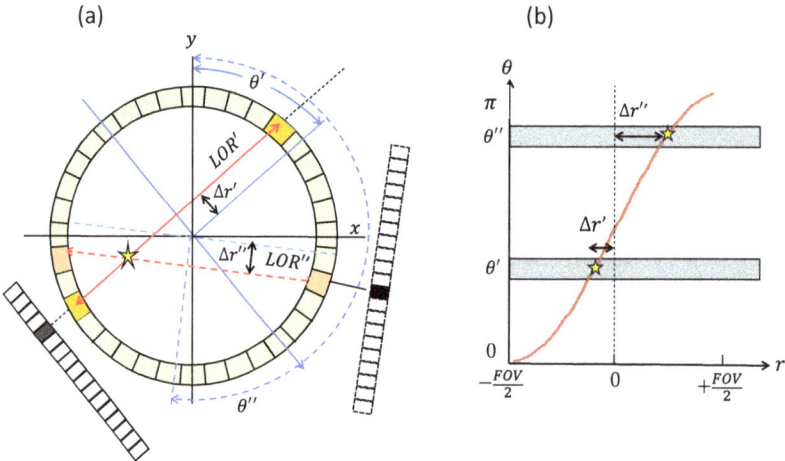

Fig. 10.7: (a) Projection of the lines of response into 1D slices as a function of polar angle and radial distance from the center. Two coincidence events are shown. Their respective coordinates are indicated by bars and double bars. The geometric projections are shown by solid and dashed lines, respectively. (b) Sinogram of events assembled from the 1D slices. The two events from panel (a) are emphasized.

10.3.4 Standard uptake value

After PET images have been generated, medical professionals usually analyze and evaluate the images with the naked eye. Color-coded areas of higher activity indicate diseased tissue and lesions. Low activity signals healthy tissue. But, whether the lesions are benign or malignant cannot be judged by eye. Only a quantitative evaluation of the images helps. Various evaluation methods for quantitative image processing are described in the literature. However, these sometimes require complex calibrations for which there is usually no time in clinical practice. Therefore, an easy-to-use semiquantitative method has become established, the so-called *standard uptake value* method (abbreviation SUV) [17, 18].

SUV is expressed as a ratio of measured activity r within an ROI to the administered activity a normalized by the patient's body weight w:

$$\text{SUV} = \frac{r}{a/w} \tag{10.5}$$

Here r is an activity density measured in Bq/ml (in the literature often falsely called activity concentration). With PET only activities can be measured. In order to obtain an activity density, the volume of higher activity has to be estimated. The surface area is determined from the image taken and the thickness is given by the PET slice thickness Δz.

The activity a is the total activity administered to the patient. As the lifetime of PET isotopes is short and the scans are usually taken 1 h after administering, the SUV should be evaluated by considering the actual activity at the time t of the scan:

$$\text{SUV} = \frac{r(t)}{a(t)/w} \tag{10.6}$$

w is simply the body weight in units of kg. Thus, the unit of SUV is $[\text{SUV}] = \text{kg/ml}$. To further simplify the SUV, it is often assumed that the body density is roughly 1 kg/l. Then the SUV becomes a dimensionless number.

Equal distribution of the administered activity in the body yields an $\text{SUV} = 1$. Experience shows that an $\text{SUV} \geq 2$ is considered to be suggestive of malignancy, whereas lesions with SUVs less than 2 are considered to be benign.

The SUV evaluation outlined above is primarily used for ^{18}F-FDG-PET. Here, it quantitatively describes the glucose metabolism of a tumor when using the tracer FDG. However, it can be used for another PET isotope as well, and actually is also used for a quick evaluation of SPECT images.

10.4 PET isotopes

10.4.1 General aspects

In PET, mainly positron emitters of light and biocompatible elements are used such as C, N, O, and F. Fortunately, these light elements can be transmuted to positron-emitting radioisotopes by proton bombardment in an accelerator. This contrasts SPECT, which employs radioisotopes, whose biocompatibility is questionable. The PET radiopharmaceuticals are authentic pharmaceuticals with identical biochemical properties to those without radioisotopes. In contrast, SPECT works with analogous but not identical radiopharmaceuticals. The availability of authentic radiopharmaceuticals is a considerable advantage of PET over SPECT. After producing β^+-emitting radioisotopes, radiochemistry is required to synthesize PET tracers bound in molecules similar to glucose. The synthesized tracer is then administered either by intravenous injection or by inhalation. Subsequently, the PET scanner detects γ-rays and registers coincidence events. Figure 10.8 illustrates the processing steps from isotope production to image recording. Table 10.2 lists the biocompatible radioisotopes for PET scans, including nuclear reaction, maximum β^+-emission energy, and β^+-straggling range in water. Two heavy β^+-emitters are also listed, which require generators like the one for ^{99m}Tc.

Tab. 10.2: Positron-emitting radioisotopes used for PET CT.

Isotope	$T_{1/2+}$ (min)	Production	E_{max} (MeV)	Range (mm)
C-11	20.5	^{14}N(p,α)^{11}C	0.96	1.1
N-13	10.0	^{16}O(p,α)^{13}N	1.19	1.4
O-15	2.1	^{14}N(d,n)^{15}O	1.72	1.5
F-18	110	^{18}O(p,n)^{18}F	0.9	1.0
Ga-68	68	Ge-68 Generator	1.9	8.1
Rb-82	1.27	Sr-82 Generator	3.15	15.0

Replacing a stable isotope in a biomolecule with a radioisotope such as ^{11}C, ^{13}N, or ^{15}O does not change the biochemistry of the tracer. Therefore, radioactive tracers permit undisturbed imaging of the metabolic process. The exemption is ^{18}F as fluorine is not part of the body biochemistry. Nevertheless, PET-CT with F-18 is particularly revealing; therefore, this procedure is presented in more detail in the next section. Other tracers and their clinical indicators are listed in Tab. 10.3. A detailed discussion of clinical PET procedures can be found in [19].

10.4.2 ^{18}F-decay

^{18}F-PET is of particular importance for PET diagnostics. Before going into further details, the most important properties of this radioisotope will be briefly outlined. Fluorine has only one stable isotope: $^{19}_{9}$F. When oxygen $^{18}_{8}$O is bombarded with protons in a cyclotron, the reaction $^{18}_{8}$O(p, n)$^{18}_{9}$F takes place, producing the unstable isotope $^{18}_{9}$F, which decays by β^+-emission back into $^{18}_{8}$O. The decay scheme is shown in Fig. 10.9. As target, usually pure or enriched ^{18}O-water is used, and the proton beam energy is typically 10 MeV, although the cross section has a maximum at 5 MeV

(see Section 5.6.4 for further information). The significance of the ^{18}F β^+ -emitting radioisotope in nuclear medicine is due to three favorable facts:
1. ^{18}F has a short half-life of only 110 min
2. ^{18}F has a short straggling length of only 1 mm
3. ^{18}F offers the possibility of incorporation into authentic radiopharmaceuticals

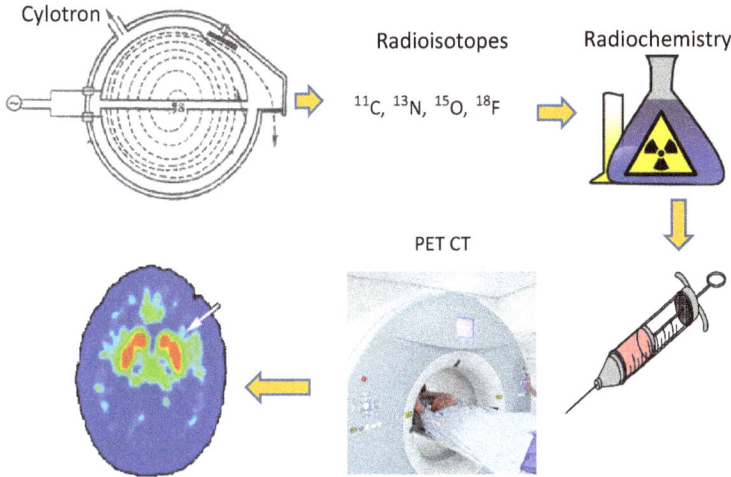

Fig. 10.8: Procedures required for PET scans with positron-emitting radioisotopes. First, suitable radioisotopes are produced with a cyclotron by proton bombardment of light elements. Usually, the cyclotron is located in the radiation clinic because of the short lifetime of the β^+-emitting radioisotopes. Then the isotopes are bound in authentic radiopharmaceuticals for administration to the patient. PET scans, including data acquisition and fast data processing, provide cross-sectional slices that map areas with enhanced activity indicative of metabolic changes symptomatic of diseases.

Tab. 10.3: Positron tracers and their applications in nuclear medicine.

Tracer	Indicator	Application
^{18}F-FDG	Glucose metabolism	Oncology, cardiology, neurology
^{18}F-FET	Glucose metabolism in the brain	Glioma, neurology, psychiatry
^{68}Ga-Dotatoc	Somatostatin receptors	Oncology
^{68}Ga-PSMA HBED-CC	Prostate-specific membrane antigen	Oncology
Na^{18}F	Bone formation	Bone metastasis
^{11}C-Acetate	Cholesterol metabolism	Cardiology
^{11}C-Methionin	Amino acid metabolism	Oncology
^{13}N-NH$_3$	Perfusion	Cardiology
H$_2$^{15}O	Perfusion	Cardiology, neurology

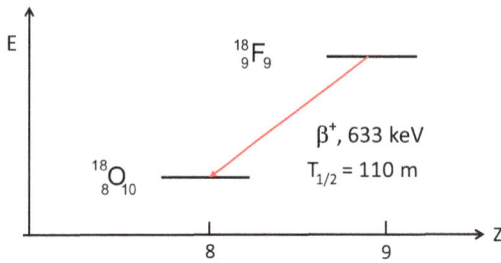

Fig. 10.9: Decay scheme of F-18 via β^+-decay.

> ! ^{18}F is the most frequently applied radioisotope in PET diagnosis of tumors.

10.5 Clinical applications of PET

10.5.1 FDG-PET

In radiochemistry laboratories, F-18 is primarily synthesized into fluorodeoxyglucose (^{18}F-FDG) for use in PET scans. ^{18}F-FDG is a glucose analog in which the positron-emitting ^{18}F isotope substitutes for a hydroxyl (-OH) group (see Fig. 10.10). ^{18}F-FDG is then used as a tracer in PET scans. ^{18}F-FDG follows a metabolic pathway similar to glucose in the body, except that it is not metabolized to CO_2 and water but remains trapped within the tissue (see Fig. 10.11). It is well known that proliferating cancer cells have a higher than normal rate of glucose metabolism, which in the oncology literature is referred to as *upregulation of glucose metabolism* in cancer cells. Because of this upregulation, FDG is well suited as a marker for cancer cells.

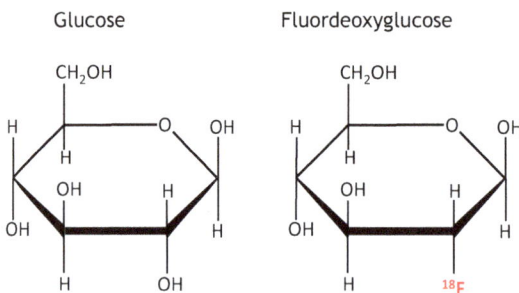

Fig. 10.10: Chemical structure of glucose and fluorodeoxyglucose ^{18}F-FDG. The fluor isotope ^{18}F is substituted for a hydroxyl group.

In standard metabolism, glucose reacts with ATP to form glucose-6-phosphate and finally fructose-6-phospate by isomerization. The first part of the FDG metabolism is identical to the one of glucose. However, in case of FDG, the metabolism stops after phosphorylation to FDG-6-phosphate and FDG becomes enriched by metabolic trapping [20] (Fig. 10.11(b)). This is of advantage for an early tumor recognition. Thus, enhanced γ -radiation indicates enrichment of radioactive ^{18}F-FDG revealing tumor cells in PET images [20, 21]. An example of a bronchial carcinoma is shown in Fig. 10.12 in comparison with different imaging modalities. On the CT image in the left panel, the carcinoma is barely visible but stands out in the PET image of the middle panel. The right panel shows a combination of CT and PET with convincing quality and a clear message of how valuable PET is for cancer detection.

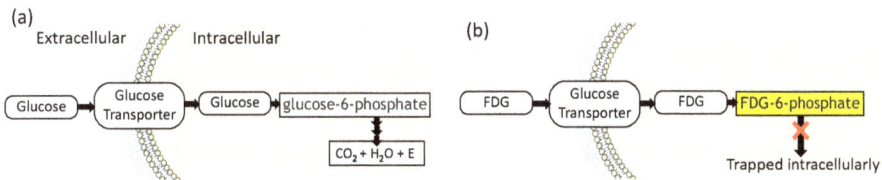

Fig. 10.11: Simplified schematics of the metabolism of glucose and fluorodeoxyglucose (FDG). Glucose is a polar compound that can only diffuse through the nonpolar lipid bilayer of the cell membrane with the help of a glucose transporter (adapted from [16]).

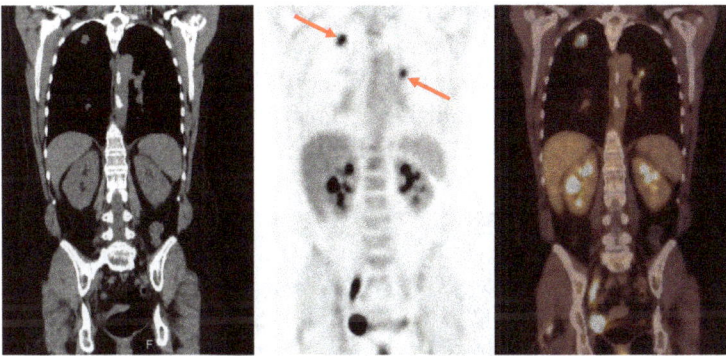

Fig. 10.12: Images of a bronchial carcinoma recorded with different techniques. Left panel: CT; middle panel: ^{18}F-FDG PET scan demonstrating intense uptake in lung cancer of right upper lobe (arrow) and in lymph nodes on the lower left lobe (arrows); right panel shows a combined CT/PET image (reproduced with permission of Dr. Margot Jonas, Clinic Knappschaft, Bochum, Germany).

10.5.2 FET-PET

FDG is a successful marker of carcinoma in the body. However, for detecting carcinoma in the brain, called glioma, FDG turns out to lack specificity. This is because the normal functioning brain already features a high glucose metabolism. Low- or high-grade gliomas do not generate sufficient contrast by FDG metabolism to the background. Furthermore, it was noticed that FDG is also taken up in inflammatory benign lesions, which could lead to false diagnosis. Therefore, it was necessary to synthesize a more specific marker for gliomas in the brain and more generally in the CNS. As the diffusivity of amino acids is upregulated in gliomas, radiolabeled amino acids are promising candidates for glioma scanning. Indeed, an artificial and radioactive amino acid ^{18}F-fluoro-ethyl-tyrosine (FET), synthesized and optimized by a research group at the national research center FZ Jülich (Germany) in the mid-1990s, turned out to provide the required glioma specificity [21]. Simultaneously, the effective dose for a PET scan could be drastically lowered compared to FDG-PET. Alternative radiolabeled amino acid tracers were developed for the same purpose such as ^{11}C-methionine (MET). However, the much longer half-life of ^{18}F (110 min) compared to ^{11}C (20 min) has a big advantage as it leaves more time for handling the marker, including radiochemistry, shipment, administering to the patient, and finally scanning. Thus, ^{18}F-FET-PET is presently the most valuable imaging modality because of a better tumor volume definition for glioma diagnosis, for guiding biopsy, and for image-guided radiotherapy. In a recent comparative meta-study of published patient data, the authors reported a strong advantage of FET-PET over FDG-PET for the diagnosis of brain tumors [22]. According to their analysis, FDG-PET carries little informational value for the differentiation of brain tumors versus nontumoral lesions, whereas FET-PET helps excluding/confirming the diagnosis of brain tumors with much higher accuracy [22]. Figure 10.13 shows a comparison of MRI, FDG-PET, and FET-PET scans of a high-grade cerebral glioma. MRI

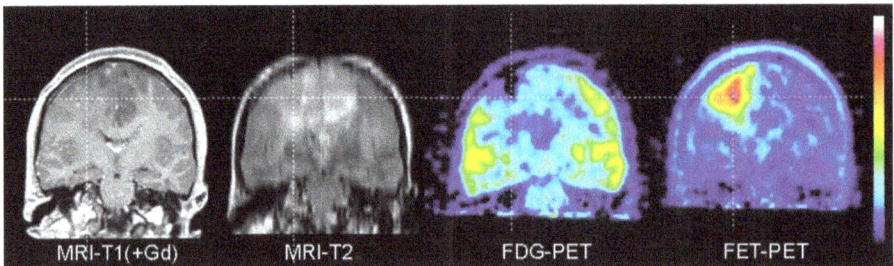

MRI-T1(+Gd)   MRI-T2   FDG-PET   FET-PET

Fig. 10.13: Brain glioma of grade III scanned with different imaging modalities: T1-weighted MRI after application of Gd-DTPA contrast enhancement (left) and T2-weighted MRI shows widespread abnormalities but does not allow a delineation of the tumor. FET-PET (right) clearly depicts the tumor while glucose metabolism (FDG-PET) is decreased in the tumor area (reproduced with permission of Institute of Neuroscience and Medicine Medical Imaging Physics (INM-4), FZ Jülich, Germany).

yielded a sensitivity[4] of 96% for the detection of tumor but a specificity[5] of only 53%. MRI has a notoriously low specificity for glioma, even if using multiple parameter MRI (see Section 3.6.6). In contrast, the combined use of MRI and FET-PET increased specificity to 94% with a sensitivity of 93%. This example demonstrates a significant improvement of diagnostic accuracy with the use of ^{18}F-FET-PET [22].

10.5.3 Prostate-specific membrane antigen PET

The prostate-specific membrane antigen (PSMA)-PET is a novel imaging modality for detecting prostate cancer (PCa). Because of the intriguing targeting method, the main idea of the PSMA–PET method is briefly presented here.

An enhanced concentration of prostate-specific antigens (PSA) measured in blood tests is an early warning sign for PCa. Normal PSA concentration increases with age but should be <1 ng/ml. Concentrations between 4 and 10 ng/ml indicate a probability for PCa of 25–35%; for values above 10 ng/ml, the probability is 50–80%. PSA is an enzyme that is naturally present in the ejaculate of males for increasing the fluidity of semen. Only through leaky vasculature in the prostate glands, some PSA can escape into the blood circulation. Here it may be free-floating in the hematocrit or bound to macroglobulin. Free PSA is excreted by the kidneys within 2–3 h, whereas bound PSA is metabolized in the liver within 2–3 days. Therefore, the PSA concentration measured at any time is in dynamic equilibrium between production rate in the prostate and excretion rate in the liver and kidneys.

Another test for early signs of PCa is a digital rectal exam that can tell an enhanced prostate volume but is unspecific to the exact tumor location. Confirmation via transrectal or perineal biopsies not only causes discomfort but also lacks diagnostic accuracy. Multiparameter MRI (mpMRI) combining $T2$-weighted and diffusion-weighted imaging shows promising results for localizing carcinoma in the prostate gland that should then be confirmed via biopsy. Nevertheless, screening PCa with mpMRI remains a problematic procedure because of contrast problems.

PSMA is a transmembrane protein that is present in all parts of the prostate gland but is significantly overexpressed in PCa cells [23]. A schematics of the protein is shown in Fig. 10.14(a). The main part of the protein is in the extracellular space, and a much smaller part reaches into the intracellular space. The function of PSMA is not entirely clear, but it is frequently found in PCa cells and other types of tumors. PMSA has gained much interest recently as a target protein for imaging via molecular contrast enhancement.

---

4 Sensitivity is the probability of finding cancers when present (see also Section 8.15).
5 Specificity is the probability of a negative result when cancer is not present (see also Section 8.15).

PSMA

(a)

^{68}Ga inhibitor

PSMA

(b)

^{177}Lu-
antibody

Extracellular

Extracellular

Intracellular

Intracellular

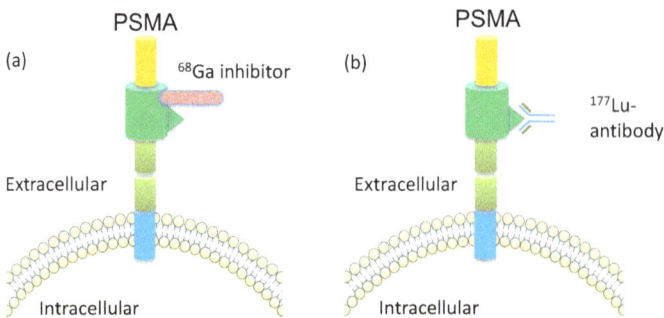

Fig. 10.14: Prostate-specific membrane antigen (PSMA) is a transmembrane protein with the head sticking in the extracellular space and the tail reaching into the intracellular region: (a) docking site for the PSMA inhibitor and (b) antibody docking to the antigen of PSMA (adapted from [20]).

PSMA-HBED-CC[6] is an extracellular PSMA inhibitor docking to the inhibitor center of PSMA. Using this molecule and attaching the positron emitter ^{68}Ga to it, (^{68}Ga)-PSMA HBED-CC becomes a labeled inhibitor enabling PET imaging. The isotope ^{68}Ga has a half-life of 67.63 min and emits positrons with a maximum energy of 1.9 MeV. It can be extracted from a generator similar to the procedures that are used for the generation of ^{99m}Tc discussed in Section 9.3. It has been demonstrated that (^{68}Ga)-PSMA HBED-CC shows a high specificity for PSMA-expressing PCa cells [23]. A comparative example of MRI and ^{68}Ga-PET is shown in Fig. 10.15, reproduced from [24]. The PET map in panel (C) appears much clearer than the MRI maps in panels (A) and (B) for identifying the location of tumor cells. The fused MRI/PET map in panel (D) leaves no doubt about the exact location and size of the tumor. Hence, PSMA-based imaging is a promising new modality for improved PCa screening [25].

Agents that target PSMA for diagnostic imaging can also be used for radiotherapy by delivering radioactive emitters to tumor sites. For PCa treatment, often ^{177}Lu-labeled agents are used. Utilization of the same agent for both diagnostic imaging and therapy is referred to as theranostics. The only distinction between agents for PET diagnostics (β^+-emitter) and those for therapeutics (γ or β^--emitter) is the radioisotope attached to the ligand, schematically compared in panels (a) and (b) of Fig. 10.14. Theranostic concepts are also implemented in nanomedical procedures using nanoparticles for diagnostics and therapeutics as discussed in Section 7.5 in Volume 3. ^{177}Lu is a β^--emitter with maximum kinetic energy of 497 keV, decaying to the ground state of ^{177}Hf with a half-life of 6.65 days. After binding to a PSMA inhibitor (antibody), researchers

---

6 The full name of this acronym is: prostate-specific membrane antigen (PSMA) Glu-NH-CO-NH-Lys-(Ahx).

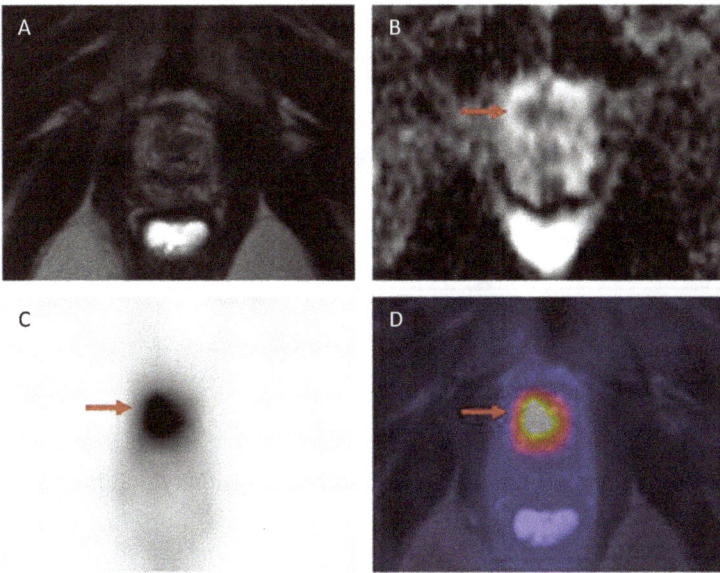

Fig. 10.15: Biopsy-proven prostate cancer of a patient with PSA of 5.4 ng/ml. (A) *T*2-weighted MRI map; (B) diffusion-weighted *T*2-MRI map; (C) ^{68}Ga-PET map; (D) fused *T*2-weighted/PET map shows intense local ^{68}Ga uptake in the direction ventrally to the urethra (white area in the lower part of the map). The red arrows in the panels (B), (C), and (D) indicate the location of the tumor volume in the prostate (reproduced with permission from [24]).

have found that the radioactive labeled agent shows high specific and rapid uptake in PCa metastases [26]. Therefore, this PCa treatment holds much promise for patients who have already developed widespread PCa metastasis.

10.6 Conclusion

Concluding, we want to compare both methods PET and SPECT that use radioisotopes for detecting various diseases. A noncomprehensive comparison is given in Tab. 10.4. PET has the decisive advantage of using light isotopes in connection with authentic radiopharmaceuticals. However, this advantage has a high price, as the production of the isotopes is rather costly. And because of their very short half-life, these isotopes have to be produced by a proton accelerator (cyclotron) in the clinic or in a nearby facility. Nevertheless, with the advent of smaller and more lightweight scanners, PET imaging finds increasing interest in neurological studies for the early recognition of brain diseases. On the other hand, clinics tend to go back to SPECT despite lower resolution and higher SNR because of easier isotope handling and overall lower treatment cost.

Tab. 10.4: Comparison of PET and SPECT.

PET	SPECT
Isotopes emit positrons	Isotopes emit gamma radiation
Higher spatial resolution	Lower spatial resolution
Short half-life radioisotopes	Longer lived radioisotopes
Requires on-site cyclotron	Radioisotopes from external laboratory
Use of authentic radiopharmaceuticals	Use of analog radiopharmaceuticals
Costlier scanner	Less capital intensive scanner

10.7 Summary

S10.1   PET is based on coincidence detection with high sensitivity and high spatial resolution of 3–6 mm.

S10.2   The spatial resolution of PET is limited by the number and size of the detectors, non-collinearities of the 2γ-emission process, and by the straggling length of the positrons.

S10.3   Coincidence measurements do not require collimator blades in front of the detectors.

S10.4   Fast scintillation counters with high light output and low dead time are essential for high sensitivity.

S10.5   PET uses positrons emitted from light isotopes with short half-life.

S10.6   Application of PET requires an isotope laboratory in close reach.

S10.7   The radioisotopes are integrated in biologically active and authentic radiopharmaceuticals.

S10.8 ^{18}F-FDG is the most frequently used tracer for PET.

S10.9 ^{18}F-FDG takes part in the metabolism of glucose, but stops after phosphorylation. The trapped ^{18}F-FDG reveals areas with enhanced metabolism.

S10.10   Among the imaging methods available, PET is the most expensive one because of the combination of accelerator, radiochemistry, and scanning unit.

Questions

Q10.1 PET requires specific radioisotopes. What are the requirements?
Q10.2 How are PET radioisotopes produced?
Q10.3 Positron annihilation is the reverse process of which process?
Q10.4 Why do the emitted gamma photons fly apart at 180°?
Q10.5 What is the typical straggling distance from positron decay to positron annihilation?
Q10.6 How is the gamma radiation from positron annihilation detected?
Q10.7 How does coincidence detection work?
Q10.8 What are possible artifacts of coincidence counting?
Q10.9 What is PET mainly used for?
Q10.10 What is the difference between perfusion and diffusion?
Q10.11 How does the F-18 metabolism work?
Q10.12 What is the significance of using ^{18}F-isotopes in PET imaging?
Q10.13 Which method has higher spatial resolution, SPECT or PET?
Q10.14 What is the main disadvantage of PET?
Q10.15 What do you understand by TOF-PET?

Attained competence checker   +   0   –

I know how positron emitting radioisotopes are generated.

I know what the term "straggling length" means and how long it is in tissue.

I know how long it takes for a γ-quant to cross the PET ring.

I am aware of artifacts in PET imaging.

I appreciate the main idea of coincidence counting.

I can distinguish between PET and SPECT imaging.

I can distinguish between LOR and FOV.

I can identify at least three clinical applications of PET.

I recognize the importance of TOF methods for higher resolution PET.

I appreciate the differences in FDG and FET and their applications.

Exercises

E10.1 **Compton scattering of 511 keV photons:** The 511 keV γ-photons produced during positron annihilation have a certain probability for Compton scattering in the tissue. Assume that the γ-photons are scattered at an angle of 30°.

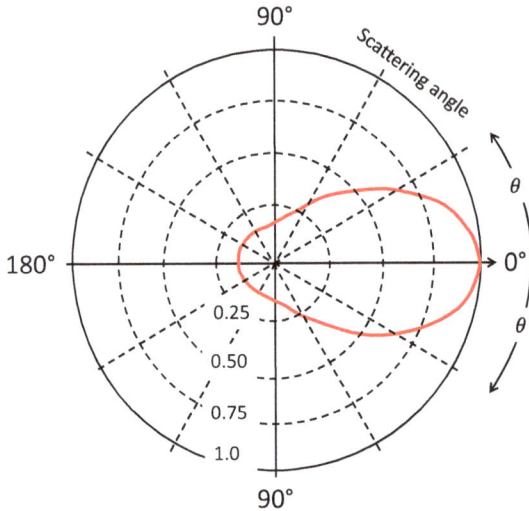

a. What is the wavelength shift of the photons?
b. What is the energy transfer to the electrons in the tissue?
c. What is the probability for Compton scattering of 511 keV photons at 30°? Give a rough estimate. The normalized angular dependence of the Compton scattering cross section is shown by the red line in the adjoining polar graph.

E10.2 **Production time for a 100 MBq ^{18}F source:** ^{18}F-isotopes are generated by proton bombardment in a cyclotron with 5 MeV proton energy via the reaction: $^{18}O(p, n)^{18}F$. This reaction has a cross section of 500 mbarn. The target is a 1 g of isotopically clean $H_2^{18}O$ water sample a cross section of 1 cm^2; the beam has a matching cross section of 1 cm^2 and a flux of 10^{16} protons/ cm^2 s. The aimed activity is 100 MBq. For how long is proton irradiation required to produce the ^{18}F source?

E10.3 **Improved precision for isotope production:** In E10.2, we have determined the production time for a 100 MBq ^{18}F source. The calculated production time is, however, rather insensitive to the actual activity. The activity could be 1 MBq or 1000 MBq without difference in the production time. How can you improve the precision of the aimed activity?

E10.4 **Angular correlation:** The positron annihilation process is described by the emission of two γ-photons flying in opposite directions. Assume that we have constructed an instrument that can measure with high precision the angular correlation between the two γ-quants, much better than 1°. In such an experiment, it is observed that there is a small angle θ between the trajectories of both γ-quants. What could be the possible reasons for the deviation from 180°?

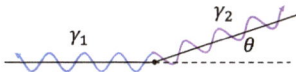

The measurement of the small angle θ is a method known as perturbed $\gamma - \gamma$ angular correlation spectroscopy. It plays an important role in condensed matter physics, but less so in the field of soft matter.

References

[1]   Rich DA. Brief history of positron emission tomography. J Nucl Med Techno. 1997; 25: 4–11.
[2]   Vandenberghe S, Moskal P, Karp JS. State of the art in total body PET. EJNMMI Phys. 2020; 25: 7–35.
[3]   Andrew W. Introduction to Biomedical Imaging. Wiley-Interscience: John Wiley & Sons, Inc; 2003.
[4]   Lecoq P, Gundacker S. SiPM applications in positron emission tomography: Toward ultimate PET time-of-flight resolution. Eur Phys J Plus. 2021; 136: 292.
[5]   Vandenberghe S, Mikhaylova E, D'Hoe E, Mollet P, Karp JS. Recent developments in time-of-flight PET. EJNMMI Phys. 2016; 3: 3.
[6]   Ahmed AM, Tashima H, Yoshida E, Yamaya T. Investigation of the optimal detector arrangement for the helmet-chin PET – a simulation study. Nucl Instr Meth. 2017; 858: 96–100.
[7]   Tashima H, Yoshida E, Iwao Y, Wakizaka H, Maeda T, Seki C, Kimura Y, Takado Y, Higuchi M, Suhara T, Yamashita T, Yamaya T. PET Helmet Hideaki – First prototyping of a dedicated PET system with the hemisphere detector arrangement. Phys Med Biol. 2019; 64: 065004.
[8]   Ritt P, Quick H. Multimodale SPECT- und PET-Bildgebung. Chap. 16 In: "Medizinische Physik". eds. Schlegel W, Karger CP, Jäkel O, Heidelberg, Berlin Springer Sektrum; 2018. p. 365–376.
[9]   Slomka PJ, Pan T, Berman DS, Germano G. Advances in SPECT and PET hardware. Prog Cardiovasc Dis. 2015; 57: 566–578.
[10] Hueso-González F, Biegun AK, Dendooven P, Enghardt W, Fiedler F, Golnik C, Heidel K, Kormoll T, Petzoldt J, Römer KE. Comparison of LSO and BGO block detectors for prompt gamma imaging in ion beam therapy. J Instrum. 2015; 10: P09015.
[11] Kim S et al., High resolution GSO block detectors using PMT-quadrant-sharing design for human whole body and breast/brain PET applications, IEEE Nuclear Science Symposium Conference Record, 2005, 2005, 2863–2867
[12] Pierce L, Miyaoka R, Lewellen T, Alessio A, Kinahan P. Determining block detector positions for PET scanners. IEEE Nucl Sci Symp Conf Rec. 1997; 2009: 2976–2980.
[13] Gundacker S, Heering A. The silicon photomultiplier: Fundamentals and applications of a modern solid-state photon detector. Phys Med Biol. 2020; 65: 17TR01.
[14] Mix M. Positronen-Emissions-Tomographie Medizinische Physik. Chap. 15 eds. Schlegel W, Karger CP, Jäkel O, Heidelberg, Berlin Springer Sektrum; 2018. p. 349–364
[15] Boudjelal A, El Moataz A, Messali Z. A new method of image reconstruction for PET using a combined regularization algorithm. In: El Moataz A, Mammass D, Mansouri A, Nouboud F eds., Image and Signal Processing. Springer Nature, Switzerland AG. ICISP 2020. Lecture Notes in Computer Science. 2020; 12119: 178–185.
[16] Galve P, Udías JM, López-Montes A, Arias-Valcayo F. Super-Iterative Image Reconstruction in PET. IEEE Trans Comput Imaging. 2021; 7: 248–257.
[17] Kinahan PE, Fletcher JW. Positron emission tomography-computed tomography standardized uptake values in clinical practice and assessing response to therapy. Semin Ultrasound CT MR. 2010; 31: 496–505.

[18]   Bertoldo A, Rizzo G, Veronese M. Deriving physiological information from PET images: From SUV to compartmental modelling. Clin Transl Imaging. 2014; 2: 239–251.
[19]   Workman RB, Coleman RE. Fundamentals of PET and PET/CT Imaging. PET/CT Essentials for Clinical Practice. Heidelberg, Berlin: Springer-Verlag GmbH; 2006.
[20]   Acker MR, Burrell SC. Utility of 18F-FDG PET in evaluating cancers of lung. J Nucl Med Technol. 2005; 33: 69–74.
[21]   Pauleit D, Floeth F, Hamacher K, Riemenschneider MJ, Reifenberger G, Müller HW, Zilles K, Coenen HH, Langen KJ. O-(2-[18F]fluoroethyl)-L-tyrosine PET combined with MRI improves the diagnostic assessment of cerebral gliomas. Brain. 2005; 128: 678–687.
[22]   Dunet V, Pomoni A, Hottinger A, Nicod-Lalonde M, Prior JO. Performance of 18F-FET versus 18F-FDG-PET for the diagnosis and grading of brain tumors: Systematic review and meta-analysis. Neuro-Oncology. 2016; 18: 426–434.
[23]   Silver DA, Pellicer I, Fair WR, Heston WD, Cordon-Cardo C. Prostate-specific membrane antigen expression in normal and malignant human tissues. Clin Cancer Res. 1997; 3: 81–85.
[24]   Eiber M, Weirich G, Holzapfel K, Souvatzoglou M, Haller B, Rauscher I, Beer AJ, Wester HJ, Gschwend JE, Schwaiger M, Maurer T. Simultaneous 68Ga-PSMA HBED-CC PET/MRI Improves the Localization of Primary Prostate Cancer. Eur Urol. 2016; 7(0): 829–836.
[25]   Maurer T, Eiber M, Schwaiger M, Gschwend JE. Current use of PSMA–PET in prostate cancer management. Nat Rev Urol. 2016; 13: 226–235.
[26]   Kwekkeboom D. Perspective on 177Lu-PSMA therapy for metastatic castration-resistant prostate cancer. J Nucl Med. 2016; 57: 1002–1003.

Useful websites

http://www.people.vcu.edu/~mhcrosthwait/PETW/index.html

Further reading

Bushberg JT, Seibert JA, Leidholdt EM, Jr, Boone JM. The essential physics of medical imaging. 3rd. Philadelphia, Baltimore, New York, London: Wolters Kluwer, Lippincott Williams & Wilkins; 2012.
Maier A, Steidl S, Christlein V, Honegger J. Medical Imaging Systems. Berlin, Heidelberg: Springer Open; 2018.
Saha GB. Basics of PET Imaging: Physics, Chemistry, and Regulations. 3rd. Berlin, Heidelberg: Springer Verlag; 2016.
Salditt T, Aspelmeier T, Aeffner S. Biomedical Imaging. Principles of Radiography, Tomography, and Medical Physics. Berlin, Boston: De Gruyter; Graduate Text 2017.
Santiago JFY. Positron Emission Tomography with Computed Tomography (PET/CT). Berlin, Heidelberg: Springer Verlag; 2015.
Schlegel W, Karger CP, Jökel O. Medizinische Physik. Grundlagen-Bildgebung-Therapie-Technik. Heidelberg, Berlin: Springer Lehrbuch: Springer Spektrum; 2018.
Webb A. Introduction to Biomedical Imaging. New York, London, Sydney, Toronto: Wiley-Interscience, John Wiley & Sons, Inc; 2003.
Wernick MN, Aarsvold JN. Emission Tomography: The Fundamentals of PET and SPECT. Elsevier; 2004.

Appendix

11 Answers to questions

Chapter 1

A1.1   Materials science, engineering, geosciences, bonding, and surface cleaning.

A1.2   Static and dynamic imaging, fragmentation of bladder stones, and lens fragmentation during cataract surgery.

A1.3   2–20 MHz.

A1.4   Acoustic impedance is defined as the product of density and sound velocity.

A1.5   Two materials with very different impedance values that share a common interface.

A1.6   Viscous damping, thermal conduction, and scattering.

A1.7   Reflected, transmitted, or scattered.

A1.8   A large impedance mismatch at the interface between two different materials and normal incidence to a smooth and flat interface.

A1.9   A transducer is a converter of one form of energy into another.

A1.10   By the use of a piezoelectrical transducer.

A1.11   Essential parts are a piezoelectrical crystal, a damping material in the back, and a quarter wave plate in the front.

A1.12   For impedance matching, i.e., to avoid reflections at the air/skin interface.

A1.13   Because imaging is achieved by detecting the echo signal, which is only possible when the US wave is pulsed.

A1.14   TGC is an amplification of echo signals arriving later than the first echo compensating for the decayed echo amplitude.

A1.15   In the near-field region, the wavefront is nearly a plane wave; in the far-field region, diffraction effects occur.

A1.16   At the border between near field and far field.

A1.17   By shaping the transducer or using a plastic lens in front of the transducer, or controlling the signal arrival time electronically.

A1.18   The depth resolution depends on the pulse length of the emitted sound wave. At higher frequencies, the pulse is shorter, and therefore the resolution increases.

A1.19   B-scan is a brightness mode scan. It consists of a sequence of A-scans that combined yield an image of a slice of the body.

A1.20   Sector scanner, array scanner, phase array scanner.

A1.21   The PRF is the reciprocal time separation between two US pulses emitted. The minimum PRF is given by the depth of view. The larger the DOV, the lower the PRF.

A1.22   A C-scan probes a plane at a certain depth taken from a sequence of B-scans.

A1.23   Double reflection, shadowing effects, refraction effects.

https://doi.org/10.1515/9783110757095-011

A1.24   a. Strongly reflecting organ in front of the ROI (bones, lungs, etc.); b. Stronger reflection than expected due to an organ in front with lower absorption; c. Soundwaves, which are reflected back and forth at flat parts of an organ.

A1.25   By US scattering from erythrocytes, which act as moving receiver and emitter for sound waves that cause a frequency shift proportional to the speed, also called Doppler shift.

A1.26   For high-resolution velocity profile determination, the probing sound source is continuous.

A1.27   For good imaging, short pulses are required, whereas high-resolution velocity determination requires a continuous wave. The compromise are long pulses.

A1.28   Imaging the depth of object sets an upper limit on PRF that depends on the reciprocal echo time. Determination of the velocity sets a lower limit on the PRF in order to avoid aliasing effects.

A1.29   RI is defined by the difference between the peak systolic velocity and end diastolic velocity normalized by the peak systolic velocity. PI is defined by the same velocity difference, however, normalized by the mean velocity.

Chapter 2

A2.1   A flexible tube that shines light into hollow spaces and channels out an image from the illuminated area.

A2.2   A white light source, a fiberglass bundle, and receiving optics of the reflected light.

A2.3   Total reflection at the inner walls of the glass fiber and constructive interference of the head and tail of a traveling light wavetrain.

A2.4   Gastroscopy and colonoscopy.

A2.5   100 μm.

A2.6   Depends on the wavelength and penetration of light into tissue, which can be as much as a few mm.

A2.7   Confocal endoscopes are scanning endoscopes. A fine light spot is scanned via a mirror scanner across the field of interest.

A2.8   Main features of confocal microscopes are 1. Formation of the focal point of the objective lens on a pinhole to decrease "noise." 2. Increase in the optical resolution and contrast of the image. 3. Ability to reconstruct a 3-D image of the specimen. 4. Ability to collect serial optical sections from thick specimens.

A2.9   Optical coherence tomography endoscope is an instrument that carries depth information by using interference of laser light with short coherence length.

A2.10   By a capsule endoscope.

A2.11   A capsule endoscope connected to a string containing optical fibers for data transfer that can be swallowed and pulled back out again.

Chapter 3

A3.1   Light nuclei with an odd number of nucleons.

A3.2   Proton-based MRI.

A3.3   The Larmor frequency is the precessional frequency of spins in an external magnetic field. In the case of nuclei, it is the product of the nuclear gyromagnetic ratio y and the magnetic induction B.

A3.4   About 50 to 150 MHz, depending on the field.

A3.5   The precession is damped by the longitudinal relaxation time $T1$ and the transverse relaxation time $T2$. $T1$ is the spin-lattice relaxation process, and T2 is the spin-spin relaxation process.

A3.6   FID is essential for recording an induced voltage in pick-up coils from precessing protons.

A3.7   FID occurs only upon relaxation of the in-plane magnetization M_{xy}.

A3.8   The magnetization can be turned by a resonating Larmor precession over a finite time that turns the magnetization by 90°.

A3.9   The transverse relaxation time $T2$.

A3.10   The longitudinal relaxation time $T1$.

A3.11   Spin-echo implies that the dephasing of the m_{xy} spins and the decay of the M_{xy} magnetization is partially recovered by a refocusing 180° field pulse.

A3.12   First, a 90° pulse turns the spins into the plane, followed by 180° pulse for spin reversal after a time span of t_0.

A3.13   At $2t_0$.

A3.14   For $T1$, short TR and short SE. For $T2$ long TR and long SE.

A3.15   Long TR and short TE.

A3.16   IR is applied when the $T1$ weighting procedure is not sufficient. Then first, a 180° pulse is applied, and a spin-echo is followed as soon as the fast system has passed the zero line of M_z, and the relaxation of the slower system is at $M_z=0$.

A3.17   Because if the relaxation time of the spin system matches the characteristic frequencies of the embedding system, the spin-lattice coupling is strongest, and the relaxation times are shortest.

A3.18   Because the Larmor frequency is zero in the rotating frame and therefore also the field in the z-direction.

A3.19   Slice selection is achieved by a magnetic field gradient in the z-direction.

A3.20   z-direction by field gradient, x-direction by a frequency gradient, and the y-direction by a phase encoding gradient.

A3.21   The FOV is mainly determined by the field gradient in the x-direction and the bandwidth of the receiving coil.

A3.22   A K-map is a two-dimensional map with 256 × 256 pixels, each one containing the frequency in the x-direction and the phase in the y-direction.

A3.23   By Fourier transformation of the K-map.

A3.24   Just before the 180° refocusing pulse.

A3.25   1.5 T and 3 T.

A3.26   By a solenoid with superconducting wires.

A3.27   Gradient in the z-direction is generated by anti-Helmholtz coils, gradients in the x- and y-direction by half circle Golay coils.

A3.28   About 20 kW.

A3.29   CAs shorten the relaxation time, preferentially either $T1$ or $T2$.

A3.30   They always contain a magnetic ion, such as Gd^{3+}.

A3.31   In hyperpolarization, MRI isotopes are used with an odd number of nuclei and unpaired nuclear spin that can be polarized externally before administering for imaging.

A3.32   With polarized ^3He, respiration can be studied; with ^{17}O and ^{19}F metabolic processes and pharmacokinetics can be investigated. The main disadvantage of ^3He and ^{17}O application is the rarity of the isotopes and, therefore, the high cost.

A3.33   With diffusion-weighted MRI, the flow of liquids can be studied using two opposing short field gradients applied just before and just after the 180° refocusing pulse.

A3.34   Investigations of neural activity, mainly in the brain.

A3.35   Functional MRI is based on high-spin deoxyhemoglobin and low-spin oxyhemoglobin. In the high-spin state, the $T1$ relaxation time is shorter than in the low-spin state. Active areas in the brain consume more oxygen than neighboring areas, which can be used for locating active areas.

A3.36   Moving organs can be imaged with the use of turbo MRI. Turbo MRI uses a sequence of 180° refocusing pulses together with phase encoding pulses within one TR.

A3.37   About 20 frames/s.

A3.38   Use of smaller scanners with conventional magnets or magnets with wires of high-temperature superconductors. Use of new and more sensitive detectors compared to induction coils. Further development of multiparameter MRI to replace radiographic imaging with x-rays and γ-rays like PET and SPECT.

A3.39   (1) The high applied magnetic fields exclude the use of any metals nearby. (2) The power consumption has to be adjusted not to heat the body by more than one centigrade. (3) Contrast agents such as Gd-chelates may be incompatible with some patients.

Chapter 4

A4.1 Energies are inversely related to wavelengths: $E \sim 1/\lambda$.

A4.2 Bremsstrahlung, characteristic radiation, and synchrotron radiation.

A4.3 X-rays can be produced by either x-ray tubes, linear accelerators, or electron synchrotrons.

A4.4 Bremsstrahlung occurs upon deaccelerating fast electrons in nuclei's Coulomb field.

A4.5 Characteristic x-ray radiation is generated by creating an electron-hole in a core-shell, which is filled by an electron from shells with less binding energy. The energy difference is emitted as x-ray radiation.

A4.6 Both are highly energetic electromagnetic waves. The distinction is only possible by the generation: γ-radiation is a byproduct of radioactive decay stemming from nuclei. X-rays are artificially generated by accelerators and electronic excitations.

A4.7 For radiography, acceleration voltages up to 150 pKV are used, for radiotherapy, the accelerating voltages are in the range of 2–20 MV.

A4.8 A cathode for the production of electrons, a high voltage supply for accelerating electrons in a potential gradient, and a target (anode) for stopping the electrons and production of bremsstrahlung.

A4.9 The x-ray intensity is proportional to the difference of the high voltage applied (E_{max}) and the energy of the emitted photons (E_{photon}): $I \sim E_{max} - E_{photon}$.

A4.10 The cut-off energy is determined by the accelerating potential U_{acc}: $E_{max} = eU_{acc}$.

A4.11 The purpose of the Al-filter is to reduce soft x-ray radiation by absorption that would otherwise increase the dose but would not contribute to the image quality of radiographs.

A4.12 Characteristic radiation has sharp spectral lines, which are characteristic of the element that emits this radiation. Therefore, by analyzing the energy of the characteristic radiation, the element that emitted the radiation can be identified.

A4.13 In forensic medicine.

A4.14 The transition from the $2p_{1/2}$ (L2) to $1s$ ($K_{\alpha1}$) energy level.

A4.15 No. The W-K_α radiation requires an excitation energy of 69.5 keV.

A4.16 The excitation energy is always higher than the corresponding x-ray characteristic radiation. For Mo, the excitation energy of the Mo-K_α radiation is 20 keV, the energy of the characteristic Mo-$K_{\alpha1}$ radiation has an energy of 17.48 keV.

A4.17 The Mosely law is similar to Bohr's model for the energy levels in hydrogen atoms modified by the charge of the nucleus and the hole in the K-shell after excitation.

A4.18   The cathode is offset, and the electrons hit the rim of a rotating anode. Cathode and anode sit in an evacuated glass tube. The anode is driven by an outside stator that couples to a rotor inside the glass tube.

A4.19   No, they are only cooled by radiation.

A4.20   The Straton tube is continuously cooled. The focus can be controlled by a magnetic field bending the electron beam, and the x-ray radiation is automatically filtered by the Al housing.

A4.21   Linear accelerators require an electron source (klystron), a pulse generator a modulator, and microwave cavities for accelerating the electrons.

A4.22   No. The electrons lose only part of their kinetic energy by radiation. For instance, an electron with 1 GeV kinetic energy may emit a 1 keV photon, which is only 1 ppm of the kinetic energy. Nevertheless, the lost energy is boosted up in the ring, so that the electrons stay in the orbit.

A4.23   Synchrotron radiation is mainly used for the analysis of materials via scattering and spectroscopy. The materials to be analyzed range from hard to soft matter, including biomaterials.

Chapter 5

A5.1   Nuclei are composed of the nucleons: protons and neutrons.

A5.2   Three. Two up and one down in protons, two down and one up in neutrons.

A5.3   A is the sum of protons and neutrons.

A5.4   Isotopes are nuclei with the same number of protons but different number of neutrons.

A5.5   Isotopes with the right number of neutrons are stable. Too many neutrons or too few neutrons make isotopes unstable.

A5.6   For light atoms, stable isotopes have roughly the same number of protons and neutrons. There are more neutrons than protons in heavy atoms.

A5.7   Natural isotopes and artificial isotopes.

A5.8   The atomic mass unit is defined as 1/12 of the mass of the isotope ^{12}C.

A5.9   There are two reasons. One is the binding energy of nucleons in nuclei, which causes a mass deficiency. The other reason is the isotope mixture that is usually present in chemical elements. The atomic weight is then an average weighted by the natural abundance.

A5.10 α, β^-, β^+-decay, EC, and γ-emission.

A5.11   Lifetime is the time T after which 1/e or 63% of radioisotopes have decayed. Half-life $T_{1/2}$ is the time after which half of the radioactive isotopes or 50% have decayed.

A5.12   Activity is the number of decays (events) per second.

A5.13   Becquerel; 1 Bq is one event per second.

A5.14   The activity of the daughter radioisotope depends on the production rate of the daughter nucleus via decay of the parent nuclei minus the decay rate of the daughter nuclei.

A5.15   Proton bombardment, nuclear fission with slow neutrons, activation with slow neutrons.

A5.16   Cyclotron for proton activation. Nuclear reactor for fission production and for thermal neutron activation.

A5.17   The most efficient way is by fission of 235-uranium. ^{99}Mo is a fission product from which ^{99m}Tc can be extracted.

A5.18 β-decay is a two-particle process. The spectral energy is divided between β-particle and a neutrino. All other decays are single-particle decays and therefore feature sharp spectral lines.

Chapter 6

A6.1   Absorption and scattering.

A6.2   Number of particles per area.

A6.3   Intensity is fluence per time.

A6.4   Photoelectric effect, Compton effect, pair production, coherent scattering.

A6.5   Thomson scattering for elastic scattering at bound electrons; Rayleigh scattering from electron clouds of atoms; x-ray scattering from crystal structures; Mie scattering from nano- to microsize objects. Small-angle x-ray scattering is the same as Mie scattering.

A6.6   First step: x-ray photon is absorbed and all photon energy is transferred to a core electron. Second step: the core hole is filled and another fluorescence photon is emitted.

A6.7   Generating characteristic x-ray radiation by electrons: there is one electron going in, two electrons and one photon coming out. Generating characteristic x-ray radiation via the PE effect: one photon goes in, one electron and one photon come out.

A6.8   In photoelectric effect, the photon emitted is characteristic for the atom from which it is emitted. The photoelectron has no directional preference. In Compton scattering, the scattered photon has no relation to the atom it is scattered from. The Compton electron has a directional preference in the forward direction.

A6.9   Coherent scattering is elastic, i.e., the photon energy does not change during the scattering process. Compton scattering is an inelastic scattering process. The photon transfers energy to the Compton electron.

A6.10   Pair production does not occur below 1 MeV. However, for photon energies above 1 MeV, pair production becomes increasingly important.

A6.11 High-energy alpha particles loose little energy per collision. But at the end of their straight path, they lose their remaining kinetic energy over a very short distance.

A6.12 5 cm.

A6.13 The Bragg peak is a peak in linear energy transfer just before the end of the range of alpha particles. This is due to the fact that the energy loss per collision is largest at the end of the track.

A6.14 No. Photons have a half thickness where half of the incident intensity has diminished.

A6.15 The range of beta particles depends on the energy and is typically meters in air.

A6.16 Fast neutrons interact via collisions, slow neutrons may be absorbed by target nuclei.

A6.17 Light and heavy water.

A6.18 In the energy region of a few eV.

Chapter 7

A7.1 Dosimetry provides precise radiation measurements, defines radiation limits at workplaces, and provides protection against unsafe radiation levels.

A7.2 Dose is defined as deposited radiation energy per mass unit.

A7.3 w_R is a weighting factor for different types of radiation (formerly called Q-factor); the weighting factor w_T characterizes sensitivities of tissues and organs towards radiation (formerly called RBE).

A7.4 Dose is the deposited radiation energy per mass unit (unit Gray), equivalent dose is radiation energy weighted by the factor w_T (unit Sievert).

A7.5 Both are weighting factors: w_R for the ionization effectiveness of radiation, and w_T for the sensitivity of body parts to radiation.

A7.6 Fluence is the number of particles per area. Energy fluence considers only fluence from photons multiplied by their energy.

A7.7 Kerma stands for kinetic energy released in matter. It is defined as the product of energy fluence and mass energy transfer coefficient.

A7.8 No. It is a definition that is solely applied for electromagnetic radiation, i.e., x-rays and γ-radiation that transfers kinetic energy to charged particles.

A7.9 The mass attenuation coefficient accounts for all kinds of interactions of photons with matter, including those which do not transfer kinetic energy to the matter. The mass-energy transfer coefficient (Kerma) considers only those attenuation processes which result in the kinetic energy of charged particles.

A7.10 Film dosimeters, finger dosimeters, pocket dosimeters, and radiation monitors.

A7.11 The design of these counters is identical, but the operational voltage is different. Proportional counters operate at a lower voltage than Geiger-Mueller counters. The output signal of proportional counters is proportional to the energy of incident particles. Geiger-Mueller counters are event counters without energy resolution.

A7.12 Dead-time is the time when a counter is insensitive to radiation. The measured count rate is lower than the true count rate because of detector dead-time.

A7.13 Cosmic radiation, terrestrial radiation, and radiation from industry and medical applications.

A7.14 5 mSv/a.

A7.15 ALARA stands for "as low as reasonably achievable".

A7.16 Shorten the exposure time and increase the distance to radioactive sources.

Chapter 8

A8.1 Contrast is defined as the difference intensity of spatially separated regions divided by the sum of the intensities.

A8.2 An attenuation profile is a lateral profile of intensity variations due to spatially varying mass absorption coefficients.

A8.3 Reasons are an extended x-ray source and Compton scattering.

A8.4 By the use of an oscillating collimator (grid).

A8.5 Beam hardening can be achieved by two methods: either by using a filter that absorbs the low energy end of the spectrum or by increasing the accelerating voltage.

A8.6 X-ray filters have two effects: they reduce the primary flux preferentially in the soft x-ray regime and shift the average energy of the bremsstrahlung spectrum to higher energies.

A8.7 X-ray films have been replaced by flat panel thin film transistors.

A8.8 No, the resolution of flat panel displays is slightly lower.

A8.9 Sharing x-ray radiographs instantly worldwide and recording life videos of surgical procedures.

A8.10 Mammography uses softer x-rays than in standard x-ray radiography, because of the low x-ray absorption of the female breast for hard x-rays.

A8.11 Injection of contrast agents for imaging the GI tract or blood vessels; dual-energy x-ray imaging; phase contrast imaging.

A8.12 The real part.

A8.13 Phase contrast imaging emphasizes topological differences even though absorption contrast may be lacking. Therefore, edges are more pronounced in phase contrast imaging than in absorption imaging.

A8.14 CT imaging can localize organs in a slice that would not be visible in projection radiography.

A8.15 The Hounsfield scale is a relative scale of x-ray absorption with respect to water.

A8.16 Air appears black, and bones appear white.

A8.17 At present 256 detectors are mounted in parallel to cover a slice in the z-direction of about 120 mm.

A8.18 $0.5 \times 0.5 \times 0.5$ mm^3.

A8.19 About 2–3 mSv.

A8.20 For recognizing bone fractures, lung diseases, and early cancer developments.

A8.21 1. Penumbra (extended source); 2. Compton scattering; 3. Too small $\Delta\mu$.

A8.22 1. Attenuation contrast; 2. Spectroscopic contrast; 3. Phase contrast.

A8.23 1. Use of two x-ray generators mounted on the gantry rotating with two detector banks; 2. Periodic switching between two anode voltages; 3. Energy dispersive recording of two parts of the bremsspectrum.

A8.24 A sinogram is the recorded attenuation profile of a rotating object.

A8.25 Any 2D function projected by a line integral into a 1D profile constitutes a Radon-transformation.

Chapter 9

A9.1 By injection.

A9.2 Gamma emitters.

A9.3 Because light radioisotopes do not emit gamma radiation.

A9.4 By nuclear fission and proton irradiation.

A9.5 Because Cs can substitute for K and Na and would disturb the electrolyte household of the body.

A9.6 ^{99m}Tc and ^{123}I.

A9.7 140 keV.

A9.8 They are analogous pharmaceuticals.

A9.9 ^{99}MoO$_4{}^{2-}$ is doubly charged and strongly bonded to alumina substrate. When decaying to ^{99m}Tc, single-charged ^{99}TcO$_4{}^-$ is less bonded to alumina, and can therefore be washed out.

A9.10 After 23 h.

A9.11 Most frequently, scintillation counters combined with photomultiplier tubes. But semiconductor counters (CdZnTe) are becoming more popular.

A9.12 Collimators are necessary to trace back the origin of the radiation. Collimators are part of the SPE camera.

A9.13 A scintillation counter consists of a scintillation crystal and a photomultiplier tube. In the scintillation crystals, high-energy γ-photons are down-converted into visible photons. The visible photons hit a photocathode and are converted to electrons. The number of photoelectrons increases in a cascade of dynodes and yields a current pulse at the exit of the photomultiplier tube, which is then converted into a voltage pulse.

A9.14 Single-photon emission computed tomography.

A9.15 Two to three detector banks.

A9.16 X-ray CT has a higher spatial resolution and does not require isotopes. SPECT is better than x-ray CT in terms of functional studies, metabolism, and perfusional investigations.

A9.17 For a myocardial stress test, bone metabolism, and brain activity.

A9.18 About 0.5 mm for x-ray radiography, about 5 mm for SPE.

Chapter 10

A10.1 The radioisotopes must be positron emitters, and they should exhibit short half-lifes.

A10.2 They are produced by proton bombardment of light elements.

A10.3 Pair production.

A10.4 Because of momentum conservation.

A10.5 A few millimeters.

A10.6 By coincidence detection.

A10.7 If one gamma quant is detected in one counter, a second one should be detected in a counter positioned 180° apart and within a narrow time window.

A10.8 Coincidence by scattering and random accidental coincidence.

A10.9 Cancer recognition and perfusion studies.

A10.10 Perfusion is the flow of gases or liquids through the body driven by a concentration gradient, pressure gradient, or by active transport; diffusion is particle transport driven by a concentration gradient only.

A10.11 The positron emitter 18F is combined with glucose-like molecules yielding FDG. In the body, FDG follows the same metabolic pathway as true glucose but is not metabolized to CO_2 and water. Instead remains in tissue with high glucose demand. Those parts of the tissue indicate potential cancerous cells.

A10.12 The significance of 18F in nuclear medicine is due to the short half-life and the possibility of incorporating it in authentic radiopharmaceuticals.

A10.13 PET.

A10.14 High cost for isotope production and for radiochemistry.

A10.15 TOF-PET stands for time-of-flight PET. The TOF method determines the arrival time of the γ-rays, and from the time difference the location of the positron along the LOR is estimated.

12 Solutions to exercises

Chapter 1

E1.1 **Solution: Liver screening**

An air-filled volume has a large impedance mismatch with respect to the surrounding tissue. Then the incoming sound wave is completely reflected at
the first wall of the air-filled intestine and cannot reach the liver.

E1.2 **Solution: Half-value thickness**

A 50% reduction corresponds to a loss by 3 dB. The half-value thickness is
then $t_{1/2}(1\,\text{MHz}) = 3\,\text{dB}/\mu \cdot 1\,\text{MHZ} = 3\,\text{dB}/0.75\,\text{dB}/\text{cm}/1\,\text{MHz} \times 1\,\text{MHz} = 4\,\text{cm}$. For a
5 MHz US wave, $t_{1/2}(5\,\text{MHz}) = 4/5 = 0.8\,\text{cm}$.

E1.3 **Solution: Attenuation**

The $1/e$ intensity loss corresponds to $10\log(1/e) = 4.3\,\text{dB}$. This loss is reached
at a distance of $4.3\,\text{dB}/0.75 = 5.7\,\text{cm}$ in soft tissue.

E1.4 **Solution: Reflection and transmission**

a. We calculate the reflection and transmission ratios at different interfaces.
The arrows in the graph are inclined for clarity. There are three reflected
and three transmitted beams, if all reflections and transmissions beyond
R_3 are neglected. The transducer will receive first intensity from R_1 and
with some time delay from T_3.

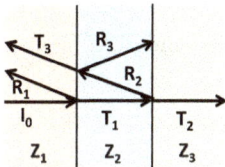

b. The reflected and transmission ratios are as follows:

$$R_1 = \left(\tfrac{Z_1 - Z_2}{Z_1 + Z_2}\right)^2, \; T_1 = 1 - R_1;$$

$$R_2 = T_1 \left(\tfrac{Z_2 - Z_3}{Z_2 + Z_3}\right)^2, \; T_2 = 1 - R_2 = 1 - T_1 \left(\tfrac{Z_2 - Z_3}{Z_2 + Z_3}\right)^2;$$

$$R_3 = R_2 \left(\tfrac{Z_2 - Z_1}{Z_2 + Z_1}\right)^2, \; T_3 = 1 - R_3 = 1 - R_2 \left(\tfrac{Z_2 - Z_1}{Z_2 + Z_1}\right)^2.$$

c. Using the impedance values, we obtain

$R_1 = 0.0039$, $T_1 = 0.996$; $R_2 = 0.443$, $T_2 = 0.5513$;
$R_3 = 0.0017$, $T_3 = 0.9983$.

The transducer receives the intensities from $R_1 = 0.0039$ and $T_3 = 0.9983$,
but mainly from T_3.

https://doi.org/10.1515/9783110757095-012

    d. The intensity received by the transducer is dominated by the reflected and transmitted waves at the water/lung interface.

E1.5 **Solution: Examination at high resolution**
    a. The field of interest should be positioned in focus within the near-field area.
    b. For 10 cm depth, a frequency of 5 MHz should be chosen according to Tab. 1.3.
    c. The lateral resolution is then 1.2 mm.
    d. The axial resolution is 0.3 mm.
    e. The PRT can be determined from the maximal depth to be probed because the PRT should be larger than the echo expected. For a 10 cm depth, the PRT should be larger than $\Delta t > 0.20$ m/1540 m/s = 130 µs. Therefore, the PRF should be lower than 7.7 kHz.

E1.6 **Solution: Bulk modulus**
    a. The sound velocity in the tissue is $v_{\text{tissue}} = 2L/\Delta t = 0.1\,\text{m}/68\,\mu\text{s} = 1470\,\text{m/s}$.
    b. According to Tab. 1.2, the velocity corresponds to fatty tissue.
    c. The sound velocity is $v = \sqrt{B/\rho}$. Since the impedance is $Z = \rho v$, the bulk modulus is $B = vZ$. Using the values for fatty tissue, the bulk modulus is $B = 1470$ m/s $\times$ 1.38 10^6 Ns/m^3 = 2028 $\times$ 10^6 N/m^2 = 2 GPa.

E1.7 **Solution: Doppler sonography**
$\text{RI} = 1 - \text{EDV}/\text{PSV}$, and $\text{EDV} = (1 - \text{RI})\text{PSV} = 0.45\,\text{m/s} = 2\,\text{m/s}$.
The Doppler shift is $(\Delta f)_{\text{PSV}} = 2 \cdot 7\,\text{MHz} \cdot \frac{5\,\text{m/s}}{1540\,\text{m/s}} = 45\,\text{kHz}$, $(\Delta f)_{\text{EDV}} = 2 \cdot 7\,\text{MHz} \cdot \frac{2\,\text{m/s}}{1540\,\text{m/s}} = 18\,\text{kHz}$.
The difference is, therefore, 27 kHz between PSV and EDV.

Chapter 2

E2.1 **Solution: Optic fiber**
The condition for total reflection is.

$$\cos\beta_c = \sin\gamma_c = n_0/n_1$$

Using Snell's law, we find:

$$n_0 \sin\alpha_c = n_1 \sin\beta_c$$

or

$$n_0^2 \sin^2\alpha_c = n_1^2 (1 - \cos^2\beta_c),$$

which is

$$\sin^2\alpha_c = \frac{1}{n_0^2}\left(n_1^2 - n_0^2\right).$$

Taking the root and solving for α_c yields eq. (2.3).

E2.2   **Solution: Refractive index**
The maximum angle $\alpha_{max} = 90°$. For this angle,

$$\sin\alpha_{max} = \sqrt{n_1^2 - n_0^2} = 1.$$

Therefore,

$n_1^2 - n_0^2 = 1$ or $n_1 = \sqrt{2} = 1.41$.

Fiberglass with a higher refractive index would not be helpful unless using gladding material with a refractive index n_2, which fulfills the condition $n_1^2 - n_2^2 < 1$.

E2.3   **Solution: Order of interference**
Using eq. (2.8), we find for $\lambda = 500$ mm:

$$m \le \frac{2 \times 1\,\mu m}{0.5\,\mu m}\sqrt{1.5^2 - 1.2^2} = 4 \times 0.9 = 3.6.$$

Therefore, $m = 3$.
For a 1000 nm light source, $m = 1$.

E2.4   **Solution: Longitudinal coherence length**
The condition for destructive interference is $N\lambda = (N - 1/2)(\lambda + \Delta\lambda)$. Solving for N, we obtain: $N = (\lambda + \Delta\lambda)/(2\Delta\lambda)$. Neglecting $\Delta\lambda$ in the numerator, the coherence length is therefore:

$$L_c = N\lambda = \frac{\lambda^2}{2\Delta\lambda},$$

which is similar to the result in eq. (2.10), derived by calculating the Gaussian width of the wavenumber distribution in a short wave train. $\Delta\lambda$ is the spectral bandwidth of a wave train, not the length of the wave train. The length of the wave train is determined by the coherence length.

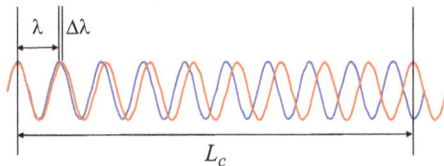

In our case, the coherence length is

$$L_c = \frac{\lambda^2}{2\Delta\lambda} = \frac{\lambda^2}{2 \times 0.1\lambda} = \frac{\lambda}{0.2} = 5 \times \lambda = 5000\text{ nm.}$$

E2.5 **Solution: OCT scanner**

The coherence length should be equal to the depth resolution: $L_c = dz$. For the depth resolution, we request $dz = 0.05 \times 200$ μm $= 10$ μm. A wave train of 10 μm length comprises 10 wavelengths with wavelength $\lambda = 1$ μm. With a coherence length that comprises 10 wave trains, 20 fringes can be scanned, yielding a total scanning depth of $z = 20 \times 10$ μm $= 200$ μm.

Chapter 3

E3.1 **Solution: Magnetization**

Equation (3.19) is

$$N_1/N_2 = \exp(+\gamma\hbar B_z/k_B T) = \exp(x).$$

The magnetization is defined as

$$M = \frac{\mu_B}{V}(N_1 - N_2) = \frac{\mu_B N}{V}\left(\frac{N_1 - N_2}{N_1 + N_2}\right).$$

Therefore, we obtain for

$$\frac{N_1 - N_2}{N_1 + N_2} = \frac{\exp(x/2) - \exp(-x/2)}{\exp(x/2) + \exp(-x/2)} = \tanh(x/2)$$

and for

$$M = \frac{\mu_B N}{V}\tanh(x/2) = \frac{\mu_B N}{V}\tanh(\gamma\hbar B_z/2k_B T).$$

E3.2 **Solution: Pulse sequence**

The trivial solution is that TR must always be longer than TE because TE takes place during TR. Let us assume that $TE < TR$ and that TR is short. Then the situation is similar to the $T1$ contrast. However, by waiting too long for TE, the $T1$ contrast is lost again. Therefore, the pulse sequence, short TR and long TE, does not yield useful images.

E3.3 **Solution: contrast elimination**

By selecting a short TE, short TE eliminates $T2$ contrast.

E3.4   **Solution: Time of inversion**

For the inversion time, we have the condition:

$$M_z(t = t_i) = 0 = M_{z,\,sat}(1 - 2\exp(-t_i/T1)),$$

see eq. (3.46). This condition can be fulfilled, if

$$1/2 = \exp(-t_i/T1).$$

Therefore, $\ln 2 = t_i/T1$, or $t_i = T1\ln 2$.

E3.5   **Solution: Liquid He consumption**

The daily He loss is 2.25 l. In a year, 821 l He are vaporized. Refill is necessary after 333 days.

E3.6   **Solution: G_z-gradient**

According to eq. (3.50), the field gradient is determined by: $G_z = \frac{2\pi \times df}{\gamma \times FOV}$.
In our case, we have df=100 Hz, FOV=1 mm, $\gamma = 42.6$ MHz/T. This yields a field gradient of 14.7 mT/m.

E3.7   **Solution: $T1$ relaxation**

Both water and fat have a high proton concentration. However, the protons in water are much more mobile than in fat, where they are bound to large molecules. Fat is more solid-like than liquid-like; the diffusivity of protons in fat is low. Therefore, it is reasonable that the relaxation time for protons in fat is much shorter.

E3.8   **Solution: Water contrast**

In $T1$-weighted images, shorter $T1$ appears brighter. Protons in water have a very long $T1$ of 3000 s. Therefore, within TR, protons in water do not produce sufficient FID signal and appear therefore black.

E3.9   **Solution: Power consumption of solenoid**

The magnetic induction at the center of a circular solenoid is $B = \mu_0 H = \mu_0 NI/2\pi r$. N is the number of windings and r is the radius of the coil. Therefore, the current is $I = B2\pi r/\mu_0 N$. Inserting numbers, the current is 600 A. The resistance of the wire is $R = \rho L/A = \rho N2\pi r/A$. Here, A is the cross section of the wire. Inserting numbers, we find that the wire is about 250 m long and has a resistance of about 0.16 Ω. Therefore, voltage is about 100 V and power consumption is 60 kW. The energy consumption during the 5 s turn-on is 3×10^5 J.

E3.10 **Solution: Frequency gradient**

The frequency gradient per meter is $\Delta f/m = (42.8\,\text{MHz}/1\,\text{T}) \times 10\,\text{mT}/m =$ $42.8\,\text{MHz} \times 10^{-2}/m = 0.428\,\text{MHz}/m$. The frequency gradient per millimeter is then 428 Hz.

Chapter 4

E4.1 **Solution: Threshold energy**

The threshold energy is the Mo Kα absorption energy, which is 20.00 keV (taken from x-ray tables; for instance, x-ray data booklet: http://xdb.lbl.gov/). The Mo K_α characteristic radiation has an energy of only 17.4 keV. The difference of 2.6 keV is due to the fact that absorption brings the electron from the K-shell to the vacuum level, while the characteristic radiation corresponds to the energy difference between the L- and the K-shell, which is a smaller energy difference. Excitation energy is always larger than characteristic x-ray emission photon energy.

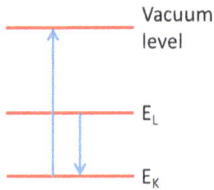

E4.2 **Solution: Energy resolution**

Mo $K_{\alpha1}$ radiation has an energy of 17.48 keV and a wavelength $\lambda_{\alpha1}$ = 709 pm. Mo $K_{\alpha2}$ radiation has an energy of 17.37 keV and a wavelength $\lambda_{\alpha2}$ = 713.8 pm. Therefore, the absolute energy difference $\Delta E = = 0.11\,\text{keV}$, and the relative energy difference is $\Delta E/E = 0.11/17.4 = 0.0063$ or 0.63%. The wavelength difference is $\Delta\lambda = 4.8\,\text{pm}$ and $\Delta\lambda/\lambda = 0.6\%$. This is the same relative value. As the photon energy is $E = hc/\lambda$ (h = Planck's constant, c = speed of light), therefore, $\Delta\lambda/\lambda = \Delta E/E$.

E4.3 **Solution: Cut-off energy**

For an accelerating voltage of 40 kV, the maximum photon energy in the bremsspectrum is E_{max} = 40 keV. Since $E_{max} = hc/\lambda_{min}$, the cut-off wavelength is

$$\lambda_{min} = hc/E_{max} = 1.239\,\text{keV nm}/40\,\text{keV} = 0.03097\,\text{nm}.$$

E4.4 **Solution: Absorption edge energy**
The Mo $K_{\alpha 1}$ radiation has an energy of $E_{K_{\alpha 1}} = E_K - E_L = 17.48\,\text{keV}$. The absorption edge energy of the Mo $K_{\alpha 1}$ radiation is 20 keV. Therefore, the absorption edge energy of the Mo $L_{\alpha 1}$ radiation is $E_L = E_K - (E_K - E_L) = 20\,\text{keV} - 17.48\,\text{keV} = 2.52\,\text{keV}$.

Chapter 5

E5.1 **Solution: Activity**
The initial activity is A_0. Activity after time t is A_1. Activity after time $2t$ is $A_2 = A_1\ (A_1/A_0) = 900\ (900/1000) = 810$.

E5.2 **Solution: Biological half-life**
It is required that after n half-lives, the activity is less than $0.01\,A_0$, i.e., $(1/2)^n < 0.01(1/2)^n$. With $n = 6$, the activity is 1.5%; with $n = 7$, the activity is 0.8%. The patient has to wait between approx. 6 and 7 half-lives, i.e., approx. $7 \times 6\,\text{h} = 42\,\text{h}$ or approx. 2 days. You can also calculate n exactly: $n = \ln(0.01)/\ln(0.5) = 6.64$. This results in a waiting time of 39.86 h = 1.66 days.

E5.3 **Solution: Radioactive oxygen isotope**
First solution:

$$\frac{A_n}{A_0} = \left(\frac{1}{2}\right)^n = 0.001$$

Solving for n yields

$$n = 3\frac{\ln 10}{\ln 2} = 9.96.$$

As 20 min correspond to n half-lives, one half-life is $20\,\text{min}/9.96 = 2\,\text{min}$. Alternatively, we may write.

$$\frac{A_n}{A_0} = \exp\left(-\frac{t \cdot \ln 2}{T_{1/2}}\right) = 0.001$$

Solving for $T_{1/2}$ yields.

$$T_{1/2} = -\frac{20\ \text{min} \cdot \ln 2}{\ln(0.001)} = 2\,\text{min}$$

E5.4 **Solution: Remaining isotopes**

After 5 h, $2 \times 10^3 \times 5 \times 3600$ s $= 3.6 \times 10^7$. Then the amount of decayable isotopes left is $10^{11} - 3.6 \times 10^7 \cong 10^{11}$. The number of decayed isotopes is negligible in comparison to those still decayable.

E5.5 **Solution: Reactor accident**

Let us first consider only the physical half-life. After one half-life, only half of the ^{134}Cs is available, after two half-lives only 1/4, after three half-lives only 1/8, etc. If you only consider the physical half-life, then 8 years would be the correct solution. Together with the biological half-life, we first calculate the effective half-life with the expression stated in eq. (5.5). Inserting the time gives

$$T_{\text{eff}} = \frac{730 \times 140}{730 + 140} = 117.5 \text{ days.}$$

After three effective half-lives, i.e., after approx. 351 days ($\approx$1 year), the radiation exposure of the body has fallen to 1/8.

E5.6 **Solution: Radioactive decay**

a. The number of nuclides produced per unit time is given by the production rate minus the decay rate: $dN = gdt - \lambda N(t)dt$ or $dN/dt = g - \lambda N(t)$. This is an inhomogeneous differential equation. In the homogeneous case, $N(t) = N_0 \exp(-\lambda t)$. However, the initial number of nuclides is not N_0, but 0. With the solution approach $N(t) = g/\lambda \cdot (1 - \exp(-\lambda t))$, all conditions are fulfilled. The activity is then $A(t) = \lambda N(t) = g(1 - \exp(-\lambda t))$.

b. The maximum number is reached at t toward infinity and corresponds to $N_{\max}(\lambda) = g/\lambda$.

E5.7 **Solution: Activation of ^{60}Co**

In 1 g sample, there are $(1/59) \times 6 \times 10^{23} = 10^{22}$ ^{59}Co nuclei. Thus, $N = 10^{22}$.
The cross section for thermal neutrons with energy of about 25 meV is about 40 barn.
According to eq. (5.21) and the plot of the neutron capture cross section, the reaction rate is

$$R_B = N \sigma_n J_n = 10^{22} \times 40 \times 10^{-24} \text{ cm}^2 \times 10^9 \frac{\text{neutrons}}{\text{cm}^2 \text{s}} = 4 \times 10^8 \frac{1}{s}.$$

According to eq. (5.25),
$A_B(t) = R_B(1 - e^{-\lambda_B t})$, where $A_B(t)$ is the targeted activity and R_B is the activity at full saturation. Therefore, the activation time is

$$t = -\frac{1}{\lambda_B} \ln\left(1 - \frac{A}{R}\right) = -\frac{T_{1/2}}{\ln 2} \ln\left(1 - \frac{A}{R}\right) = -\frac{5.26 \text{ y}}{0.692} \ln\left(1 - \frac{1}{4}\right) = 2.18 \text{ years}$$

for reaching an activity of 100 MBq.

E5.8 **Solution: Cyclotron production rate**

a. The nuclear reaction is a proton capture $^{18}O(p, n)^{18}F$ reaction, in which high-energy neutrons are released together with positrons.

b. The highest cross section occurs at about 5 MeV protons.

c. The activity of 2000 mCi corresponds to 7.4×10^{10} Bq. The half-life of ^{18}F is 110 min. The activity and the reaction rate are related as follows:

$$A_B(t) = R_B\left(1 - e^{-\lambda_B t}\right).$$

Therefore, the reaction rate is

$$R_B = \frac{A_B(t)}{\left(1 - e^{-t\ln 2/T_{1/2}}\right)} = \frac{7.4 \times 10^{10} \text{Bq}}{(1 - e^{-120\ln 2/110})} = 1.4 \times 10^{11} \text{Bq}.$$

d. The beam current is quoted as 2 mA/cm². This is equivalent to $N_p \times q/t$. With the charge $q = 1.6 \times 10^{-19}$ As, the proton intensity is
$I_p = 1.25 \times 10^{16}$ cm²/s.
Thus the ratio $R_B/I_p = 1.12 \times 10^{-5}$.
This means that roughly 1 out of 10^5 protons generates a ^{18}F isotope.

E5.9 **Solution: Isotope production in a neutron reactor**

129.6 g contain 6×10^{23} $NiCl_2$ molecules and 12×10^{23} Cl^- ions. Therefore, in 1 g, there are $12 \times 10^{23}/129.6$ Cl^- ions, multiplied with the natural abundance yields 7×10^{21} ions of $^{35}Cl^-$.

The production rate of $^{36}Cl^-$ by neutron capture is according to eq. (5.20):

$$R = N_A\sigma_a J_a = 7 \times 10^{21} \times 43 \times 10^{-24} \text{cm}^2 \times 10^{14} \text{cm}^{-2} \text{s}^{-1} = 3 \times 10^{13} \text{ s}^{-1}.$$

The targeted activity is $A = 100$ µCi $= 3.7 \times 10^6$ Bq. The decay of ^{36}Cl can be neglected.

To reach an activity of $A = N_{Cl}\lambda$, we need to activate
$N_{Cl} = A/\lambda = AT_{1/2}/\ln 2 = 3.7 \times 10^6 \text{Bq} \times 9.5 \times 10^{12} \text{s}/\ln 2 = 5 \times 10^{19}$ $^{35}Cl^-$ ions.
To activate 5×10^{19} $^{35}Cl^-$ ions with a rate of 3×10^{13} s^{-1} will take 1.7×10^6 s $=$ 19.6 *days*.

E5.10 **Solution: Double decay**

According to eq. (5.15):

$$\frac{N_B(t)}{N_A(0)} = \frac{\lambda_A}{2\lambda_A - \lambda_A}\left(e^{-\lambda_A t} - e^{-2\lambda_A t}\right).$$

The maximum of N_B follows from setting the first derivative with respect to t equal to zero:

$$-\lambda_A e^{-\lambda_A t} + 2\lambda_A e^{-2\lambda_A t} = 0$$

or

$$e^{\lambda_A t} = 2.$$

Therefore, N_B has a maximum at $t = \ln 2/\lambda_A$, or $\lambda_A t = \ln 2$. The plot shows the time dependence of N_B in units of $\lambda_A t$.

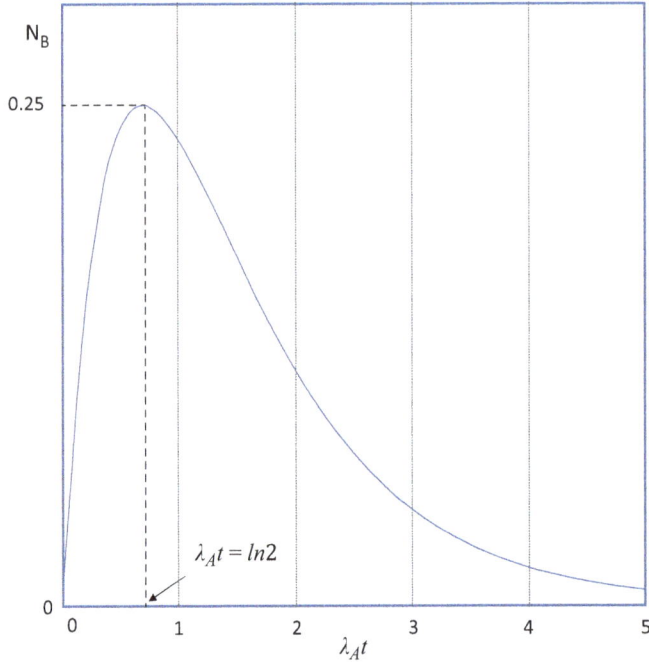

E5.11 **Solution: Prolonged time activity**

According to eq. (5.15):

$$N_B(t) = \frac{\lambda_A}{\lambda_B - \lambda_A} N_{A0}(\exp(-\lambda_A t) - \exp(-\lambda_B t)),$$

which can be rewritten as

$$N_B(t) = \frac{\lambda_A}{\lambda_B - \lambda_A} N_A(t)(1 - \exp(-(\lambda_B - \lambda_A)t)).$$

If $\lambda_B \approx \lambda_A$ and $t \gg \tau$, the exponential function approaches zero and we have

$$\frac{N_B(t)}{N_A(t)} = \frac{\lambda_A}{\lambda_B - \lambda_A}.$$

E5.12 Solution: Multiple decays of radioactive substances

a. The number of B isotopes is the result of the production rate and the decay rate of B. Therefore, the following rate equation applies at all times:

$$dN_B(t) = \lambda_A N_A dt - \lambda_B N_B dt.$$

The activity of B is, therefore:

$$\frac{dN_B(t)}{dt} = \lambda_A N_A - \lambda_B N_B,$$

where

$$N_A = N_{A0}\exp(-\lambda_A t).$$

b. Inserting it into the equation for the activity gives the inhomogeneous differential equation:

$$\frac{dN_B(t)}{dt} + \lambda_B N_B - \lambda_A N_{A0}\exp(-\lambda_A t) = 0.$$

c. Now we show that equation:

$$N_B(t) = \frac{\lambda_A}{\lambda_B - \lambda_A} N_{A0}(\exp(-\lambda_A t) - \exp(-\lambda_B t)) + N_{B0}\exp(-\lambda_B t)$$

is a solution of the inhomogeneous differential equation. We take the first derivative with respect to the time:

$$\frac{dN_B(t)}{dt} = -\frac{\lambda_A \lambda_A}{\lambda_B - \lambda_A} N_{A0}\exp(-\lambda_A t) + \frac{\lambda_A \lambda_B}{\lambda_B - \lambda_A} N_{A0}\exp(-\lambda_B t) - \lambda_B N_{B0}\exp(-\lambda_B t).$$

Now we insert this expression into the equation of the inhomogeneous differential equation and obtain

$$-\frac{\lambda_A \lambda_A}{\lambda_B - \lambda_A} N_{A0}\exp(-\lambda_A t) + \frac{\lambda_A \lambda_B}{\lambda_B - \lambda_A} N_{A0}\exp(-\lambda_B t) - \lambda_B N_{B0}\exp(-\lambda_B t)$$

$$+ \lambda_B \left[\frac{\lambda_A}{\lambda_B - \lambda_A} N_{A0}(\exp(-\lambda_A t) - \exp(-\lambda_B t))\right.$$

$$\left. + N_{B0}\exp(-\lambda_B t)\right] - \lambda_A N_{A0}\exp(-\lambda_A t) = 0,$$

which is

$$-\frac{\lambda_A \lambda_A}{\lambda_B - \lambda_A} N_{A0}\exp(-\lambda_A t) + \frac{\lambda_A \lambda_B}{\lambda_B - \lambda_A} N_{A0}\exp(-\lambda_B t) - \lambda_B N_{B0}\exp(-\lambda_B t)$$

$$+ \frac{\lambda_A \lambda_B}{\lambda_B - \lambda_A} N_{A0}\exp(-\lambda_A t) - \frac{\lambda_A \lambda_B}{\lambda_B - \lambda_A} N_{A0}\exp(-\lambda_B t)$$

$$+ \lambda_B N_{B0}\exp(-\lambda_B t) - \lambda_A N_{A0}\exp(-\lambda_A t) = 0.$$

Collecting terms, we obtain

$$-\frac{\lambda_A\lambda_A}{\lambda_B-\lambda_A}N_A + \frac{\lambda_A\lambda_B}{\lambda_B-\lambda_A}N_{A0}\exp(-\lambda_B t) - \lambda_B N_B + \frac{\lambda_A\lambda_B}{\lambda_B-\lambda_A}N_A$$

$$-\frac{\lambda_A\lambda_B}{\lambda_B-\lambda_A}N_{A0}\exp(-\lambda_B t) + \lambda_B N_B - \lambda_A N_A = 0,$$

which is

$$-\frac{\lambda_A\lambda_A}{\lambda_B-\lambda_A}N_A + \frac{\lambda_A\lambda_B}{\lambda_B-\lambda_A}N_A - \lambda_A N_A = 0.$$

If the bracket is zero, then the condition is fulfilled and the above equation is a solution to the inhomogeneous differential equation:

$$\frac{N_A}{\lambda_B-\lambda_A}[-\lambda_A\lambda_A + \lambda_A\lambda_B - \lambda_A(\lambda_B-\lambda_A)] = \frac{N_A}{\lambda_B-\lambda_A}[-\lambda_A\lambda_A + \lambda_A\lambda_B - \lambda_A\lambda_B + \lambda_A\lambda_A] = 0,$$

which was to be shown.

d. For $N_A(t) = N_{A0}\exp(-t/1h)$; $N_{B0} = 0$;

$$N_B(t) = \frac{1/1h}{1/5h - 1/1h}N_{A0}(\exp(-t/1h) - \exp(-t/5h));$$

$$\frac{N_B(t)}{N_{A0}} = 1.25(-\exp(-t/h) + \exp(-0.2t/h)).$$

At $t=0$, $N_{B0}=0$, as demanded. At $t>0$, $N_B(t)$ has a maximum at $t=2h$.

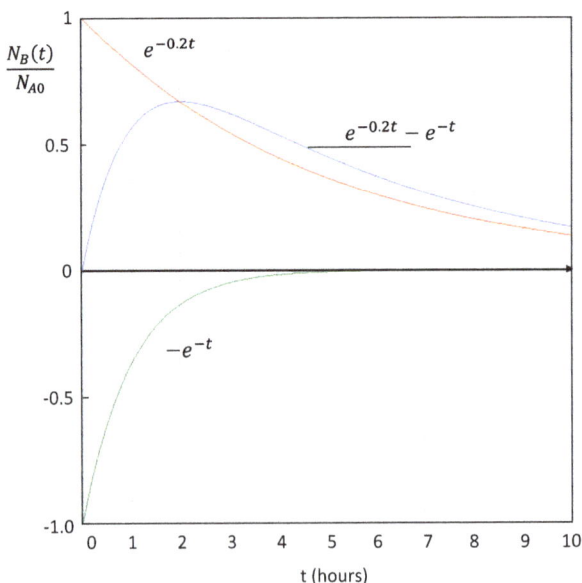

Chapter 6

E6.1 **Solution: Half-value thickness**

The half-value thickness is $x_{1/2} = \ln 2/\mu$. According to Fig. 6.9, the mass attenuation coefficient μ/ρ is 1 cm^2/g for 100 keV photons and 0.1 cm^2/g for
500 keV photons. With $\mu = (\mu/\rho)\rho$ and $\rho = 1$ g/cm^3, we find
$x_{1/2}(100\,\text{keV}) = 1\,\text{cm}; \; x_{1/2}(500\,\text{keV}) = 7\,\text{cm}.$

E6.2 **Solution: Photoelectric effect and the Compton effect**

Compton effect of a free electron at rest:
Momentum conservation: $\vec{p}_{\lambda_i} = \vec{p}_{\lambda_f} + \vec{p}_e.$
Energy conservation: $hf_i + E_{0,e} = hf_f + E_e.$
Photoelectric effect assuming an unbound free electron at rest:
Momentum conservation: $\vec{p}_{\lambda_i} = \vec{p}_e.$
Energy conservation: $hf_i + E_{0,e} = E_e.$
$E_{0,e}$ is the rest energy of the electron and E_e is the kinetic energy plus rest
energy of the electron after collision.

E6.3 **Solution: Conservation of momentum and energy**

Momentum conservation yields:

$$\vec{p}_e = \vec{p}_{\lambda_f} - \vec{p}_{\lambda_i}; \; (\vec{p}_e c)^2 = \left(\vec{p}_{\lambda_f} c - \vec{p}_{\lambda_i} c\right)^2;$$

$$(\vec{p}_e c)^2 = \left(p_{\lambda_f} c\right)^2 + \left(p_{\lambda_i} c\right)^2 - 2\left(p_{\lambda_f} c\right)\left(p_{\lambda_i} c\right)\cos\theta.$$

Since for photons holds $p_\lambda c = hf$, then we follow from the last equation:

$$(\vec{p}_e c)^2 = (hf_f)^2 + (hf_i)^2 - 2(hf_f)(hf_i)\cos\theta.$$

On the other hand, for electrons, the equation holds: $(\vec{p}_e c)^2 = E_e^2 - E_{e,0}^2.$
Therefore, for the momentum conservation we can write:

$$E_e^2 - E_{e,0}^2 = (hf_f)^2 + (hf_i)^2 - 2(hf_f)(hf_i)\cos\theta. \tag{a}$$

Now we consider the energy conservation:

$$E_e - E_{0,e} = hf_f - hf_i.$$

Squaring both sides yields

$$E_e^2 + E_{e,0}^2 - 2E_e E_{0,e} = (hf_f)^2 + (hf_i)^2 - 2(hf_f)(hf_i). \tag{b}$$

Taking the difference of eqs. (b) and (a):

$$2E_{e,0}^2 - 2E_e E_{0,e} = 2(hf_f)(hf_i)(\cos\theta - 1),$$

$$2E_{0,e}(E_e - E_{0,e}) = 2E_{0,e}(hf_i - hf_f) = 2(hf_f)(hf_i)(\cos\theta - 1).$$

Therefore,

$$f_i - f_f = \frac{h f_f f_i}{E_{0,e}} (\cos\theta - 1),$$

which is

$$\frac{f_i - f_f}{f_f f_i} = \frac{h}{m_0 c^2} (\cos\theta - 1),$$

with the final result:

$$\Delta\lambda = \frac{h}{m_0 c} (1 - \cos\theta),$$

which is the wavelength shift of the Compton photon.

E6.4 **Solution: Violation of momentum and energy conservation**
Energy conservation for the photoelectric effect of a free electron:
$h f_i + E_{0,e} = E_e$ or $h f_i = E_e - E_{0,e}$.
Momentum conservation:

$$\frac{h f_i}{c} = p_e = \frac{\sqrt{E_e^2 - E_{0,e}^2}}{c} = \frac{\sqrt{(E_e - E_{0,e})(E_e + E_{0,e})}}{c}.$$

Comparing both by multiplying the last equation with c:

$$h f_i = E_e - E_{0,e} = \sqrt{(E_e - E_{0,e})(E_e + E_{0,e})},$$

which results in

$$\sqrt{(E_e - E_{0,e})} = \sqrt{(E_e + E_{0,e})}.$$

The last equation is only fulfilled for $E_{0,e} = 0$. However, there is no particle with $E_{0,e} = 0$, and therefore, photoelectric effect of a free particle is not possible.

E6.5 **Solution: Comparison of PE and Compton effect**
Compton effect is a scattering process. The energy and momentum are partitioned during the scattering, and therefore both are conserved. Photoelectric effect implies that the total photon energy is transferred to the free electron. This is equivalent to the Compton effect for a final photon energy in the limit $E_f \to 0$ or $\lambda_f \to \infty$. Then $\Delta\lambda = \frac{h}{m_e c}(1 - \cos\theta)$ would also go to infinite, violating the limiting value $\Delta\lambda = 2\lambda_C$.

E6.6   **Solution: Photon momentum**

The rest mass of an electron has the energy $E_{0,e} = m_0 c^2 = 511\,\text{keV}$. The energy can be expressed as photon wavelength:

$$\lambda = \frac{hc}{E} = \frac{1.24\ \text{keV nm}}{511\ \text{keV}} = 0.0024\,\text{nm} = 2.4\,\text{pm}.$$

Note that this is the Compton wavelength: $\lambda_c = hc/m_0 c^2 = hc/m_0 c$. Frequency:

$$f = E/h = 511\ \text{keV} \times 1.6 \times 10^{-19}\text{J}/6.62 \times 10^{-34}\text{J s} = 1.23 \times 10^{20}\text{s}^{-1}.$$

Momentum:

$$p = h/\lambda = 6.6 \times 10^{-34}\text{J s}/2.4\,\text{pm} = 2.75 \times 10^{-22}\text{Js/m}.$$

E6.7   **Solution: Compton effect**

a. γ-Beam of ^{137}Cs has an energy of 662 keV. This corresponds to a wavelength of

$$\lambda_\gamma = \frac{hc}{E_\gamma} = \frac{1.240\ \text{keV} \cdot \text{nm}}{662\ \text{keV}} = 1.87 \times 10^{-3}\text{nm}.$$

b.   The Compton wavelength shift at an angle of $\theta = 90°$ is

$$\Delta\lambda = \lambda_C(1 - \cos 90°) = \lambda_C = 0.00243\,\text{nm}.$$

c.   Energy transfer to the electron

$$\frac{E_e}{E_i} = \frac{\alpha(1 - \cos\theta)}{1 + \alpha(1 - \cos\theta)} = \frac{\alpha}{1 + \alpha}$$

with

$$\alpha = \frac{E_\gamma}{m_0 c^2} = \frac{662\ \text{keV}}{511\ \text{keV}} = 1.29 \text{ and } \frac{\alpha}{1 + \alpha} = 0.562.$$

Therefore, energy transfer to the recoiling electron is

$$E_e = \frac{\alpha}{1 + \alpha} E_i = 0.562 \times 662\,\text{keV} = 372\,\text{keV}.$$

The relative energy transfer is 56.2%, and the absolute transfer is 372 keV.

d.   The energy loss of the electron must be

$$\Delta E_\gamma = 662\,\text{keV} - 372\,\text{keV} = 290\,\text{keV}$$

due to energy conservation. The relative energy loss is 290/662 = 43.8%.

e. The scattering angle ϕ of the recoiling electron is

$$\tan\phi = \frac{\cot(\theta/2)}{1+\alpha} = \frac{\cot 45°}{1+1.29} = \frac{1}{2.29}$$

and

$$\phi = \tan^{-1}\left(\frac{1}{2.29}\right) = 23.6°.$$

f. The total angle $\theta + \phi = 113.6°$.

E6.8 **Solution: Range of α-particles**

a. According to eq. (6.22), the range of 5 MeV α-particles in air is

$$R_\alpha^{air}(\text{cm}) = 0.325(E)^{1.5}\text{cm} = 0.325 \times 5^{1.5}\text{cm} = 0.325 \times 11.18\,\text{cm} = 3.63\,\text{cm}.$$

b. The range of 5 MeV α-particles in water is according to the scaling equation (6.23):

$$R_{\alpha,1} = \frac{\rho_{air}}{\rho_{water}}\frac{\sqrt{A_{water}}}{\sqrt{A_{air}}}R_{\alpha,air} = \frac{0.00127}{1} \times \frac{\sqrt{18}}{\sqrt{14.6}}\langle 3.63\ \text{cm}\rangle = 51\ \mu\text{m}$$

and in tissue, assuming an average density and atomic number:

$$R_{\alpha,1} = \frac{\rho_{air}}{\rho_{tissue}}\frac{\sqrt{A_{tissue}}}{\sqrt{A_{air}}}R_{\alpha,air} = \frac{0.0012}{1} \times \frac{\sqrt{9}}{\sqrt{14.6}}\langle 3.63\ \text{cm}\rangle = 34\ \mu\text{m}.$$

According to the tables in Ref. [6.10], .the range of 5 MeV α-particles in water is 3.613 g/cm². The range (in cm) is multiplied by the density of water (1 g/cm³). Therefore, we have to divide the result by the density in order to get the range in cm: 3.613 cm. This is in good agreement with the scaling result.

E6.9 **Solution: Confirm eqs. (6.51) and (6.52)**

For a central elastic collision, energy and momentum is conserved:

Energy conservation: $m_1v_1^2 = m_1v'^2_1 + m_2v'^2_2$.

Momentum conservation: $m_1v_1 = -m_1v'_1 + m_2v'_2$.

Eliminating $v'_2 = \frac{m_1}{m_2}(v_1 - v'_1)$, yields for v'_1:

$$v'_1 = \left(\frac{m_2 - m_1}{m_2 + m_1}\right)v_1 \text{ and}$$

$$E'_1 = E_1\left(\frac{m_2 - m_1}{m_2 + m_1}\right)^2. \text{ Setting } m_2 = A \text{ and } E'_1 = E_1\left(\frac{m_2 - m_1}{m_2 + m_1}\right)^2 \text{ confirms eq. (6.51).}$$

The energy transfer per collision follows from

$$\frac{\Delta E_1'}{E_1} = \left(\frac{m_2 - m_1}{m_2 + m_1}\right)^2 - 1 = \frac{4m_1 m_2}{(m_1 + m_2)^2}$$

Or

$$\frac{\Delta E_1'}{E_1} = \frac{4A}{(A+1)^2}$$

which confirms eq. (6.52).

Chapter 7

E7.1   **Solution: Dose**

$$D_{x-ray} = I \times \Delta t \times E \times w_R \times \left(\frac{\mu}{\rho}\right) = \Phi \times E \times w_R \times \left(\frac{\mu}{\rho}\right)$$

$$= 10^{10}\,\frac{\text{photons}}{\text{cm}^2\text{s}} \times 50 \times 10^{-3}\,\text{s} \times 60 \times 10^3\,\text{eV} \times 5\frac{\text{cm}^2}{\text{g}} \times 1.6 \times 10^{-19}\,\text{J/eV}$$

$$= 24 \times 10^{-6}\,\frac{\text{J}}{10^{-3}\,\text{kg}} = 24\ \text{mSv}.$$

E7.2   **Solution: Radiation exposure through ^{131}I therapy**

a.   Dose rate is defined by the deposited energy of a single decay per mass, times the activity: Dose rate $\dot{D} = A(E/m)$.
The activity is 1.5×10^8 Bq, the mass is 0.02 kg, and the energy is 89%×610 keV and 11%×350 keV. The energy is thus: 581 keV = 9.3×10^{-14}J. This results in a dose rate of
$1.5\times10^8\,\text{s}^{-1}\times9.3\times10^{-14}\text{J}/0.02\text{kg} = 7\times10^{-4}\text{J/kgs} \cong 0.7\,\text{mGy/s}$. With a quality factor of 1 for β-radiation, the dose rate is approx. 1 mSv/s.

b.   The effective half-life follows from

$$T_{\text{eff}} = \frac{T_{\text{bio}} \times T_{\text{phys}}}{T_{\text{bio}} + T_{\text{phys}}} = 7.3\,\text{days}.$$

c.   The initial activity is

$$A = \lambda N_0 = \left(\ln 2/T_{1/2}\right)N_0.$$

The initial number of radioactive isotopes is therefore:

$$N_0 = \left(AT_{1/2}\right)/\ln 2.$$

Inserting the numbers gives

N_0 = 1.5 x 10^8 Bq × 7.3 days × 86400 s/day/ln 2 = 1.36 × 10^{14} radioactive isotopes.

After a half-life, half of it disintegrated:

$$N_0/2 = 0.68 \times 10^{14}$$

and this half delivers an integral dose to the gland:

$$D = 0.68 \times 10^{14} \times 9.310^{-14} \text{J}/0.02 \, \text{kg} \ = \ 316 \, \text{Gy}.$$

E7.3 Solution: Radiation exposure of bystanders

After a half-life, the activity is

$$A_{1/2} = A_0/2 = \lambda N_0/2 = \ln 2 N_0/2T_{1/2} = 6 \times \ 10^{12} \, \text{Bq}.$$

The activity goes in 4π. At a distance of 1 m, a spherical surface has an area of 12.56 m^2. A person has a front surface of approx. 1 m^2 through which the radiation passes. The activity spread over the surface of the person is:

$$A_{1/2}/(12.56)^2 = 6 \times \ 10^{12} \, \text{Bq}/1.5 \times 10^2 = 4 \times 10^{10} \text{Bq}.$$

The absorbed dose rate is then

$$\dot{D} = AE/m = 4 \times 10^{10} \text{Bq} \times 580 \, \text{keV} \times 1.6 \times 10^{-19} \text{J}/70 \, \text{kg} \ = \ 46 \, \mu\text{Gy/s}.$$

The total accumulated dose within 1 h is

$$D = 46 \ \mu\text{Gy/s} \times \ 3600 \, \text{s} = 0.16 \ \text{Gy}.$$

That's quite a lot. Today, ^{99}Tc is preferred for thyroid treatment because the biological half-life is much shorter: 6 h instead of 8 days.

E7.4 Solution: Count rate

a. The true corrected count rate is $\dot{N}/(1 - \dot{N}\tau)$ = 5 × 10^4/0.9 = 5.55 × 10^4 counts/s.

b. For small differences, eq. (7.20) can approximately be expressed as $\Delta\dot{N}/\dot{N}^2 = \tau$. The count rate difference is $\Delta\dot{N} = 1 \times 10^5$. Then $\Delta\dot{N}/\dot{N}^2 = \frac{1 \times 10^5}{16 \times 10^{10}} = 6.2 \times 10^{-7}$ s. The dead time is about 0.6 μs.

E7.5 Solution: Dead time

Using eq. (7.26) and solving for the dead time τ, we obtain

$$\tau = \frac{\dot{N}_t - \dot{N}_m}{\dot{N}_t \dot{N}_m} = \frac{(1000 - 900)\,\text{s}^{-1}}{(1000 \times 900)\,\text{s}^{-2}} = \frac{100}{9 \times 10^5}\,\text{s} = 1.11 \times 10^{-4} \approx 11\,\mu\text{s}.$$

E7.6 Solution: Dip in the μ_{tr}/ρ_m curve

Two effects contribute to the dip: First, the mass attenuation coefficient μ/ρ_m is always larger than mass energy transfer coefficient μ_{tr}/ρ_m because μ/ρ_m considers more contributions to the attenuation. When disregarding the attenuation due to coherent x-ray scattering, a dip occurs in μ_{tr}/ρ_m as seen around 0.1 MeV. So, the dip is mainly due to coherent scattering.

Chapter 8

E8.1 Solution: Magnification and penumbra

a. The magnification is for geometrical reasons:

$$\frac{D_2}{D_1} = \frac{L_2}{L_1} = \frac{1\,\text{m}}{0.9\,\text{m}} = 1.11.$$

The magnification is 11%.

b. The penumbra extension is
$\frac{L_1}{L_3} = \frac{A/2}{U}$. Therefore, $U = \frac{A}{2}\frac{L_3}{L_1} = 0.01\,\text{m}\,\frac{0.1\,\text{m}}{0.9\,\text{m}} = 1.1\,\text{mm}.$
Whether the full extension of the focal spot A enters the equation or only half depends on the setting of the first collimator.

E8.2   **Solution: Mass attenuation coefficient and contrast**
The transmitted intensity is

$$I(t_i) = I_0 \exp\left(-\left(\frac{\mu_i}{\rho_i}\right)\rho_i t_i \right).$$

First step: we multiply: (μ/ρ) and ρ The result is in the fourth column.
Second step: We calculate the transmitted intensity for a thickness of 1 cm of
the respective material. The result is listed in the fifth column. Taking the
difference of the recorded intensities, we obtain:

C(tissue-bone)	+ 0.25
C(tissue-air)	− 0.2
C(muscle-blood)	− 0.003

The results confirm the values listed in Tab. 8.2.

E8.3   **Solution: Dual-energy radiography**
First, we make a table with the necessary attenuation values $(\mu/\rho)\rho t$ at two
different energies and two different thicknesses:

	60 keV	100 keV
Bones	1.743	1.03
Tissue	1.6	1.23

In order to make the bone contrast vanish, the weighting factor for bones has
to be
$w_1 = \frac{b_{60}}{b_{100}} = \frac{1.743}{1.03} = 1.69$. Then the bone contrast vanishes: $(b_{60} - w_1 b_{100}) = 0$.
The image taken at 100 keV has to be multiplied with the factor 1.69 before
subtracting from the image taken at 60 keV in order to make the bone con-
trast vanish.
The tissue contrast is then: $(t_{60} - t_{100} w_1) = 1.6 - 1.23 \times 1.69 = -0.48$.
Vice versa, for emphasizing the bones, the tissue contrast should vanish:
$(t_{60} - t_{100} w_2) = 0$.

$$w_2 = \frac{t_{60}}{t_{100}} = \frac{1.6}{1.23} = 1.3.$$

The bone contrast is then $(b_{60} - w_2 b_{100}) = 1.743 - 1.3 \times 1.03 = +0.404$.

E8.4   **Solution: Bone contrast**
At lower energies, the photoelectric effect is mainly responsible for the x-ray
attenuation. In this region, the attenuation effect scales with Z^3 (see eq.
(6.20)). Bones contain large amount of Ca (39%) with an atomic number

$Z = 20$, which is much higher than the Z-values of the surrounding tissue, which contains mainly carbon ($Z = 6$), oxygen (8), nitrogen (9), phosphorus ($Z = 15$). At higher x-ray energies, the photoelectric effect diminishes, and the Compton effect becomes more important. In the range from 100 keV to 1 MeV the mass absorption coefficient is small (~ 0.1 cm^2/g), constant, and independent of the atomic number Z. Small differences in mass density are not important. Therefore, the x-ray attenuation coefficient at high energies is less different for bones and tissue.

E8.5 **Solution: Attenuation profile**

a. $I(x) = I_0 \exp(-\mu x)$, $\frac{I(x)}{I_0} = \exp(-1\ \text{cm}^{-1}x(\text{cm})) = \exp(-1\ \text{cm}^{-1} \times N \times 0.1(\text{cm}))$, where N is the number of voxels in the beam.

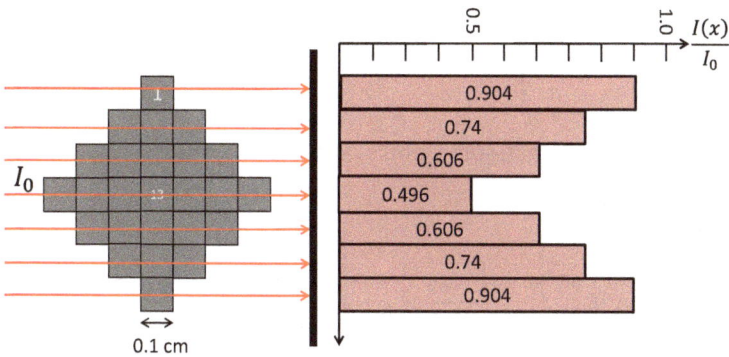

b. Intensity distribution in voxels.

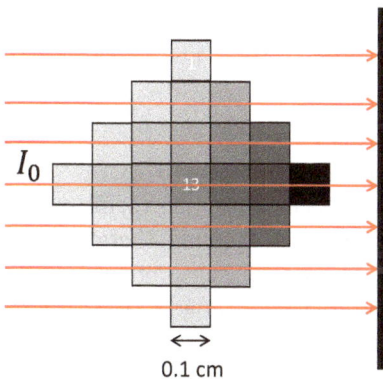

c. The Weber contrast is

$$C = \frac{I_1 - I_{13}}{I_1} = 1 - \frac{I_{13}}{I_1} = 1 - \frac{\exp(-0.4\mu)}{\exp(-0.1\mu)} = 1 - \exp(-0.3\mu) = 0.26.$$

The contrast between voxels 1 and 13 is 26%.

E8.6 **Solution: Backprojection**

First, we fill in the 3×3 matrix with the result of the projections (line integrals) for all four orientations without discriminating between different voxels along the projected line:

20	20	20
35	35	35
20	20	20

25	20	5
20	25	20
5	20	25

20	35	20
20	35	20
20	35	20

5	20	25
20	25	20
25	20	5

In the second step, we add up all four matrices, voxel by voxel. This procedure corresponds to the backprojection. The result is

70	95	70
95	120	95
70	95	70

We may subtract some noise and rescale the result with factor of 25, yielding:

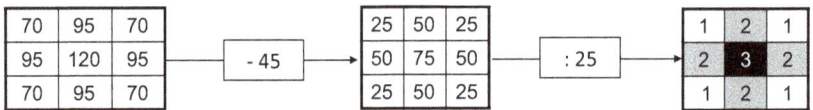

70	95	70
95	120	95
70	95	70

- 45 →

25	50	25
50	75	50
25	50	25

: 25 →

1	2	1
2	3	2
1	2	1

The final result shows that the voxel with the highest attenuation is sitting at the center. The attenuation values of all other voxels are now determined as well.

Chapter 9

E9.1 **Solution: Wall thickness for 1% transmission**

According to the NIST table, the mass absorption coefficient of lead at 140 keV is 2 cm^2/g. The density of lead is 11.3 g/cm^3. Therefore, the absorption coefficient $(\mu/\rho)\rho = \mu = 22.6\,\mathrm{cm}^{-1}$. For a 99% absorption, the transmission is 1%. According to the exponential law, it follows that the wall thickness is $t = 2\ln 10/22.6$ cm $= 0.2\,\mathrm{cm} = 2\,\mathrm{mm}$.

E9.2 **Solution: Effective Wall thickness**

a. $t_{eff} = \frac{t}{\sin(\alpha + \Delta\alpha)}$; $\alpha = \tan^{-1}\left(\frac{d}{L}\right)$. Therefore, we obtain for t_{eff} the expression:

$$t_{eff} = \frac{t}{\sin\left(\tan^{-1}\left(\frac{d}{L}\right) + \Delta\alpha\right)} \cong \frac{t}{\sin\left(\frac{d}{L} + \Delta\alpha\right)}.$$

b. $t = t_{eff} \times \sin\left(\frac{d}{L} + \Delta\alpha\right) = 2\,\text{mm} \times \sin(°(0.1) + 10°) = 2\,\text{mm} \times \sin(5.7° + 10°) = 0.54\,\text{mm}$.

A septal thickness of 0.54 mm provides the same attenuation for 140 keV γ-photons at incident angle of 15.7° as a wall thickness of 2 mm for 140 keV γ-photons at normal incidence.

E9.3 **Solution: Detector quantum efficiency**

The energy of 500 nm photon is $E = hc/\lambda = 1240\,\text{eV nm}/500\,\text{nm} = 2.48\,\text{eV} \cong 2.5\,\text{eV}$. If the quantum efficiency were 100%, 140 000/2.5 = 56 000 photons would be generated. However, only 5000 are actually produced, which corresponds to an efficiency of 8.9%.

E9.4 **Solution: Activity uptake**

For a homogeneous distribution, the activity per area is 500 MBq/1.2 m² = $5 \times 10^8\,\text{s}^{-1}/1.2 \times 10^4\,\text{cm}^2 = 4.2 \times 10^4\,\text{s}^{-1}\,\text{cm}^{-2}$.

At a counting efficiency of 1%, the count rate per pixel is about 400 counts/s.

E9.5 **Solution: SNR**

The present situation is as follows: at a count rate of 50/s and a counting time of 50 s, the total accumulated counts are 2500. This yields an SNR of $\sqrt{2500} = 50$.

In order to reach an SNR of 100, a total count of 10 000 per pixel is required. This can be achieved by counting four times longer (200 s) at the same rate or by increasing the activity to 200/s at a counting time of 50 s.

Increasing the counting time has the advantage that the administered dose to the patient remains constant and has the disadvantage of more patient movement during the recording time.

Increasing the count rate has the advantage of shortening the recording time and fewer effects due to patient movement. However, it has the disadvantage of a four times higher dose administered to the patient, which continues long after the recording time is over.

The first option should be chosen from a patient safety perspective.

E9.6   **Solution: Compton peak**

The peak is due to the Compton scattering cut-off: the maximum energy that can be transferred to electrons in the target is at a scattering angle $\theta = 180°$. At this scattering angle, the final γ-photon energy is

$$E_{f,\,max}^{photon} = E_i^{photon} \frac{1}{1 + \frac{E_i^{photon}}{m_e c^2}(1 - \cos\theta)} = 140 \text{ keV} \frac{1}{1 + \frac{140}{511} \times 2} = 90.4 \text{ keV}.$$

This energy corresponds to the Compton peak.

E9.7   **Solution: Compton peak – continued**

The Compton peak is actually not a real peak defined by an event that is more likely and therefore accumulates a higher count rate. The peak is an artifact. It occurs because scattering events at lower energies are less likely. Hence, the Compton peak occurs because of the valley adjacent to it. For energies less than 90 keV, scattered intensity can still be observed. This is due to double scattering. The peak at 66 keV is the cut-off peak due to the Compton scattering of the 90 keV photons.

Chapter 10

E10.1   **Solution: Compton scattering of 511 keV photons**

a.   The wavelength shift is $\Delta\lambda = \lambda_c(1 - \cos\theta) = 2.43 \text{ pm} \times 0.13 = 0.32 \text{ pm}$.

b.   The energy gain of the Compton scattered electron is according to eq. (6.26):

$$E_e = E_i \frac{\alpha(1 - \cos\theta)}{1 + \alpha(1 - \cos\theta)},$$

where E_i=511 keV is the initial photon energy and $\alpha = E_i/m_e c^2 = 1$. Therefore, $E_e = 511 \text{ keV} \times (0.13/1.13) = 58.8 \text{ keV}$.

c.   The cross section for Compton scattering at 30° is 0.65 or 65% of the Compton scattered photons are emitted in an angular cone of opening 2×30°.

E10.2 **Solution: Production time for a 100 MBq ^{18}F source**

According to Tab. 10.1, the half-life of ^{18}F is $T_{1/2} = 110$ min.

According to eq. (5.23), the activity as a function of time for the production of ^{18}F is

$$A(t) = R(1 - e^{-\lambda t}),$$

where R is the production rate.

Therefore, the time to reach an activity A is

$$t = -\frac{T_{1/2}}{\ln 2}\left(1 - \frac{A}{R}\right).$$

We first determine the production rate $R = N_\sigma \sigma_p J_p$ (see eq. (5.20)). Here, N_A is the total number of ^{18}O isotopes in the target. Here NA is not the Avogadro number. One mole of $H_2^{18}O$ weighs 18 g + 2 g = 20 g. These 20 g of $H_2^{18}O$ contain 6×10^{23} ^{18}O nuclei and 12×10^{23} protons. One gram of $H_2^{18}O$ contains $6 \times 10^{23}/20$ ^{18}O isotopes. Therefore, $N_A = 3 \times 10^{22}$.

The cross section for ^{18}F production is 500 mbarn = 5×10^{-25} cm^2. The proton beam flux is $J_p = 10^{16}$ cm^{-2} s^{-1}.

Inserting these numbers into the production rate yields:

$$R = N_A \sigma_p J_p = 3 \times 10^{22} \times 5 \times 10^{-25}\text{cm}^2 \times 10^{16}\text{cm}^{-2}\text{s}^{-1} = 1.5 \times 10^{14}\text{ s}^{-1}.$$

The production time is therefore

$$t = \frac{110 \text{ min}}{\ln 2}\left(1 - \frac{10^8}{1.5 \times 10^{14}}\right) = 159 \text{ min}(1 - 6.6 \times 10^{-7}) \cong 160 \text{ min}.$$

The production time is determined by the half-life of the ^{18}F isotope.

E10.3 **Solution: Improved precision for isotope production**

The lack of precision is due to the ratio A/R, which is too small in E10.2. Therefore, we have to reduce the production R, and this can only be achieved in two ways:

a. reducing the ^{16}O/^{18}O ratio in the target to decrease N_A;
b. decreasing the proton flux J_p in the cyclotron.

In both cases, the ratio A/R should be about 0.5 for improved precision of isotope production.

E10.4 **Solution: Angular correlation**

The finite angle θ immediately indicates that there must be an additional momentum in the scattering process, which is compensated by the angular perturbation. Therefore, we are led to the discussion of the origin of such an additional momentum. In metals, the electrons exhibit a high momentum, particularly when they have energies close to the Fermi surface. Angular perturbations may also occur in the presence of a magnetic field and nuclear hyperfine fields.

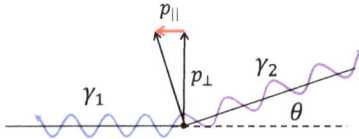

13 List of acronyms (used in all three volumes)

ACD	annihilation coincidence detection	DBT	digital breast tomosynthesis
ADC	analog digital converter	DCE	dynamic contrast enhancement
ADC	apparent diffusional constant	DES	drug elusion stent
ADP	adenosine diphosphate	DNA	deoxyribonucleic acid
ALARA	as low as reasonably achievable	DNP	dynamic nuclear polarization
amu	atomic mass units	DOV	depth of view
ANF	auditory nerve fiber	DSA	digital subtraction angiography
ATP	adenosine triphosphate	DSB	double strand break
AV	atrioventricular node	DTI	diffusion tensor imaging
AVV	atrioventricular valve	DTL	drift tube linac
BBB	blood-brain barrier	DWI	diffusion weighted imaging
BF	breath frequency	EBRT	external beam radiotherapy
BMD	bone mineral density	ECC	extra-corporal circulation
BMR	basal metabolic rate	ECG	electrocardiography
BMS	bare metal stents	EDP	end diastolic pressure
BNCT	boron neutron capture therapy	EDV	end diastolic volume
BOLD fMRI	blood oxygen level dependent fMRI	EEG	electroencephalography
		EF	ejection fraction
		EM	electromagnetic
BPS	biodegradable polymeric stents	EMG	electromyography
BSA	beam shaping assembly	EPI	echo planar imaging
BSA	body surface area	EPP	end plate potential
BW	body weight	EPR	enhanced permeation and retention
CA	contrast agent		
CAP	cardiac action potential	ERBT	external radiation beam therapy
CBF	cerebral blood flow	ERV	expirational rest volume
CCC	continuous curvilinear capsulorhexis	ESP	end systolic pressure
		ESV	end systolic volume
CCD	caput-collum-diaphyseal angle	ETL	echo train length
CCD	charge coupled device	FCRM	fiber optic confocal reflectance microscope
CD	coincidence detector		
CIRT	carbon ion radiation therapy	FDG	18F-fluoro-deoxy-glucose
CK	cyber knife	FE	fractional excretion
CLE	confocal laser endoscopy	FEG	frequency encoding gradient
CNR	contrast to noise ratio	FET	18F-fluoro-ethyl-L-tyrosine
CO	cardiac output	FF	flattening filter
COE	caloric oxygen equivalent	FF	filtration fraction
c.m.	center of mass	FFDM	full-field digital mammography
CPA	charge particle activation	FFF	flattening filter free
CPB	cardiopulmonary bypass	FFT	fast Fourier transform
CS	Compton scattering	FID	free induction decay
CSF	cerebrospinal fluid	FLACS	femtosecond laser-assisted cataract surgery
CT	computed tomography		
CTV	clinical Target Volume	FLAIR	fluid-attenuation inversion recovery
cw	continuous wave		
CZT	CdZnTe	fMRI	functional magnetic resonance imaging
dB	decibel		

https://doi.org/10.1515/9783110757095-013

FOV	field of view	MHR	metabolic heat production
FRC	fractional rest volume	MHT	magnetic hyperthermia
FSE	fast spin echo	MHz	megahertz
FT	Fourier transform	MI	mechanical index
GFR	glomerular filtration rate	MLC	multi-leaf collimator
GTV	gross tumor volume	MNP	magnetic nanoparticle
Hct	hematocrit value	mpMRI	multiparameter MRI
HD	hydrodynamic diameter	MRI	magnetic resonance imaging
HDR	high dose rate	MRSI	magnetic resonance
HEP	hemi endo prosthesis		spectroscopy imaging
hMRI	hyperpolarization magnetic	MRgRT	magnetic resonance-guided
	resonance imaging		radiation therapy
HSLS	high spin–low spin	MRT	magnetic resonance
IAEA	International Atomic Energy		tomography
	Agency	MSFP	mean systemic filling pressure
ICRP	International Commission for	MSO	medial superior olive
	Radiological Protection	MUAP	motor unit action potential
IGRT	image guided radiotherapy	MV	minute ventilation
IHC	inner hair cell	NBI	narrow band imaging
ILD	intensity level differences	NCT	neutron capture therapy
IMRT	intensity modulated radiation	NIR	near infrared
	therapy	NIRS	near infrared spectroscopy
IRT	internal radiation therapy	NMJ	neuromuscular junction
IRV	inspirational rest volume	NP	nanoparticle
ITD	interaural time difference	NRT	neutron radiation treatment
Kerma	kinetic energy release in matter	NTD	non-target dose
kHz	kilohertz	OAR	organ at risk
kV	kilovolt	OCT	optical coherence tomography
kVp	peak kilovoltage	OER	oxygen enhancement ratio
kW	kilowatt	OHC	outer hair cell
LASEK	laser epithelial keratomileusis	PAH	para-aminohippuric acid
LASER	light amplification by	PCI	percutaneous coronary
	stimulated emission of		intervention
	radiation	PCI	phase contrast imaging
LASIK	laser-assisted interstitial	PCV	packed cell volume
	keratomileusis	PD	proton density
LCI	low-coherence interferometry	PDD	percent depth dose
LDR	low dose rate	PDR	proliferative diabetic
LED	light-emitting device		retinopathy
LET	linear energy transfer	PDR	pulse dose rate
Linac	linear accelerator	PDT	photodynamic therapy
LOR	line of response	PE	photoelectric effect
LQ	linear-quadratic	PEG	phase encoding gradient
LSO	lateral superior olive	PEG	polyethylene glycol
MC	Monte Carlo	PES	photoelectron emission
MEG	magnetoencephalography		spectroscopy
MEI	middle ear implants	PET	positron emission tomography
MET	mechanoelectric transduction	PHIP	parahydrogen-induced
MeV	mega-electron volt		polarization

PI	pulsatility index	SPE	single photon emission
PMT	photomultiplier tube	SPECT	single-photon emission
PPD	percentage photon dose		computed tomography
PRF	pulse repeat frequency	SPIO	superparamagnetic iron oxide
PRK	photorefractive keratectomy	SPR	surface plasmon resonance
PRP	pan-retinal photocoagulation	SSB	single strand break
PRT	proton radiotherapy	SSD	source-to-surface distance
PRT	pulse repeat time	SSG	slice selection gradient
PSA	prostate-specific antigen	SUV	standard uptake value
PSMA	prostate-specific membrane	SV	stroke volume
	antigen	TCE	transient charged particle
PTV	planning target volume		equilibrium
PWV	pulse wave velocity	TE	time of (spin, acoustic) echo
PZT	$PbZrTiO_3$	TEP	total endo prosthesis
Q	quality factor	TERMA	total energy released per unit
QA	quality assurance		mass
RBC	red blood cell	TGC	time gain compensation
RBE	relative biological effectiveness	THz	terahertz
RBF	renal blood flow	TIPPB	transperineal interstitial
RC	respiratory coefficient		permanent prostate
Re	Reynolds number		brachytherapy
RES	reticuloendothelial system	TLC	total lung capacity
RF	radio frequency	TMR	targeted muscle reinnervation
RF	respiratory fraction	ToF	time of flight
RMR	resting metabolic rate	TPFR	total peripheral flow resistance
RPF	renal plasma flow	TPS	treatment planning system
RT	radiotherapy	TR	time of repetition
RTR	real time radiography	TRIGA	Training, Research, and
RV	residual volume		Isotopes, General Atomics
SA	sinoatrial node	TT	transfer time
SAD	source to axis distance	TV	tidal volume
SATP	standard ambient temperature	TV	target volume
	and pressure	UHMWPE	ultra-high molecular weight
SAXS	small angle x-ray scattering		polyethylene
SBRT	stereotactic body radiation	ULFMRI	ultra-low field magnetic
	therapy		resonance imaging
SCI	spinal cord injury	US	ultrasound
SE	spin echo	VC	vital capacity
SID	source to image distance	VEGF	vascular endothelial growth
SERS	surface-enhanced Raman		factor
	scattering	VOR	vestibulo ocular reflex
SGRT	surface-guided radiotherapy	XRR	x-ray radiography
SLAC	Stanford linear accelerator	XRT	x-ray radiotherapy
SNR	signal-to-noise ratio	YAG	yttrium-aluminum garnet
SOBP	spread out Bragg peak	ZFC	zero field cooling

14 Selection of fundamental physical constants, conversions, and relationships

Speed of light c	$299792458 \text{ m/s} \sim 3 \times 10^8 \text{ m/s}$
Gravitational acceleration of the earth g	9.81 m/s^2
Planck constant h	$6.623 \times 10^{-34} \text{ Js}$
Compton wavelength λ_c	2.426 pm
Elementary charge e	$1.602 \times 10^{-19} \text{ As}$
Atomic mass unit u	$1.66 \times 10^{-27} \text{ kg}$
Electron mass m_e	$9.109 \times 10^{-31} \text{ kg}$
Avogadro number N_A	$6.022 \times 10^{23} \text{ mol}^{-1}$
Boltzmann constant k_B	$1.38 \times 10^{-23} \text{ JK}^{-1}$
Dielectric constant of the vacuum ε_0	$8.854 \times 10^{-12} \text{ As/Vm}$
Magnetic permeability μ_0	$4\pi \times 10^{-7} \text{Vs/Am} = 1.256 \times 10^{-6} \text{ Vs/Am}$
Faraday constant F	$9.65 \times 10^4 \text{ C/mol}$
General gas constant R	8.314 J/Kmol
Stefan-Boltzmann constant σ_{SB}	$5.66 \times 10^{-8} \text{ AV/m}^2\text{K}^4$
Bohr magneton μ_B	$9.274 \times 10^{-24} \text{ J/T}$
Proton gyromagnetic ratio γ_p	$2.675 \times 10^8 \text{ T}^{-1}\text{rad s}^{-1}$
Proton mass m_p	$1.673 \times 10^{-27} \text{ kg}$
Neutron mass m_n	$1.675 \times 10^{-27} \text{ kg}$
Nuclear magneton μ_N	$5.05783 \times 10^{-27} \text{J/T.}$

Conversions

1 eV	$= 1.602 \times 10^{-19} \text{ J}$
1 Joule	$= 6.242 \times 10^{18} \text{ eV}$
1 Tesla	$= 1 \text{ N/Am} = 1 \text{ Vs/m}^2 = 10^4 \text{ Gauss}$
1 Pascal	$= 1 \text{ N/m}^2$
1000 hPa	$= 1000 \text{ mbar}$
1 A/m	$= 10^{-3} \text{ emu/cm}^3$
$\text{emu/cm}^3\rho$	$= \text{emu/g}$

Relationships

Avogadro number $N_A = 1 \text{ } g/1u$	$= 1 \text{ g}/1.66 \times 10^{-24} \text{ g}$
Faraday constant $F = N_A \times e$	$= 6.022 \times 10^{23} \text{ mol}^{-1} \times 1.602 \times 10^{-19} \text{ As}$
General gas constant $R = N_A \times k_B$	$= 6.022 \times 10^{23} \text{ mol}^{-1} \times 1.38 \times 10^{-23} \text{ JK}^{-1}$
Speed of light in vacuum $c = 1/\sqrt{\mu_0\varepsilon_0}$	$= \left(\sqrt{1.25 \times 10^{-6} \text{ Vs/Am} \times 8.85 \text{ } 10^{-12} \text{ As/Vm}} \right)^{-1}$

https://doi.org/10.1515/9783110757095-014

15 List of scientists named in this volume

Abbé, Ernst Karl 56
Anderson, Carl 350
Anger, Hal 331
Avogadro, Lorenzo R.A.C. 183, 217

Bateman, Harry 194
Becquerel, Antoine Henri 192
Beer, August 215
Bloch, Felix 77
Bohr, Niels 163
Brown, Robert 97

Cormack, Allan M. 306
Coulomb, Charles Augustin de 157
Cowan, Clyde 186
Curie, Marie 193
Curie, Paul Jacques 4
Curie, Pierre 4, 193

Dirac, Paul 78, 350
Doppler, Christian Andreas 33

Einstein, Albert 97, 219

Faraday, Michael 89
Fermi, Enrico 78
Fourier, Joseph 13
Fraunhofer, Joseph von 16
Fresnel, Jean 9

Gauss, Johann Carl Friedrich 288
Geiger, Johannes Wilhelm 258
Golay, Marcel J.E. 122
Gray, Harald 245

Hahn 95
Helmholtz, Hermann von 121
Hounsfield, Godfey 304

Kapany, Nadrinder Singh 50
Kirchhoff, Gustav 12
Kramers, Hendrik Anthony 158

Lambert, Johann Heinrich 215
Langevin, Paul 4
Larmor, Joseph 81
Lauterbur, Paul Christian 77
Lawrence, Ernest 201
Lorentz, Hendrikus Albertus 200

Mansfield, Peter 77
Michelson, Albeert A. 64
Michelson, Albert A. 275
Morley, Edward W. 64
Moseley, Henry Gwyn Jeffreys 163
Müller, Walther 258

Nyquist, Harry 39

Pauli, Wolfgang 186
Planck, Max 79
Purcell, Eduard Mills 77

Rabi, Isodor Isaac 77
Radon, Johann 308
Ramachandran, Ramachandran 314
Rayleigh, John William 56
Reines,Frederick 186
Richardson, Owen Willans 157
Röntgen, Wilhelm Conrad 155, 269
Rutherford, Ernest 199
Rydberg, Johannes Robert 163

Schottky, Walter H. 157
Segrè, Emilio Gino 180
sievert, Rolf Maximillian 247
Slichter, Charles P. 80
Smolukowski, Marian von 97
Stejskal, Eward O. 131

Talbot, Henry Fox 301
Tanner, John E. 131

Weber, Ernst Heinrich 275
Wehnelt, Arthur Rudoph B. 156

Zeeman, Pieter 80

https://doi.org/10.1515/9783110757095-015

16 Glossary

Acoustic impedance: Defined as the product of density and sound velocity.

A-mode scan: Sonographic single line scan through the body.

Attenuation contrast enhancement: Use of contrast agents with high Z molecules to highlight structures in soft matter, gas-filled volumes, and liquids.

Attenuation: Weakening of a particle beam as it passes through matter due to absorption and scattering.

B-mode scan: Sonographic scan providing two-dimensional images of tissues within one section.

Bragg peak: Peak in linear energy transfer (LET) near the end of particle tracks also defines the charged particle range.

Bremsstrahl spectrum: Broad x-ray spectrum emitted by electrons hitting a solid target.

Capsule endoscopy: Encapsulated miniaturized optics for illumination and imaging, swallowed and traveling through the gastrointestinal tract.

Characteristic x-ray spectrum: Ejection of electrons from the K-shell by electron or photon impact causing electrons from the L-, M-, etc. shell to fill the hole in the K-shell, thereby emitting the energy difference as x-rays with sharp spectral lines.

Chemical shift: MRI resonance frequency shift due to the local diamagnetic environment.

C-mode scan: Collection of B-mode scans for reconstructing two-dimensional images at a fixed depth perpendicular to B-mode scans.

Colonoscopy: Endoscopic imaging of the colon through the anus.

Compton effect: Inelastic scattering of x-ray photons by free electrons that transfers kinetic energy to electrons.

Computed tomography (CT): Projection radiography taken from many different angles with an x-ray generator and opposing detector bank rotating about the stationary patient at the center.

Confocal laser endoscopy: Scanning imaging technique with a small focal spot and narrow ocular aperture to suppress scattered light.

Contrast enhancement agents (CA): Magnetic substances that change the $T1$ and/or $T2$ relaxation times.

Doppler method: Sonographic method used to determine the blood flow velocity.

Dual-energy x-ray absorptiometry: Weighted difference radiography with two different mean x-ray energies for contrast enhancement.

Duplex mode: Combines B-mode scans with pulsed Doppler mode for determining blood flow velocity and direction.

Duplexer: Electronic device allowing bi-directional communication via one joined signal path.

Endoscopy: Using fiber optics, endoscopes provide images of hollow spaces in the body.

Far field (acoustic): Area at some distance from the transducer where interference effects are minimal and sound waves diverge.

Fluid-attenuated inversion recovery (FLAIR): MRI pulse sequence that nullifies signals from the slower $T1$ relaxing systems, such as the cerebrospinal fluid.

Fluoroscope: Real-time radiography that allows monitoring surgical procedures in vivo.

Free induction decay (FID): Damped oscillatory voltage signal induced by precessing transverse magnetization M_{xy}, recorded by a pick-up coil.

https://doi.org/10.1515/9783110757095-016

Frequency encoding gradient (FEG): Linear magnetic field gradient along the x-direction, defining columns in the XY-slice.

Functional MRI (fMRI): Neural activity that promotes oxygenated blood flow that shortens $T2$ relaxation time. $T2$-weighted maps show the centers of cerebral activity during stimulation.

Gastroscopy: Endoscopic imaging of the stomach through the esophagus.

Hounsfield scale: Normalized x-ray attenuation difference coefficients compared to water as zero point.

Inversion recovery: Pulse sequence that separates two different $T1$ relaxation times, emphasizing the faster relaxation.

Isotopes: Atoms with a fixed number of protons but varying number of neutrons.

Kerma: Mean photon energy fluence converted into kinetic energy of charged particles.

K-map: Fourier representation of recorded frequencies and phase angles for each voxel in a slice. Back transformation generates real space MRI images.

Linear energy transfer (LET): Rate at which a particular type of radiation deposits energy by passing through matter.

Mechanical index: Estimate of the maximum power that can be safely administered during sonographic examination.

MRI contrast: Contrast achieved by differences in $T1$ and $T2$ relaxation times or differences in proton density.

Narrow band imaging: Endoscopic imaging using a narrow spectral band of light to highlight abnormal or cancerous tissues.

Near field (acoustic): Area close to the transducer with main propagation direction in the forward direction, while sideways spreading is suppressed by destructive interference.

Optical coherence tomography (OCT): Using a short coherent wave train, OCT scans capture depth-depended images below the skin surface.

Phase contrast imaging: Image modality that results from phase shifts in the x-ray radiation wavefront that is visualized with a diffraction grating in front of the image plate.

Phase encoding gradient (PEG): Short field gradient pulse in the X-direction with different slopes from top to bottom, effectively causing a phase gradient in the Y-direction.

Photon beam attenuation: Occurs through photoelectric effect, Compton scattering, coherent scattering, and pair production.

Piezoelectric effect: Strain applied to special dielectric crystals creates electrical voltage or vice versa electrical voltage produces strain.

Piezoelectric materials: Mainly $Pb(Zr_{1-x}Ti_x)O_3$ is used for generating sound waves or serves as sensor for detecting sound waves.

Positron emission: Emission of positive or negative electrons together with neutrinos, where the decay energy is shared between both.

Positron emission tomography (PET): Annihilation of electron-positron pairs and the subsequent emission of two γ photons are used for coincidence counting, which generates images for the diagnosis of cardiac, neurological, and oncological issues.

Radiation dose: Mean energy of ionizing radiation acting on a mass.

Radioactive decay: Stochastic process of nuclear decay with an exponential time dependence of the number of remaining isotopes.

Radioisotope production methods: Fission of nuclei, proton bombardment, or neutron capture.

Radioisotopes: Instable isotopes which decay in time by emitting α-, β-, and γ-radiation.

Radon transformation: Back transformation of sinograms into real space tomographies.

Range: Photon beams have no range, only attenuation; α-particles stop after 5 cm in the air, and β-particles reach 1–2 m before they stop.

Scintigraphy: Imaging modality that uses the γ-emission of radioisotopes accumulated in malignant tissue to detect tumors or examines the perfusion of fluids in the heart and kidneys.

Short time inversion recovery: MRI pulse sequence emphasizing the slower $T1$ relaxing system.

Sinogram: Attenuation profiles as function of rotation angle.

Slice encoding gradient: Linear magnetic field gradient along the Z-direction superposed on a constant field, defining an XY-slice in the body.

$T1$ relaxation: Recovery time of the magnetization M_Z, following a 90° field pulse. $T1$ depends on the spin–lattice interaction.

$T2$ relaxation: Decay time of the transverse magnetization M_{xy}, depending on the spin–spin interaction.

Time gain compensation: Amplification of successive echoes from deeper interfaces on a logarithmic scale, yielding similar echo amplitudes from similar interfaces.

Time to echo (TE): Time between the first initializing 90° field pulse and the 180° inversion pulse, plus the time from the inversion pulse to the echo – in total, twice the time between 90° and 180° pulses.

Time to repetition (TR): Time to repeat a specific sequence of field pulses, usually starting with an initial 90° pulse.

Transducer: Transducers are devices that convert one form of energy into another such as electrical energy in vibrational and vice versa.

X-ray radiography: Use of bremsstrahlung to create a grayscale attenuation shadow that creates a contrast between soft tissues and hard bones.

17 Index of terms

^{18}F-FDG 203
^{18}F-PET 362
^{99m}Tc 326

absorption 215
acoustic impedances 9
activity unit
– Bequerel 192
– curie 193
activity
– nuclear 192
afterglow 171
ALARA 260
aliasing effect 39
allowed x-ray transitions 161
aluminum-filter 159
Anger logic 331
angiography 294
angio-MRI 130
annihilation coincidence detection 352
anode current 157
anti-Helmholtz coil 122
antiparticles 185
apparent diffusional constant 131
array scanner 22
atomic cross section 217
atomic form factor 227
atomic mass 181
atomic number 179
atomic weight 183
attenuation 215
– via absorption 12
– via scattering 11
attenuation coefficient 217
attenuation profile 304
attenuation slice 278
authentic radiopharmaceuticals 361
autocorrelation function 98
average lifetime 190
Avogadro number 183

Bateman equations 194
beam hardening 271
Becquerel 193
beta-minus decay 184
Bethe–Bloch equation 232
binomial distribution 287

binomial function 287
biological clock 191
biological effectiveness 246
biological half-life 190
Bloch equations 87
blood–brain barrier 128
Bohr model 163
BOLD fMRI 134
boron neutron capture therapy 207
Bragg peak 232
Bragg–Kleeman rule 234
branching 188
breast tomosynthesis 292
bremsstrahlung 158
Brownian motion 97
bucky factor 274
butterfly coils 122

capsule endoscopy 68
carotid artery
– Doppler test 37
cathode current 157
characteristic x-ray radiation 160
charge particle activation 198
chemical shift 101
Cherenkov radiation 235
chromoendoscopy 60
cladding 52
classical electron radius 227
CMOS detector 285
coherence length 64
coherent x-ray scattering 218
coincidence
– counting 352
– detectors 352
– electronics 352
– event 352
collimators
– field of view 327
– hexagonal arrangement 328
– magnification 328
– mimification 328
– resolution 327
– sensitivity 327
colonoscopy 49
Compton coupling strength 224
Compton scattering 218, 221, 274

https://doi.org/10.1515/9783110757095-017

Compton wavelength 224
computed tomography 303
confocal laser endoscopy 60
confocal reflectance endoscope 63
contrast
– Michelson 276
– Weber 276
– x-ray 275
contrast-enhancing agent 293
contrast-to-noise ratio 288
cosmic radiation 259
cosmic radioisotopes 181
counting statistics 359
cross section 197
CT scanner 303
CT voxel size 308
cutoff energy 159
cyclotron 201
cyclotron isotope production 204
CZT
– detector 331
– sensor 341

data acquisition
– MRI 119
daughter nucleus 194
dead time correction 258
decay constant 190
dephasing time 93
depth of view 30
detective quantum efficiency 289
detector assessment 286
detector bank 306
diamagnetic screening 101
diffusion-weighted MRI 130
digital recording 284
digital subtraction angiography 295
direct detectors 285
Doppler effect 33
Doppler method
– laminar flow 36
– turbulent flow 36
dose 251
– charge particle radiation 254
– flat panel x-ray radiography 253
– radioactive source 253
– scintigraphic imaging 253
dose calculation 253
dose rate 248

dose weighting factor 245
dual-energy CT 309
dual-energy x-ray absorptiometry 295
duplex mode 38
duplexer 8, 14
dynamic contrast enhancement MRI 127
dynodes 329

echo planar MRI 136
echo train length 136
effective dose 247
elastic scattering 222
electron anti-neutrino 179
electron capture decay 186
electron synchrotron 171
elution, radioisotopes 337
emission energy 161
emission radiography 325
endomicroscopy 60
endoscopes 49
endothermic reaction 196
energy fluence 249
energy loss
– electrons 235
– neutrons 237
energy-dispersive x-ray detection 310
equation of motion 86
equilibrium magnetization 85
equivalent dose 246
equivalent dose rate 248
excitation energy 161
exothermic reaction 196
exponential decay law 190
extremity MRI scanner 124

fast spin-echo MRI 136
fiber optics 50
fiber-optic confocal reflectance microscope 62
fiber-optic endoscopes 54
field of view
– endoscopy 50
– MRI 113–114
film badge 255
film blackening 282
filtered backprojection 315
fission process 205
fluence 215, 249
fluorine decay 362
fluorodeoxyglucose 203, 364

fluoroscopic imaging 284
flux 214, 249
Fourier transform 118
Fourier transformed filter 315
Fourier transforms 311
Fraunhofer region 16
free induction decay 89, 93
frequency encoding gradient 114
full-body MRI scanner 124
functional MRI 133
fuzziness, source of 273

gadolinium chelate complex 128
gamma-radiation 187
gastroscopy 49
Geiger–Müller detectors 258
generator
– isotope 337
glass fibers 50
Golay coils 122
gyration radius 227
gyromagnetic ratio 80
gyromagnetic ratio for protons 80

half-life 189–190
half-value layer 262
handheld radiation monitors 256
head MRI scanner 124
high-definition video endoscope 57
Hounsfield scale 304
Hounsfield window 305
hyperpolarization-MRI 138
hyperpolarized gas pulmonary MRI 140

indirect detectors 285
in-line holographic imaging 300
intensity 214
inverse positron decay 187
inverse Radon transform 313
inversion recovery 110
ionization chamber 256
isobaric transitions 184
isotope production rate 199
isotopes 179

kerma 249–251
kerma contrast 281
kilovoltage peak 159
Klein–Nishina cross section 225

klystrons 169
K-map 117
Kramers equation 158
K-shell electron 160

lambda quarter plate 7
Lambert–Beer law 215
Larmor frequency 81, 92
lepton number 184
light
– total reflection 50
line of response 353
linear attenuation coefficient 215
linear electron accelerator 168
linear energy transfer 232, 246
linear gradient field 113
linear x-ray mass attenuation coefficient 219
longitudinal magnetization 85
longitudinal relaxation time 85
Lorentz correction factor 203
Lorentz force 200

magnetic flux 89
magnetic induction 80
magnetic moment
– nuclear 80
magnetic resonance spectroscopy imaging 139
magnetic resonance tomography 77
magnetization 83
mammography 292
mass attenuation coefficient 217
mass attenuation cross section
– energy dependence 230
mass deficiency 182
mass density 217
mass energy absorption coefficient 251
mass energy transfer coefficient 250
mass number 179, 182
mechanical index 18
medical linac 170
medical radiation 259
Michelson–Morley interferometer 64
microwave cavities 169
mole number 183
Moseley's law 163
MR signal strength 96
MRI scanner 120
multiparameter MRI 133

narrowband imaging endoscope 58
natural radioisotopes 181
neutron activation 206
neutron capture isotope production 205
neutron energy loss 237
neutron moderation 238
neutrons 178
– resonance region 238
noise
– x-ray radiography 282
normal distribution 288
nuclear activation analysis 206
nuclear charge number 179
nuclear fission 182
nuclear fission reaction 205
nuclear force 178
nuclear fusion 182
nuclear magnetic moment 77
nuclear reactions 196
nuclear shell 78
nuclear spin paramagnetic susceptibility 84
nuclear
– g-factor 80
nucleus
– magnetic moment 80
number density 217
Nyquist criterion 39

OCT
– spectral domain 67
– time domain 67
optical coherence tomography 63
optical density 280
orbital radius 201
oscillating grid 274

pair production 218, 229
parent nucleus 194
particle flux 196
penetration depth 30
penumbra 272
personal dosimeters 255
PET
– 18F 362, 366
– artifacts 353
– brain scanner 357
– detectors 358
– image reconstruction 359
– LOR 359

– PSMA 367
– scanner 356–357
– spatial resolution 354
– time-of-flight 355
phase contrast imaging 298
phase encoding gradient 115
photoelectric effect 218–219
photomultiplier tube 330
pick-up coil 88
piezoelectric coefficient 5
piezoelectric effect 5
piezoelectric materials 5
Planck constant 79
pocket ionization dosimeters 255
Poisson distribution 287
polarization factor 227
positron emission 185
positron emission tomography 350
precession
– nuclear 83
production rate 197
projection image 278
projection radiography 272
prompt gamma emission 351
proportional counter 256
prostate-specific membrane antigen 367
proton density 95
proton density-weighted contrast 107
protons 178
pulsatile index 35
pulse repeat frequency 8, 13
pulse repetition frequency 29
PZT crystals 6

quadrupole moment 79
quality factor 13
quantum tunneling 199
quarks 178
quarter-wave plate 7

radiation cone 172
radiation dose 245
radiation monitors 255
radiation
– cosmic 259
– industrial 259
– terrestrial 259
radio frequency coil 88
radioactivity 180

radiographs 271
radiography
– emission 326
– transmission 326
radioisotope 191
radioisotopes 180
radiopharmaceuticals
– SPE 336
Radon transformation 310, 312
Ram–Lak filter 314
range of photoelectrons 221
range
– charge particles 234
– electrons 236
reaction rate 197
real-time radiographs 282
real-time radiography 291
relaxation time 99
relaxivity 127
resistance index 35
Richardson's law 157
Rutherford scattering 199
Rydberg constant 163

scanning electron microscope 164
scattering 215
Schottky emission 157
Schottky–Richardson law 157
scintigraphy
– single photon emission computed
    tomography 326, 341
– single photon emission 326
scintillation detector 329
scintillator crystal 329
sector array scanner 21
Segre chart 180
selection rule for dipole transitions 161
self-correlation function 98
shear wave elastography 27
Shepp–Logan phantom 316
shielding 262
short time inversion recovery 111
signal-to-noise ratio 276, 288
single mode wave guide 53
sinogram 308, 312
slice selection 113
small-angle scattering 227
sonography
– important relationships 4

sound velocity 4
sound waves
– amplitude 10
– basic properties 4
– far field 16
– near field 16
– velocity 10
source-to-image distance 291
SPE scan
– full body 339
– kidneys 339
SPE
– artifacts 333
– contrast 334
– signal-to-noise ratio 334
– thyroid 340
specific activity 193
SPECT applications 343
SPECT
– cameras 341
– emission scanner 342
– line integral 345
– myocardial perfusion 343
– Radon transformation 345
– stress test 343
spectral density 97–98
spin-echo techniques 92
spin-echo time 95
spin–lattice interaction 87
spin–spin relaxation 87
standard uptake value 361
Stejskal–Tanner equation 131
stopping power 233
straggling distance 232
straggling length 351
Straton rotary x-ray tube 167
superconducting magnets 121
superconducting solenoids 121
synchrotron storage ring 171

$T1$-weighted images 105
$T2$-weighted images 107
Talbot distance 301
technetium 189, 205
terrestrial radiation 259
terristic radioisotopes 181
the nuclear g-factor 78
therapeutic radioisotope 208
thermionic emission 157

thin-film transistors 284
Thomson scattering 222, 227
threshold energy 229
time focusing 17
time gain compensation 14
total dose rate 259
toy model 87
transceiver 114
transducer 4
– resonance condition 6
transducers 6
transmission radiography 325
transmitted intensity 9
transverse magnetization 86
transverse relaxation time 86
tritium decay 180
tungsten target 165
turbo-MRI 136

ultra-low-field MRI 142
ultrasound imaging
– artifacts 32
ultrasound
– A-mode scan 19
– axial resolution 21, 31
– B-mode scan 21
– C-mode scans 24
– dynamic focusing 29
– field of view 23
– lateral resolution 31
– phase array scanner 23
– pulse mode 39

– spatial resolution 31
umbilical cord 40
unified atomic mass unit 182
upregulation of glucose metabolism 364

velocity profile 36
video endoscope 55
voxel 277

waveguide 53
Wehnelt cylinder 156
weighting factor 246
weighting filters 314

x-ray contrast 275
x-ray film
– optical density 280
– quantum efficiency 280
x-ray films 279
x-ray fluoroscope 282
x-ray intensity 158
x-ray shadow formation 272
x-ray spectroscopy 162
x-ray tube for radiography 166
x-ray tubes 156
x-rays
– refractive index 216, 299

Zeeman energy 80

α-particle decay 187
β⁺-emitters 351

www.ingramcontent.com/pod-product-compliance
Lightning Source LLC
Chambersburg PA
CBHW080136220326
41598CB00032B/5078